A Dictionary of Physical Sciences

This book is a companion to
A Dictionary of Earth Sciences
A Dictionary of Life Sciences
also published in Pan Reference Books
The series has been prepared for Pan Books Ltd
by Laurence Urdang Associates Ltd, Aylesbury

Series editor: Alan Isaacs PhD
Additional contributions by:
Ephraim Borowski BSc MA BPhil
Julia Brailsford BSc
William Gould MA
G. W. Harris BSc DPhil

edited by John Daintith BSc PhD

A Dictionary of
Physical Sciences

Pan Books in association with The Macmillan Press

key to symbols

Asterisks before words in the text denote
cross-references to entries that will provide additional
information. Entries are defined under the most
commonly used term, with synonyms shown in brackets.

Unless otherwise stated, melting and boiling points are
given for standard temperature and pressure: relative
densities of liquids are taken at 20°C relative to water
at 4°C; relative densities of gases relate to air at
standard temperature and pressure. The chemical
terminology used is that recommended by the
International Union of Pure and Applied Chemistry.

First published 1976 by The Macmillan Press Ltd
This edition published 1978 by Pan Books Ltd. Cavaye Place, London SW10 9PG
© Laurence Urdang Associates Limited 1976
ISBN 0 330 25433 2

Prepared for Automatic typesetting by Laurence Urdang Associates Ltd
Typesetting by
Oriel Computer Services Ltd, Shipston-on-Stour
Printed in Great Britain by
Richard Clay (The Chaucer Press) Ltd Bungay Suffolk

A

ab-. Prefix placed in front of the name of a practical electrical unit to name the corresponding *electromagnetic unit. Thus the e.m.u. of current is the abampere.

Abbe condenser. A simple two-lens system used as a condenser in microscopes. [After Ernst Abbe (1840-1905), German physicist.]

aberration. 1. Any of certain defects in lenses, telescopes, or other optical systems, in which a true image is not formed. The types of aberration are classified into *chromatic aberration, *spherical aberration, *astigmatism, *curvature of field, *distortion, and *coma.
2. An apparent displacement of a star's position due to the fact that light travelling at finite velocity reaches not a stationary earth but one that is moving in orbit.

A-bomb. See nuclear weapon.

abrasive. A substance used for wearing away a solid surface, as in polishing or cleaning. Common examples are emery and pumice.

abscissa. The x coordinate of a point on a two-dimensional Cartesian graph: i.e. the distance of a point from the y axis, measured along the x axis from the origin; a value of the independent variable of a function of x. Compare ordinate.

absolute alcohol. Pure alcohol: i.e. ethanol containing little or no water.

absolute expansion. See expansion.

absolute humidity. See humidity.

absolute zero. The zero value of *thermodynamic temperature, equal to 0 kelvin or -273.15°C. It may be regarded as the temperature at which molecular motion ceases. See also zero-point energy.

absorptance (absorption factor). Symbol: α. A measure of the extent to which something can absorb radiation, equal to the ratio of the absorbed *flux to the flux incident on the body. For a *black body, the absorptance is 1. The quantity was formerly called the absorptivity. See also Kirchoff's law.

absorption. 1. Solution of a gas in a solid or liquid. The gas permeates into the bulk of the material: this process is distinguished from adsorption, in which the gas in held on the surface. Absorption in solids is sometimes called sorption.
2. Conversion of the energy of electromagnetic or sound waves into some other form of energy in a medium. A beam of light, for example, passing through matter loses intensity. Part of the loss may result from excitation of atoms or molecules. This process is distinguished from diffusion, in which light is scattered out of the beam. In absorption, the process occurring depends on the way the radiation interacts with the material: infrared radiation, for instance, is converted directly into heat by exciting vibrations of the atoms.

absorption spectrum. A *spectrum produced by absorption of electromagnetic radiation by matter. In producing an absorption spectrum, a continuous source of radiation is used: i.e. one having a range of wavelengths. This is passed through the sample and the emerging beam is dispersed using a prism or grating. The electromagnetic radiation is absorbed at certain frequencies, these being capable of exciting the atoms or molecules of the sample from their ground state to an excited state. Thus, in the case of visible light, the absorption spectrum consists of dark lines or bands on the bright continuous background. The frequency (ν) at which a line occurs depends on the difference ΔE between the energy of the ground state and that of the excited state: $\Delta E = h\nu$, where h is the Planck constant. *See also* emission spectrum, Fraunhofer lines.

absorptivity. *See* absorptance.

abundance. 1. Symbol: C. The concentration of a given isotope in a mixture of isotopes, equal to the ratio of the number of atoms of the isotope to the total number of atoms; often expressed as a percentage.
2. The concentration of a specified element in the earth's crust, the universe, etc.

a.c. *See* alternating current.

acceleration. The rate of change of velocity with time. Linear acceleration (symbol: a) is measured in metres per second per second, etc. Angular acceleration (symbol: ω) is measured in radians per second per second, etc.

acceleration of free fall. Symbol: g. The acceleration of a body falling freely, i.e. with no air resistance, at a specified point on the earth's surface as a result of the gravitational attraction of the earth. This acceleration, which is often called the *acceleration due to gravity*, varies from place to place because of different distances from the earth's surface to its centre. The standard value of g is 9.806 65 m s^{-2} (32.174 ft s^{-2}).

accelerator. 1. A machine for increasing the velocity (and thus the energy) of charged elementary particles and ions. Accelerators are used in fundamental research in nuclear and particle physics for producing a high-energy beam of particles which is directed onto a target, usually within a *bubble chamber. The particles are initially produced from a hot filament (in the case of electrons) or from a ion source.
The simplest method of accelerating charged particles is to use a high potential difference, the energy being gained by the particle in the electric field. This method is used in the *Van de Graaff accelerator. Higher energies are achieved by increasing the particle velocity in stages by a series of small electric fields—the method used in *linear accelerators, in which the particles follow a straight path down an evacuated tube. A similar technique is used in the *cyclotron and *synchrocyclotron, which employ magnetic fields to cause the particles to travel in spiral paths.
Cyclic accelerators of this type have the advantage that long path lengths can be used without making the accelerator impractically large. The *betatron is a cyclic accelerator in which energy is gained by magnetic induction produced by a varying magnetic field. *Synchrotrons work by a combination of electric and varying magnetic fields. The proton synchrotron is the most powerful type of accelerator, producing protons in the GeV range.
Particle energies can be effectively increased by the use of *storage rings*, which are large evacuated toroidal rings into which the particles are injected. It is possible to build up an intense beam of particles and maintain it circulating around the ring for many months. It is also possible to produce two beams, as of electrons and positrons, travelling in opposite directions and to arrange for them to intersect in the ring. In this way particle interactions can be studied at energies of up to 1700 GeV.
2. A substance added to increase the rate of a chemical reaction: i.e. to act as a catalyst. Accelerators are used to increase the rate of vulcanization of rubber.

acceptor. 1. An atom that accepts electrons in an extrinsic *semiconductor, thus producing holes in the conduction band.
2. The atom or molecule that contributes no electrons in forming a *coordinate bond. *Compare* donor.

access time. The time necessary for information to be supplied from a computer store for processing.

accommodation. *See* eye.

accumulator (storage battery). A device designed for storing electricity, consisting of one or more secondary *cells in series. Common examples are the lead-acid accumulator and the *nife. cell.

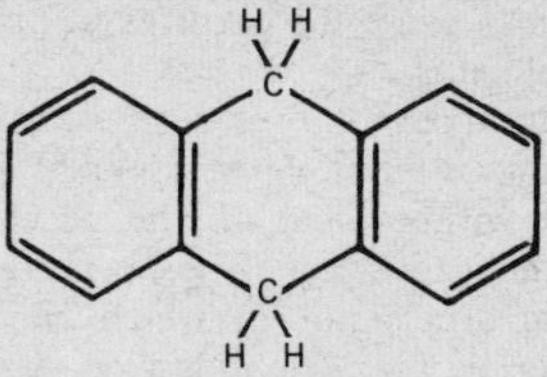

Acenaphthene

acenaphthene. A white crystalline solid, $C_{10}H_6(CH_2)_2$, obtained from coal tar and used in the manufacture of dyes and plastics. M.pt. 96.2°C; b.pt. 279°C; r.d. 1.02.

acetal. Any of a class of organic compounds with the general formula RCH-(OR')(OR''), where R, R', and R'' are organic groups. Acetals are formed by addition of an alcohol to an aldehyde. The reaction proceeds in two steps. The first is addition of one alcohol molecule to form a *hemiacetal*, with further reaction forming the full acetal. The reaction is catalysed by acid.
Ketones react similarly to yield *ketals. The term *acetal* is often applied to *1,1-diethoxyethane*, $CH_3CH(OC_2H_5)_2$. B.pt. 103.2°C; r.d. 0.83.

acetaldehyde (ethanal). A colourless liquid *aldehyde, CH_3CHO, made by the oxidation of ethanol and used in manufacturing many other compounds. M.pt. -124.6°C; b.pt. 20.8°C; r.d. 0.78.

acetaldol (aldol, 3-hydroxybutanal). A colourless syrupy liquid, CH_3C-$HOHCH_2CHO$, made by the condensation of acetaldehyde and used in medicine as a sedative and hypnotic. M.pt. below 0°C; b.pt. 83°C; r.d. 1.11.

acetamide (ethanamide). A colourless deliquescent crystalline *amide, CH_3C-ONH_2, with a mouse-like odour. It is manufactured from ethyl acetate and ammonia and used as a solvent and wetting agent. M.pt. 82.3°C; b.pt. 221.2°C; r.d. 1.0.

acetanilide An odourless white powder, $C_6H_5NH(COCH_3)$, made by the acetylation of aniline and used in the preparation of drugs, dyes, and lacquers. M.pt. 114.3°C; b.pt. 304°C; r.d. 1.22.

acetate. 1. Any salt or ester of acetic acid. **2.** *See* cellulose acetate.

acetate rayon. *See* rayon.

acetic acid (ethanoic acid). A colourless liquid carboxylic acid, CH_3COOH, manufactured by bacterial oxidation of ethanol (*see* vinegar) or by the oxidation of acetaldehyde. The anhydrous liquid is known as *glacial* acetic acid. Acetic acid is used in the manufacture of dyes, rubbers, plastics, and many other prod-

Acetal formation

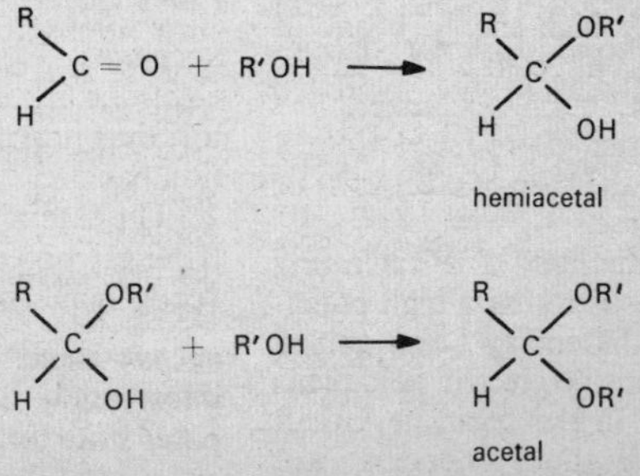

hemiacetal

acetal

ucts. M.pt. 16.6°C; b.pt. 117.9°C; r.d. 1.05.

acetic anhydride (ethanoic anhydride). A colourless liquid organic anhydride, $(CH_3CO)_2O$, used as a dehydrating and acetylating agent (see acetylation). M.pt. -73.1°C; b.pt. 139.6°C; r.d. 1.08.

acetolysis. The reaction of an organic compound with glacial acetic acid to exchange one of its groups for an acetyl group. A tertiary alkyl halide, for example, reacts to give an ester: $(CH_3)_3CCl + CH_3COOH = CH_3COOC(CH_3)_3 + HCl$. Acetolysis is a type of *solvolysis.

acetone (propanone). A colourless flammable volatile liquid *ketone, CH_3CO-CH_3, manufactured from propene and extensively used as a solvent. It is also a raw material for the production of many organic compounds. M.pt. -95.4°C; b.pt. 56.2°C; r.d. 0.79.

acetonitrile (methyl cyanide). A colourless flammable liquid *nitrile, CH_3CN, manufactured by the dehydration of acetamide and used in the synthesis of some pharmaceuticals. M.pt. -45.7°C; b.pt. 81.6°C; r.d. 0.79.

acetophenone (phenyl methyl ketone). A colourless liquid *ketone, $C_6H_5CO-CH_3$, with a sweet pungent odour and taste, used in perfumery. M.pt. 20.5°C; b.pt. 202°C; r.d. 1.03.

acetylation. The introduction of an acetyl group (CH_3CO-) into an organic compound in a chemical reaction. Acetylations are thus a special type of *acylation reaction. Most acetylations involve replacement of the hydrogen atom of a hydroxyl group (-OH) or amino group ($-NH_2$) by the acetyl group. Common acetylating agents are acetic anhydride, acetyl chloride, and ketene.

acetyl chloride (ethanoyl chloride). A colourless flammable pungent liquid *acid chloride, CH_3COCl, used as an acetylating agent (see acetylation). M.pt -112°C; b.pt. 50.9°C; r.d. 1.11.

acetylene (ethyne). 1. A colourless highly flammable gaseous *alkyne, C_2H_2, produced by the Wulff process or by the action of water on calcium carbide. It is used in the manufacture of acrylic plastics and PVC and in oxy-acetylene welding and cutting. M.pt -82°C (820 mmHg); b.pt. -84°C; r.d. 0.91.
2. See alkyne.

acetyl group. The organic group CH_3CO-.

acetylide. A *carbide that yields acetylene on hydrolysis.

acetylsalicylic acid. See aspirin.

achromat. See achromatic.

achromatic. 1. Having no colour. *Achromatic colours* are colours such as black, white, and various greys, which have no hue.
2. Having no chromatic aberration. An *achromatic lens* is one made by combining lenses of different types of glass, so that the combination has no dispersion. The simplest type, an *achromatic doublet*, is made of two lenses of flint and crown glass. The *power of the doublet is the sum of the powers of the lenses ($P = P_1 + P_2$). For the combination to be achromatic, the lenses must satisfy the relationship $P_1\omega_1 + P_2\omega_2 = 0$, where ω_1 and ω_2 are the dispersive powers of the glasses used. Achromatic lenses are called *achromats*.

acid. A substance that will react with a *base to form a salt and water. Acids turn litmus red and produce hydrogen ions if dissolved in water. They react with certain metals to give hydrogen and a metal salt. *Strong acids* are substances such as hydrochloric acid, in which the covalent molecule is fully dissociated into ions in solution: $HCl = H^+ + Cl^-$. *Weak acids* are compounds such as acetic acid, which are partially dissociated. In solution the hydrogen ion, H^+, is solvated by water and often considered to be a hydroxonium ion, H_3O^+. The idea of an acid in chemistry has been extended to include *Lewis acids.

acid chloride (acyl chloride). Any of a class of organic compounds with the general formula RCOCl, in which R is a hydrocarbon group. They are derived by replacing the hydroxyl group of a carboxylic acid by a chlorine atom using a chlorinating agent such as phosphorus trichloride. Acid chlorides, such as acetyl chloride, CH_3COCl, are easily hydrolysed: $CH_3COCl + H_2O = CH_3COOH + HCl$. They similarly react with alcohols and other hydroxy compounds and are used as acylating agents (see acylation). Other *acid halides* (or *acyl halides*), such as RCOF, and RCOBr, and RCOI, show similar behaviour.

acid dye. Any of a class of dyes that consist of salts of alkali metals containing large coloured negative ions. Acid dyes can be used on wool and silk. *Azo dyes are examples of acid dyes.

acid halide. See acid chloride.

acidic. Denoting a substance that is an acid or a solution containing an excess of hydrogen ions.

acidic hydrogen. A hydrogen atom in an acid that can be lost to form a hydrogen ion. For example, of the four hydrogen atoms in the acetic acid molecule, CH_3COOH, only the atom joined to the oxygen atom is an acid hydrogen: $CH_3COOH = CH_3COO^- + H^+$.

acidimetry. Determination of the amōunts of acids in solution by titration.

acidolysis. Hydrolysis of carboxylic *esters in acid solution to yield the parent carboxylic acid and the alcohol. The acid acts as a catalyst by protonating the carbonyl oxygen, thus facilitating attack by water.

acid radical. The group bound to the acidic hydrogen atom or atoms in an acid. For example, the acetate group, CH_3COO-, is the acid radical of acetic acid.

acid salt. See salt.

aclinic line. See magnetic equator.

acoustics. The scientific study of sound.

acre. An Imperial unit of area equal to 4840 sq. yards. It is equivalent to 4046.856 422 sq. metres.

acriflavine. A brown-orange granular solid, $C_{14}H_{14}N_3Cl$, used as an antiseptic and bacteriostat.

acrolein (propenal). A colourless or yellowish flammable liquid *aldehyde, $CH_2:CHCHO$, manufactured from propene and used for producing polyurethane and polyester resins. M.pt. $-87\,°C$; b.pt. $53\,°C$; r.d. 0.84.

acrylic resin. Any of a class of synthetic resins obtained by polymerization of acrylic acid, methacrylic acid, or their esters. The acrylics are thermoplastic materials noted for their transparency. The commonest is *polymethylmethacrylate.

acrylonitrile (vinyl cyanide). A colourless liquid *nitrile, $H_2C:CHCN$, manufactured from propene, air, and ammonia. It is copolymerized with butadiene to make synthetic rubbers and is also used in the production of acrylic fibres. M.pt. $-83.5\,°C$; r.d. 0.81.

actinic. Denoting radiation that causes chemical reaction, especially ultraviolet radiation.

actinide. Any of 15 elements from actinium (atomic number 89) to lawrencium (atomic number 103) inclusive. The actinides are similar to the *lanthanides in that they form a series in which the 5f shell is being filled. Chemically, the actinides are reactive metals forming compounds with a variety of valencies. All are radioactive and those with atomic numbers greater than 92—the *transuranic elements*—are artificial elements, made by bombardment of other nuclei with high-energy particles.

actinium. Symbol: Ac. An *actinide element found in uranium ores. Its most stable isotope, ^{227}Ac, has a half-life of 21.7 years. A.N. 89; m.pt. 1050°C; r.d. 10.07; valency 3.

actinometer. Any of various instruments for measuring the intensity of radiation.

actinon. The isotope of radon with mass number 219. Its half-life is 3.92 s.

action. The product of momentum and distance.

activate. 1. To heat a solid porous material in order to drive off absorbed water and other substances, thus producing an absorbent material. *Activated alumina*, made by heating aluminium oxide, is used for absorbing gases and for chromatography. *Activated charcoal* is also a highly effective absorbent. It is used for removing gas from vacuum systems, separating gases by selective absorption, purifying air in gas masks, decolorizing, deodorizing, and similar applications.
2. To supply sufficient energy to an atom or molecule to make it reactive.
3. To make a material radioactive, as by bombardment with neutrons or other particles.

activated complex. See transition state.

activation analysis. A technique for determining the amount of a particular element in a sample by bombarding the sample to produce a radioactive isotope, which is then identified by its gamma-ray emission spectrum. A common method of activating the sample is by placing it in a nuclear reactor, where radioisotopes are formed by neutron capture. The technique is extremely sensitive.

activation energy. The energy required to initiate a reaction. In a chemical reaction the activation energy is the energy that must be supplied in breaking and reforming chemical bonds: i.e. it is the energy required to form the *transition state. See also Arrhenius equation.

active mass. The effective concentration of a substance in a chemical reaction, as used in the law of *mass action.

activity. Symbol: A. A measure of the radioactivity of a radioactive substance, equal to the number of atoms disintegrating per unit time. It is measured in *curies.

acute. Denoting an angle that is between 0° and 90°.

acyclic. Denoting a chemical compound that is not cyclic, i.e. does not contain a ring of atoms in its molecular structure. Propane and acetic acid are common examples of acyclic compounds. *Compare* cyclic.

acylation. The introduction of an acyl group (RCO-) into an organic compound in a chemical reaction. The most important acylations are those in which an acyl group is attached to an aromatic ring by a *Friedel-Crafts reaction. See also acetylation.

acyl chloride. See acid chloride.

acyl group. Any organic group with the formula RCO-, where R is a hydrocarbon group. Examples are the acetyl group, CH_3CO- and the benzoyl group, C_6H_5CO-.

acyl halide. See acid chloride.

addend. One of a set of numbers to be added.

addition. A chemical reaction in which one molecule or atom combines with another to form a third molecule. In organic chemistry, addition is usually a reaction in which a molecule adds to an unsaturated molecule by breaking a double or triple bond. *Alkenes, *alkynes, and *aldehydes all undergo addition reactions.

addition polymerization. See polymerization.

additive process. The process in which a coloured light is produced by direct combination of lights of different colours. Most colours can be produced by a suitable mixture of primary colours. *Compare* subtractive process.

adiabatic. Denoting a process in which no heat enters or leaves the system. In general, the temperature changes when an adiabatic change occurs: for example, in the adiabatic compression of a gas work is done on the system and the gas temperature rises. *Compare* isothermal.

adiabatic demagnetization. A technique for producing very low temperatures (close to absolute zero) by demagnetizing a paramagnetic salt. The magnetized material is first cooled to the temperature of liquid helium and isolated thermally from its surroundings. When the field is removed the sample is demagnetized and cools still further.

adiabatic equation. The equation $pV^\gamma = C$, describing the relation between the pressure and volume of a gas during an adiabatic change. C is a constant and γ is the ratio of the principal *specific heat capacities of the gas.

adipic acid (hexanedioic acid). A colourless crystalline carboxylic acid, $HOOC(CH_2)_4COOH$, manufactured by oxidation of cyclohexanol. It is used in the manufacture of synthetic polymers, particularly *nylon. M.pt. $153°C$; b.pt. $267°C$ (750 mmHg); r.d. 1.4.

admittance. Symbol: Y. A measure of the ability of a circuit to pass electric current, equal to the reciprocal of its *impedance. It is a complex quantity: $Y = G + iB$, where G is the *conductance and B is the *susceptance. Admittance is measured in siemens.

adsorption. The process in which a compound, usually a gas, forms a layer on the surface of a solid: this process is distinguished from *absorption*, in which the gas permeates into the material. Two types of adsorption are distinguished: *chemisorption*, in which the gas molecules or atoms are held by covalent bonds, and *physisorption*, in which they are held by the weaker van der Waals forces. In chemisorption, a single layer of adsorbed molecules is formed: physisorption may involve the production of several layers.

aerial (antenna). A device for transmitting and receiving radio waves. A simple form is the *dipole aerial*—a long metal rod divided at the centre point with connections made at this point. The overall length of the dipole is about one half of the operating wavelength. In transmitting, the electromagnetic waves are produced by accelerating charges in the aerial. In receiving, the electromagnetic radiation induces small varying currents. Many different types of aerial exist suitable for use at various frequencies. Arrays of elements are used for reception and transmission in particular directions.

aerodynamics. The branch of science concerned with gases in motion, in particular the relative motion of solid objects and gases, as in the movement of bodies in air.

aerosol. A fine dispersion of liquid or solid particles in a gas.

aether. *See* ether.

afterdamp. Carbon dioxide produced in coal mines by an explosion of firedamp.

agate. A naturally occurring form of silica, used for ornaments and, because of its hardness, for mortars and for knife edges in chemical balances.

aggregation. A cluster or group of particles held together in a gas or liquid by *intermolecular forces.

agonic line. A line on the earth's surface joining points of zero magnetic declination: i.e. points at which a compass indicates true north. Two main agonic lines exist; one passing from north to south through America and the other having an irregular path through eastern Europe, Arabia, Asia, and Australia.

alabaster. A naturally occurring form of gypsum, $CaSO_4.2H_2O$, used for ornamental carvings.

alanine (**2-aminopropanoic acid**). A colourless crystalline nonessential *amino acid, $CH_3CH(NH_2)COOH$. M.pt. 295-297°C (decomposes).

albedo. The ratio of the amount of light scattered from a surface to the amount of incident light.

alchemy. A philosophical system that was the forerunner of modern chemistry. It dated from early Christian times, lasted until the 17th century, and involved a mixture of astrology, mysticism, magic, and practical chemistry in attempts to demonstrate the unity of the universe with the world of man. In particular, many alchemists sought the *Philosopher's stone*—a substance that could transmute base metals into gold and produce an elixir of life.

alcohol. 1. *See* ethanol.
2. Any of a class of organic compounds with general formula ROH, where R is a hydrocarbon group or substituted hydrocarbon group. Alcohols are distinguished from *phenols, in which the hydroxyl group is attached to a carbon atom in an aromatic ring. Alcohols are classified into *primary alcohols*, with general formula RCH_2OH, *secondary alcohols*, with formula RR'CHOH, and *tertiary alcohols*, with formula RR'R''OH. They are also classified as monohydric, dihydric, trihydric, etc., according to the number of hydroxyl groups they contain.
Alcohols react with electropositive metals to form *alkoxides: $2ROH + 2M = 2MOR + H_2$. They also react with acids to yield *esters and water. With oxidizing agents, primary alcohols give aldehydes: $RCH_2OH + O = RCOH + H_2O$; secondary alcohols give ketones: $RCH(OH)R' + O = RCOR' + H_2O$. Concentrated sulphuric acid and other dehydrating agents produce alkenes: $RCH_2CH_2OH - H_2O = RCH:CH_2$.

alcohol thermometer. A type of *thermometer consisting of a small glass bulb containing alcohol, connected to a fine sealed capillary tube. The temperature is measured by the expansion of the alcohol, which is usually coloured with red dye. Alcohol has a lower freezing point than mercury and the instrument is useful at low temperatures.

aldehyde. Any of a class of organic compounds containing the group -CHO (the *aldehyde group*), consisting of a carbonyl group bound to a hydrogen atom. Aldehydes are produced by oxidizing alcohols and are further oxidized to carboxylic acids: $RCO.H + O = RCO.OH$. Aldehydes are reducing agents and give a positive reaction with *Schiff's reagent, *Tollen's reagent, and *Fehling's solution. They also have a variety of addition and condensation reactions.

aldol. 1. Any organic compound containing a hydroxyl group and an aldehyde group bound to adjacent carbon atoms, i.e. containing the *aldol group*, -CH(OH)CH(CHO)-.
2. *See* acetaldol.

aldose. A *sugar that is an aldehyde in its straight-chain form.

Alfvén waves. Magnetohydrodynamic waves propagated through a plasma under certain conditions. Alfvén waves travel in the direction of the magnetic field, and the particles of plasma oscillate in a plane perpendicular to this direction. The motion of particles is communicated both by collisions and by interaction with electric and magnetic fields. [After Hannes Alfvén (b. 1908), Swedish physicist.]

algebra. 1. The branch of mathematics concerned with the general properties of numbers. Algebra is a generalization of arithmetic, and achieves its generality by investigating relations between different representations of numbers treated abstractly. Letters are employed to represent unspecified numbers.
2. Any abstract mathematical system involving formal rules for operations on and relations between entities. The entities can be vectors, matrices, sets, logical propositions, etc.

algebraic. Of or relating to algebra or the rules of algebra. The *algebraic sum* of a set of numbers is the result of their addition with due regard to sign; for example the algebraic sum of 25 and −30 is −5 (not 55).

algin. *See* alginic acid.

alginate. *See* alginic acid.

alginic acid. A polymeric gelatinous substance extracted from certain seaweeds. Alginic acid has a chain similar to that in carbohydrates with carboxyl groups attached. Its salts, the *alginates*, form viscous gelatinous solutions. Sodium alginate (*algin*) is used as an emulsifier and as a thickener in some foods. Calcium alginate can be spun into fibres.

algorithm. A procedure or formula applied mechanically in order to carry out a computation. For example, *Euclid's algorithm* is a mechanical method for calculating the highest common factor of two large numbers, by dividing the larger by the smaller, then dividing the smaller by the remainder, then dividing the first remainder by the second, and repeating this procedure until one division is exact. When this occurs, the last divisor is the highest common factor.

alicyclic. Denoting a chemical compound that is both *aliphatic and *cyclic. Cyclohexane is a common example.

alidade. *See* astrolabe.

aliphatic. Denoting an organic compound that is not *aromatic: i.e. does not have the characteristic chemical behaviour of benzene. Aliphatic compounds include the alkanes, alkenes, alkynes, cycloalkanes, and their derivatives.

alizarin (1,2-dihydroxyanthraquinone). An orange-red crystalline solid, $C_6H_4(CO)_2C_6H_2(OH)_2$, used as a dyestuff. It was formerly extracted from madder roots but is now synthesized. M.pt. 289°C; b.pt. 430°C.

Alizarin

alkali. A *base that is soluble in water, thus producing a solution of *hydroxide ions.

alkali metals. The metallic elements lithium, sodium, potassium, rubidium, caesium, and francium, belonging to group IA of the periodic table. They tend to be soft silvery metals, which are univalent, electropositive, and highly reactive. Their oxides and hydroxides are strongly alkaline and their salts are typical ionic crystalline solids containing M^+ ions.

alkalimetry. Determination of the amounts of alkalis in solution by titration.

alkaline-earth metals. The metallic elements beryllium, magnesium, calcium, strontium, and barium, belonging to group IIA of the periodic table. They are silvery-white metals, which are divalent and electropositive and are not quite as reactive as the alkali metals. Like the alkali metals they form basic oxides and hydroxides and crystalline ionic salts. Strictly speaking the **alkaline earths** are the oxides of these metals, although the term is often used for the metals themselves.

alkaloid. Any of a class of organic compounds found in plants, characterized by their physiological activity. Typically alkaloids are poisonous insoluble crystalline compounds containing oxygen and nitrogen. They are basic, forming soluble crystalline salts. Typical examples are quinine and strychnine.

alkane (paraffin). Any of the saturated hydrocarbons with general formula

C_nH_{2n+2}. The alkanes form a *homologous series starting with methane, CH_4. The first four alkanes are gases and higher members of the series are liquids or colourless waxes. Alkanes tend to be rather unreactive: they can be converted into alkyl halides by *substitution.

alkanization. Hydrogenation of an unsaturated hydrocarbon, such as an alkene, to produce an alkane.

alkene. Any one of a class of hydrocarbons characterized by the presence of double bonds between carbon atoms. The simplest example is *ethylene. Alkenes can be prepared by treatment of an alkyl halide with alcoholic potassium hydroxide: $C_2H_5Cl + KOH = CH_2:CH_2 + KCl + H_2O$. Other common methods are the dehydration of alcohols, dehalogenation of vicinal dihalides, and reduction of alkynes. The main type of reaction of alkenes are *addition reactions. For example, under acid conditions: $CH_2:CH_2 + H_2O = CH_3CH_2OH$. Under some conditions substitution can also occur: $CH_3CH_2: CH_2 + Cl_2 = CH_2ClCH_2:CH_2$. Alkenes also react with ozone to form an unstable intermediate which breaks down to give aldehydes or ketones: $RCH:CHR' + O_3 = RCOH \cdot + R'COH$. Alkenes with one double bond form a homologous series with general formula C_nH_{2n}, called the *ethylene series*.

alkyd resin. See polyester resin.

alkyl. An organometallic compound in which a metal atom is directly bound to an alkyl group. Tetraethyl lead, $Pb(C_2H_5)_4$, is a common example. The aluminium alkyls, with general formula $Al(C_nH_{2n+1})_3$, are also commercially important, being used in Ziegler-Natta catalysts. Alkyls can be made by direct reaction of a metal with an alkyl halide.

alkylation. The introduction of an alkyl group into an organic compound in a chemical reaction. Usually alkylations involve addition of an alkyl group to an alkene, or substitution of an alkyl group into an aromatic ring by a *Friedel-Crafts reaction.

alkylbenzene. Any hydrocarbon containing an aromatic group bound to an alkyl group, and having the chemical properties of both.

alkyl group. A univalent organic group derived by removing one hydrogen atom from an aliphatic hydrocarbon, usually an alkane. An example is the methyl group, CH_3-.

alkyne. Any one of a class of hydrocarbons characterized by the presence of triple bonds between carbon atoms.

allene. A colourless gaseous hydrocarbon, $CH_2:C:CH_2$.

allotropy. The existence of a solid compound or element in more than one physical form (allotropes). Diamond and graphite, for example, are allotropes of carbon.

Alloxan

alloxan. A white crystalline heterocyclic compound, $C_4H_2O_4N_2$. M.pt. $170°C$ (decomposes).

alloy. A metallic substance consisting of a mixture of two or more metals or of a metal with nonmetals. Alloys have properties that differ from those of their components. They may be solid solutions, in which the material is homogeneous, or may contain small inclusions of one phase in a matrix of another phase.

allyl alcohol (prop-2-en-1-ol). A colourless poisonous liquid, $CH_2:CHCH_2OH$, with a pungent mustard-like odour, used in the manufacture of plasticizers and synthetic resins. M.pt. $-129°C$; b.pt. $97°C$; r.d. 0.85.

allyl group. The organic group CH_2:-$CHCH_2$-.

alpha decay. A type of radioactive decay in which an unstable nucleus ejects on alpha particle. The product has a mass number that has decreased by 4 and an atomic number that has decreased by 2. Radium-226, for example, with an atomic number of 88, decays to give radon-222, which has an atomic number of 86. The radiation emitted during alpha decay follows the *Geiger-Nuttall law. *See also* radioactivity.

alpha iron. The allotropic form of iron that is stable below 906°C. It has a body-centred cubic crystal structure and is ferromagnetic up to its Curie temperature (768°C). *See also* beta iron, ferrite.

alpha particle (α-particle). A helium-atom nucleus, containing two protons and two neutrons. Alpha particles are emitted in streams *(alpha rays)* by radioactive decay.

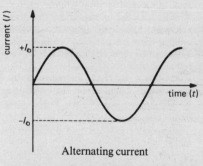

Alternating current

alternating current (a.c.). An electric current that varies in strength, periodically reversing its direction. The commonest practical form is a sinusoidal current, in which the value of the current follows the equation $I = I_0 \sin 2\pi ft$. I_0 is the peak value of the current, f is the *frequency, and t is the time.

alternating series. An infinite series whose elements are alternately positive and negative, e.g. $1 - \frac{1}{2} + \frac{1}{4} - \frac{1}{6} + \frac{1}{12} -.....$
Such a series converges if the moduli of the successive terms decrease and tend to zero.

alternator. A device for producing an alternating current. *See* generator.

altitude. 1. A celestial coordinate that determines the angular distance of a star, planet, etc., above the horizon.
2. The height of a geometric figure or solid measured perpendicular to the base.

alum. Any of a class of isomorphous hydrated crystalline double salts having the formula, $M_2SO_4.M'_2(SO_4)_3.24H_2O$, where M is a monovalent metal and M' a trivalent metal. The term is often applied to the double sulphate of potassium and aluminium, $K_2SO_4.Al_2(SO_4)_3.24H_2O$, known as *potash alum.* The potassium can be replaced by other monovalent ions as in *ammonium alum,* $(NH_4)_2SO_4.Al_2(SO_4)_3.24H_2O$. Similarly the term alum is used for double salts in which the aluminium has been replaced by a trivalent metal, as in *chrome alum,* $K_2SO_4.Cr_2(SO_4)_3.24H_2O$, or *ammonium ferric alum,* $(NH_4)_2SO_4.Fe_2(SO_4)_3.24H_2O$. Alums are prepared by crystallizing solutions of the sulphates mixed in the correct proportions. In strict chemical usage they are named according to the rules for double salts: aluminium potassium sulphate, etc.

alumina. *See* aluminium oxide.

aluminium. Symbol: Al. A silvery ductile metallic element obtained from *bauxite; the most abundant metal in the earth's crust. It is produced by converting the bauxite to aluminium oxide by addition of caustic soda and heating the precipitated aluminium hydroxide. A fused mixture of the oxide with cryolite is electrolysed to obtain the aluminium. It is extensively used in a wide range of light aluminium alloys, which also contain manganese, nickel, copper, zinc, and other metals and are used in aircraft, ships, machinery, cooking utensils, etc. It is also a good conductor of electricity. Chemically, aluminium is a reactive metal, forming a film of transparent oxide in air, which protects against further oxidation. It also reacts with other nonmetals. Aluminium is more typically metallic than boron, the first member of its group in the periodic table, and forms ionic crystalline salts such as aluminium sulphate. The halides, however, are volatile compounds, dimeric in the vapour, and seem to be covalently bonded indicating a degree of nonmetallic behaviour. A.N. 13; A.W. 26.9815; m.pt. 659.7°C; b.pt. 1800°C; r.d. 2.6; valency 3.

aluminium chloride. A white or yellow crystalline solid, prepared by reacting chlorine with molten aluminium and used in *Friedel-Crafts reactions. The vapour is dimeric, Al_2Cl_6. M.pt. 190°C (2.5 atm); sublimes at 178°C; r.d. 2.44.

aluminium hydroxide. A white odourless tasteless powder, also obtainable as a gel, $Al(OH)_3$. A hydrated *amphoteric oxide, it is prepared by adding alkali to a solution of an aluminium salt. It is used as a mordant for dyes and as a gastric antacid.

aluminium oxide (alumina). A white powder, Al_2O_3. An *amphoteric oxide that occurs naturally as corundum. Emery, ruby, and sapphire are impure crystalline forms. Aluminium oxide is obtained commercially from bauxite and its principal use is in the manufacture of aluminium. It is also used in abrasives, ceramics, and chromatography. M.pt. 2030°C; r.d. 3.4–4.0.

aluminium sulphate. A white crystalline salt, $Al_2(SO_4)_3$ or $Al_2(SO_4)_3.18H_2O$, made by treating bauxite with sulphuric acid. It is used in sizing paper, as a mordant, and in making fireproof and waterproof materials. Decomposes before melting; r.d. 2.71 (anhydrous).

alunite. A white or pink mineral consisting of a basic sulphate of aluminium and potassium, $K_2SO_4.Al_2(SO_4)_3.4Al(OH)_3$, used as a source of potash alum.

amalgam. An alloy of mercury with one or more other metals. Amalgams are either liquid or solid, some being simple mixtures and others containing definite compounds (such as Hg_2Na). Most metals will form amalgams: iron and platinum are notable exceptions.

amatol. An explosive consisting of a mixture of ammonium nitrate and TNT.

ambergris. A black or grey waxy substance found in the intestinal tract of the sperm whale. Mainly cholesterol, it contains fatty oils and steroids and has a musky scent, hence its use in perfumery.

americium. Symbol: Am. A silvery-white transuranic *actinide element made by neutron bombardment of plutonium. The most stable isotope, ^{243}Am, has a half-life of 8800 years. A.N. 95; r.d. 11.7; valency 3, 4, 5, or 6.

amide. 1. Any of a class of organic compounds with the general formula $RCONH_2$, where R is an organic group and $-CONH_2$ is the *amide group*. Amides can be prepared by reacting acid chlorides with ammonia. Unlike amines they are not basic compounds. They can be reduced to amines by some reducing agents, $RCONH_2 + 2H_2 = RCH_2NH_2 + H_2O$; dehydrated by phosphorus pentoxide to nitriles, $RCONH_2 = RCN + H_2O$; and hydrolysed to acids, $RCONH_2 + H_2O = RCOOH + NH_3$. 2. An inorganic compound containing the $-NH_2$ group or NH_2^- ion, as in sodamide, $NaNH_2$.

amination. A chemical reaction in which an aldehyde or ketone is converted into an amine by reaction with hydrogen and ammonia in the presence of a catalyst. For example, an aldehyde may be converted thus: $RCHO + NH_3 + H_2 = RCH_2NH_2 + H_2O$. Nickel is a typical catalyst for this type of reaction. *Compare* ammonolysis.

amine. Any of a class of organic compounds derived by replacing one or more of the hydrogen atoms of ammonia with an alkyl or aryl group. If one hydrogen is replaced, a *primary* amine results having a formula RNH_2. *Secondary* amines have the general formula $RR'NH$ and *tertiary* amines have the formula $RR'R''N$. Methods of preparing amines include reduction of *nitro compounds, *nitriles, and amides. They are basic substances, forming quaternary salts, and also react with nitrous acid to give *diazo compounds.

amino acid. Any of a class of organic compounds that contain both an amino group and a carboxyl group in their molecules. Glycine, H_2NCH_2COOH, is the simplest example. Amino acids are the units present in *peptides and *proteins. Most proteins contain amounts of 23 amino acids. *Essential* amino acids are those that have to be present in the diet of an animal for conversion into protein; *nonessential* amino acids can be synthesized in the body. There are eight essential amino acids in the diet of man.

aminoethane. *See* ethylamine.

amino group. The group $-NH_2$, as present in *amines.

aminoplastic. *See* amino resin.

amino resin (aminoplastic). Any of a class of synthetic resins made by copolymerizing an aldehyde, usually formaldehyde, with an amine. The most important amino resins are *melamine and *urea-formaldehyde.

ammeter. An instrument for measuring an electric current. The *moving-coil instrument is fairly sensitive but can be used only for direct current. Alternating currents are measured by the *moving-iron instrument, which is less sensitive and has a nonlinear scale. Both types have very low resistances.
At high frequencies *hot-wire ammeters* are used. The current to be measured passes along a thin resistance wire, causing a rise in temperature. This is detected by a thermocouple or by the resulting expansion of the wire.

ammine. An inorganic metal complex containing coordinated ammonia molecules, as in copper ammine, $[Cu(NH_3)_4]^{2+}$.

ammonal. An explosive consisting of a mixture of ammonium nitrate and aluminium powder.

ammonia. A colourless pungent toxic gas, NH_3, manufactured by methods that have evolved from the *Haber process of direct catalytic combination of nitrogen and hydrogen. It is used in the manufacture of fertilizers, nitric acid, explosives, and synthetic fibres. Ammonia is often used in aqueous solution, which is alkaline as a result of hydroxide ions (OH^-) being produced with ammonium ions (NH_4^+). The compound NH_4OH (*ammonium hydroxide*) is unstable and cannot be isolated. M.pt. $-78\,^\circ$C; b.pt. $-33.5\,^\circ$C.

ammoniacal. Denoting the presence of ammonia. The term is usually used to describe solutions in aqueous ammonia.

ammonia soda process. *See* Solvay process.

ammonium alum. *See* alum.

ammonium carbonate. A white powder with an odour of ammonia, made by heating ammonium chloride with calcium sulphate. It is a mixed salt of ammonium bicarbonate and ammonium carbamate, $(NH_4)HCO_3.(NH_4)CO_2NH_2$. It is used in baking powders and smelling salts. The true carbonate, $(NH_4)_2CO_3$, can be produced by treating this with ammonia. It is an unstable crystalline solid.

ammonium chloride (sal ammoniac). A white crystalline solid, NH_4Cl, manufactured as a by-product in the *Solvay process. It is used in batteries and as a pickling agent in zinc coating and tinning. Sublimes without melting; r.d. 1.54.

ammonium hydroxide. *See* ammonia.

ammonium ion. The monovalent ion NH_4^+ produced by protonation of ammonia. Ammonium salts contain these ions: ammonium chloride, for example, is NH_4^+ Cl^-.

ammonium nitrate (Norway saltpetre). A colourless odourless hygroscopic crystalline solid, NH_4NO_3, used in fertilizers and explosives. M.pt. 169°C; decomposes at 210°C; r.d. 1.73.

ammonium sulphate. A white crystalline solid, $(NH_4)_2SO_4$, used in fertilizers and in making ammonium alum. M.pt. 513°C (decomposes); r.d. 1.77.

ammonolysis. The reaction of an organic compound with liquid ammonia to exchange one of its groups for an amine group. Acyl halides, for example, produce amides: $RCOCl + 2NH_3 = RCONH_2 + NH_4Cl$. Ammonolysis is a type of *solvolysis. Compare amination.

amorphous. Denoting a solid substance with no regular *crystalline structure. The term specifically refers to the lack of order in the arrangement of atoms in the solid—many apparently amorphous substances are composed of minute crystals and are in fact crystalline. True amorphous solids are uncommon: examples are soot (carbon), flowers of sulphur, and *glass.

amount of substance. Symbol: n. A measure of quantity proportional to the number of particles of substance present. The amount of substance is not the same as the mass. It is always used with reference to a specified type of particle. Thus the amount of substance of an element is proportional to the number of atoms present; the amount of substance of light is proportional to the number of photons, etc. The constant of proportionality is the *Avogadro constant. Amount of substance is always measured in *moles.

ampere. Symbol: A. The basic *SI unit of electric current, equal to the current that, when passed through two parallel conductors of infinite length and negligible circular cross section placed one metre apart in a vacuum, produces a force of 2×10^{-7} newton per sq. metre

between them. Formerly, this was the *absolute ampere*, taken to be one tenth of the *CGS-electromagnetic unit of current. The international unit of current, the *international ampere*, was defined in 1893 as the current that, when passed through a solution of silver nitrate under specified conditions, deposits silver at a rate of 0.001 118 gram per second. In 1948 the international ampere was redefined as 0.999 850 A. [After André-Marie Ampère (1775-1836), French physicist.]

ampere-hour. A unit of quantity of electricity (charge) equal to the amount flowing in one hour as the result of a steady current of one ampere. It is equal to 3600 coulombs.

Ampère's law. The formula $dB = (\mu_o I \sin \theta. dl)/4\pi r^2$, giving the elemental magnetic flux density dB produced by a current I at a point a distance r from a conductor element dl. A line from the point to the conductor makes an angle θ with the current direction and μ_o is the *magnetic constant.

ampere-turn. Symbol: A or AT. A unit of magnetomotive force equal to the magnetomotive force produced by a current of one ampere flowing through one turn of a coil.

amphetamine. A colourless mobile liquid amine, $C_6H_5CH_2CH(NH_2)CH_3$. Its salt, amphetamine sulphate, is used as a nasal decongestant and as a stimulant. B.pt. 200-203°C; r.d. 0.91.

ampholyte ion. See zwitterion.

amphoteric. Capable of reacting both as an acidic substance and a basic substance. For example, with strong acids aluminium hydroxide acts as a base: $Al(OH)_3 + 3HCl = AlCl_3 + 3H_2O$. With strong bases it acts as an acid to form aluminates: $NaOH + Al(OH)_3 = NaAlO_2 + 2H_2O$.

amplifier. An electronic circuit for increasing the voltage of an input signal.

amplitude. 1. The maximum departure of an oscillating system from its equilibrium value. The amplitude of an alternating current, for instance, is the peak value of the current. The amplitude of a pendulum is the maximum distance it swings from its central position.
2. (of a complex number) The angle subtended with the x axis by the line joining the representation of the given point on the *Argand diagram with the origin. *See also* argument.

amplitude modulation. *See* modulation.

amu. *See* atomic mass unit.

amyl acetate. A colourless flammable liquid with a characteristic odour of pear drops. It is a mixture of esters with the formula $CH_3COOC_5H_{11}$, prepared from amyl alcohol and used chiefly as a solvent for lacquers and paints. The pure isomers can also be obtained.

amyl alcohol. Any of eight aliphatic alcohols that are structural isomers with the formula $C_5H_{11}OH$. They are distinguished as n-*amyl alcohol*, sec-*amyl alcohol*, etc. The amyl alcohols are now more usually named according to the main chain of carbon atoms: *pentan-1-ol*, *pentan-2-ol*, etc. The term is still used for commercial mixtures of alcohols.

amyl group. An organic group with the formula $C_5H_{11}-$. Several isomeric forms exist.

analog computer. A type of computer in which problems are solved by using a mechanism or electric circuit that has the same behaviour as the system investigated. Analog computers are used for solving differential equations or for investigating or simulating complex physical systems. Usually they employ electrical circuits in which variables are represented by currents or voltages input to a network of amplifiers and other components. The network is designed so that the relationship of the output signal to the input signal is given

by the equation under investigation. A trivial example would be a single fixed resistance R through which is passed a current I. The potential difference across the resistor is given by IR and in principle the circuit could be used for multiplying an input variable I by a constant factor R. Mechanical devices operate by gears, levers, etc., and variables are represented by distances. In analog computers the input is continuous rather than discrete. *See also* digital computer, hybrid computer.

analysis. Determination of substances or elements present in a sample. *Qualitative analysis* is simply the identification of a substance or of the constituents of a mixture. *Quantitative analysis* measures the amounts of each substance present. Many different chemical and physical analytical techniques are in use. *See* volumetric analysis, spectrographic analysis, colorimetric analysis, chromatography.

analytic geometry (coordinate geometry). A form of geometry using a system of *coordinates, in which lines and curves can be represented by equations and the properties of shapes can be deduced by algebraic reasoning.

aneroid barometer. *See* barometer.

anethole. A white crystalline derivative of anisole, $CH_3CH:CHC_6H_4OCH_3$, found in some essential oils. It is used in perfumery. M.pt. 22°C; b.pt. 234-7°C; r.d. 0.98.

angle. The figure formed between two intersecting lines. The amount of rotation between such lines is measured in degrees (360° to a complete turn) or in radians (2π to a circle). The angle between two plane surfaces is measured between the perpendicular lines on each to a point on the line of intersection of the surfaces.

angstrom (angstrom unit). Symbol: Å. A unit of length equal to 10^{-10} metre. It is used for expressing wavelengths of electromagnetic radiation and interatomic distances in molecules and crystals.

angular frequency. Symbol: ω. Frequency measured in radians per unit time, given by $2\pi f$, where f is the frequency in hertz.

angular momentum. *See* momentum.

angular velocity. Symbol: ω. The rate of rotation of a body expressed as the angle turned through per unit time. It is usually measured in radians per second.

anhydride. 1. A compound that gives a specified compound when dissolved in water. Sulphur trioxide, for example, is the acid anhydride of sulphuric acid: $SO_3 + H_2O = H_2SO_4$.
2. Any of a class of organic compounds containing the -CO.O.CO- group (the *anhydride group*). Anhydrides are produced by abstracting water from carboxylic acids using dehydrating agents. For instance, two molecules of acetic acid condense to give acetic anhydride: $2CH_3COOH - H_2O = CH_3CO.O.-COCH_3$. Cyclic anhydrides are also formed as in *maleic anhydride. Anhydrides react with water to give the parent acid. They also react with amines and alcohols, replacing a hydrogen atom with an acyl group (*see* acylation).

anhydrite. A naturally occurring anhydrous form of calcium sulphate, $CaSO_4$, used in making sulphuric acid.

anhydrous. Without water. The term is often used to describe salts that have no water of crystallization, as in anhydrous *copper sulphate.

aniline. A colourless oily flammable liquid amine, $C_6H_5NH_2$, prepared by the reduction of nitrobenzene. It is used as the parent substance for many dyes, plastics, and pharmaceuticals. M.pt. $-6°C$; b.pt. $184°C$; r.d. 1.02.

anilinium ion. The positive ion $[C_6H_5NH_3]^+$, obtained by protonating aniline. Anilinium salts containing this ion are produced by reaction of aniline with acids, as in the reaction yielding anilinium chloride (aniline hydro-chloride): $C_6H_5NH_2 + HCl = C_6H_5NH_3Cl$.

animal charcoal (bone black). A form of carbon obtained by heating bones. It contains large amounts of calcium phosphate. Activated animal charcoal is used as an absorbent for decolorizing, etc.

anion (negative ion). An *ion with a negative charge. Such ions are attracted to the anode during electrolysis. *Compare* cation.

anisole (methoxybenzene). A colourless liquid aromatic ether, $C_6H_5OCH_3$, used chiefly as a solvent. M.pt. $-37°C$; b.pt. $155°C$; r.d. 0.99.

anisotropic. Having different properties in different directions. For example, an anisotropic crystal has different physical properties (refractive index, modulus of elasticity, etc.) along different axes because of its crystal structure. *Compare* isotropic.

annealing. The process of heating a solid and cooling it slowly so as to remove strain and crystal imperfections.

annihilation. Collision of a particle and its antiparticle resulting in conversion of the particles into electromagnetic radiation (*annihilation radiation*). For example, an electron and a positron can collide to yield two gamma-ray photons. The total energy of the electromagnetic radiation is equivalent to the total mass of the particles according to the law of *conservation of mass-energy. *See also* pair production.

annulus. 1. A two-dimensional hollow ring; the plane figure bounded by two concentric circles. **2.** A solid equivalent to a solid cylinder with a hole along its axis.

anode. An electrode with a positive electric charge, as in a cell, thermionic valve, rectifier, etc. *Compare* cathode.

anodizing. The process of forming an oxide layer on aluminium by making it the anode in an electrolytic bath containing sulphuric or chromic acid solu-

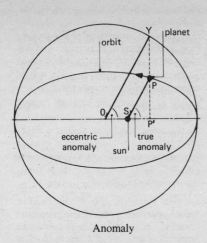

Anomaly

tion. The layer of aluminium oxide is porous and can be coloured by certain dyes.

anomaly. An angular measure used in determining the position of a planet in its orbital path about the sun. *True anomaly* is the angle between the planet's *perihelion, the sun, and the planet itself measured in the direction of the planet's motion. *Mean anomaly* is the angle between the perihelion, the sun, and an imaginary planet having the same period as the planet in question but moving at a constant velocity (*see* Kepler's laws). A third type, *eccentric anomaly*, is based on the planetary orbit conceived as a circle having the major axis of the planet's true *elliptical* orbit as its diameter. The eccentric anomaly is the angle P'OY in the illustration.

antenna. *See* aerial.

Anthracene

anthracene. A colourless crystalline solid tricyclic aromatic hydrocarbon,

Anthraquinone

$C_{14}H_{10}$, obtained from coal tar and used as an important source of dyestuffs. M.pt. 218°C; b.pt. 340°C; r.d. 1.25.

anthraquinone. A yellow crystalline solid quinone, $C_6H_4(CO)_2C_6H_4$, manufactured by a *Friedel-Crafts reaction between benzene and phthalic anhydride and used in the production of dyes. M.pt. 286°C; b.pt. 379.8°C; r.d. 1.44.

antichlor. A substance used for removing chlorine from a solution, material, etc.

antiderivative. The function whose derivative is the given function. Thus, the antiderivative of $2x$ is x^2 since the derivative of x^2 is $2x$. *See also* integration.

antiferromagnetism. A type of magnetic behaviour in which the material has a very low *susceptibility that increases with temperature. It is found in certain inorganic solids, such as MnO and FeO, having lattices in which adjacent atoms have antiparallel spins. Above a certain temperature, the *Néel temperature*, the susceptibility falls as the material becomes paramagnetic. *See also* magnetism.

antilogarithm. The number whose logarithm is the given number. Thus, if $x = 10^y$, so that $y = \log x$ to base 10, then x is the antilogarithm (written antilog, or $\log^{-1}$) of y to base 10.

antimatter. Hypothetical matter in which the atoms are composed of antiparticles. Thus, atoms of antimatter would have negatively charged nuclei containing antiprotons and antineutrons with orbiting antielectrons (positrons). If antimatter came into contact with ordinary matter, *annihilation would occur. Speculations that antimatter exists in the universe have not been substantiated.

antimonic. See antimony.

antimonous. See antimony.

antimony. Symbol: Sb. A brittle bluish-white metalloid element occurring in several minerals, especially stibnite, Sb_2S_3. It is extracted by roasting to produce the oxide, which is reduced with carbon. Besides metallic antimony there are three other nonmetallic allotropes. The element is used for increasing the hardness of lead that is used in accumulators, type metal, and sheathing cables. The chemistry of antimony is similar to that of arsenic. It forms covalent antimony III (*antimonous*) and antimony V (*antimonic*) compounds. Ionic antimony III compounds, containing the Sb^{3+} ion, also exist. In addition the element has *antimonyl* compounds, containing the SbO^+ ion. A.N. 51; A.W. 121.75; m.pt. 650°C; b.pt. 1380°C; r.d. 6.69.

antimony hydride. See stibine.

antimony oxide. Either of two amphoteric oxides of antimony. *Antimony (III) oxide* (antimony trioxide) is a white solid, Sb_2O_3, obtained by roasting antimony ores and used in flameproofing textiles, as a mordant, and as a white pigment (*antimony white*). M.pt. 450°C; r.d. 5.6. *Antimony (V) oxide* (antimony pentoxide) is a yellow solid, Sb_2O_5, made by the action of nitric acid on the metal. It decomposes to the III oxide on heating. R.d. 5.6.

antinode. A point of maximum displacement in a standing *wave. See node.

antiparallel. Designating vectors that have parallel lines of action and act in opposite directions.

antiparticle. An elementary particle that has the same mass as a given particle and a charge, baryon number, and isospin quantum number of the same magnitude but of opposite sign. Thus, the positron is the antiparticle of the electron; the antiproton, with negative charge, is the antiparticle of the proton. Collision between a particle and its antiparticle can result in *annihilation.

apatite. A naturally occurring phosphate of calcium, $Ca_5(PO_4)_3$, used as a source of phosphorus and of *superphosphate fertilizers.

aperture. The part of a mirror or lens that reflects or admits light in an optical system. The aperture of a lens or curved mirror is measured by the diameter of the refracting or reflecting surface.

aperture synthesis. A technique used in *radio astronomy to obtain observations on a large area of the sky using an array of small radio telescopes. The array of antennas is moved into a large number of different positions and observations made in each position are recorded. The data can be used to construct a map of radio activity over a wide area of the sky—the result is equivalent to using a single radio telescope with a very large aperture.

aphelion. The point of furthest recession from the sun attained by a planet, asteroid, comet, or other orbiting body. *Compare* perihelion.

aplanatic. Denoting a lens or other optical system that is free from spherical aberration.

apogee. The point at which the moon or any other orbiting satellite is furthest away from the earth. *Compare* perigee.

apothecaries' weight. A system of weights based on the grain. *See Appendix.*

Appleton layer. See ionosphere. [After Sir Edward Victor Appleton (1892-1965), English physicist.]

apsis. Either of two points (the *apsides*) at the ends of the major axis of an elliptical orbit. In the case of the moon's orbit, for example, one point is the perigee and the other the apogee.

Aquadag. ® A colloidal suspension of graphite in water, used as a lubricant and for producing conductive graphite coatings.

aqua fortis. See nitric acid.

aqua regia. A fuming yellow corrosive mixture of one part of nitric acid to three or four parts hydrochloric acid. It is used in metallurgy, since it dissolves all metals, even gold. The mixture contains chlorine and nitrosyl chloride (NOCl).

aqueous. Denoting a solution in which water is the solvent.

aqueous humour. See eye.

arc. 1. An open segment of a curve.
2. A luminous electrical *discharge between two electrodes, characterized by a high current and a low potential difference. The electrodes are heated by the discharge and their evaporation helps maintain it.
3. Denoting the inverse of a circular or *hyperbolic function. Thus the arc tangent of a number is any angle whose tangent is the given number, denoted arctan x or $\tan^{-1} x$.

Archimedes' principle. The principle that if a body is partly or completely immersed in a liquid, its apparent loss in weight is equal to the weight of liquid displaced. [After Archimedes of Syracuse (287-212 B.C.).]

are. Symbol: a. A metric unit of area equal to 100 sq. metres. It is equivalent to 119.60 sq. yards.

area. A measure of a two-dimensional surface. Units of area are the square of any unit length, e.g. square centimetres (cm^2) or square feet (ft^2).

Argand diagram. A representation of a complex number as a point on a plane (see diagram). With Cartesian coordinates, the complex number $x + iy$ is represented by the point (x,y). In *polar coordinates, this point is (r,θ), where θ is the argument of the complex number and r the modulus. [After Jean Robert

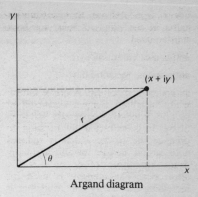

Argand diagram

Argand (1768-1822), French mathematician born in Geneva.]

argentic. See silver.

argentite (silver glance). A grey-black mineral consisting of silver sulphide, Ag_2S, used as an ore of silver.

argentous. See silver.

argol (tartar). A crude form of potassium hydrogen tartrate deposited as a reddish-brown solid on the side of wine vats.

argon. Symbol: A. A colourless odourless tasteless nonflammable gaseous element present in the atmosphere (0.94%) from which it is obtained as a by-product in the liquefaction of air. Argon is used for filling fluorescent lamps and electric-light bulbs, and in argon-arc welding. It is one of the *inert gases. A.N. 18; A.W. 39.948; m.pt. -189.3°C; b.pt. -185.8°C; r.d. 1.38 (air = 1).

argument. 1. (of a function). A value of an independent variable yielding some particular value of the function. The function $y = x^2$ takes the value 4 in the field of rational numbers for arguments +2 and -2.
2. (of a complex number). The angle between the line joining the representation of a complex number on the *Argand diagram with its origin, and the x-axis, normally measured in radians. The complex number can also be represen-

ted uniquely in terms of its argument and its *modulus. If $x + iy$ has modulus θ, then tan $\theta = y/x$.

Ariel. *See* Uranus.

arithmetic. The elementary study of the properties of and relations between the integers or the rational numbers under the operations of addition, subtraction, multiplication, and division.

arithmetic mean. The result of dividing the sum of a given set of numbers by the number of numbers in the set.

arithmetic progression. A *sequence in which each term differs from the preceding one by a fixed number. An arithmetic progression has the form a, $(a + d)$, $(a + 2d)$, ... $[a + (n - 1)d]$, where d is the *common difference* and n is the number of terms. The *arithmetic series* formed from this sequence has a sum equal to $(n/2)[2a + (n - 1)d]$. *Compare* geometric progression.

armature. 1. The part of a *generator in which the electric current is induced or the corresponding part of an *electric motor in which the driving current flows.
2. Any of various parts of electrical equipment in which an electric voltage is induced, as in a loudspeaker or record-player pick-up.
3. A piece of iron or steel held across the poles of an electromagnet so that it can be used to apply a force.
4. *See* keeper.

aromatic. Denoting a chemical compound that has the property of *aromaticity, as characterized by benzene. The term was originally used in a more literal sense to distinguish fragrant compounds from *aliphatic ("fatty" compounds).

aromaticity. The property of certain chemical compounds of having unsaturated molecules yet undergoing substitution reactions rather than addition reactions. Such compounds are said to be *aromatic*. The archetypal example is benzene, which, in the nineteenth cen-

tury, was the subject of considerable speculation. The basic problem—the *benzene problem*—was that of reconciling the formula of benzene, C_6H_6, with its chemical reactions. The empirical formula is the same as that of acetylene, C_2H_2, and it might be expected that benzene would undergo similar reactions. However, benzene does not show the usual behaviour of a compound containing double or triple bonds.

For example, acetylene adds bromine to yield CHBr:CHBr and eventually $CHBr_2CHBr_2$: benzene, with an iron III bromide catalyst, suffers displacement of one of its hydrogen atoms to yield C_6H_5Br. This type of activity indicates that benzene might be saturated and a postulated structure is that of *Ladenburg benzene* (see illustration).

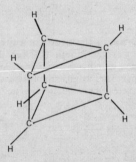

Fig. 1: Ladenburg benzene

Benzene, however, does not always act as a saturated compound. In sunlight bromine is added to give $C_6H_6Cl_6$ and hydrogen can also be added with a nickel catalyst to yield cyclohexane, C_6H_{12}. Some of the postulated structures accounting for this behaviour are shown in the first illustration.

In 1865 the German chemist August Kekulé (1829–96) suggested a structure with alternate double and single bonds in a hexagonal ring. To account for the fact that benzene has only three disubstitution products, he further proposed that the positions of the bonds oscillate, so that two molecules are in equilibrium. This structure—the *Kekulé formula*—is the one usually used in formulas of compounds containing benzene rings.

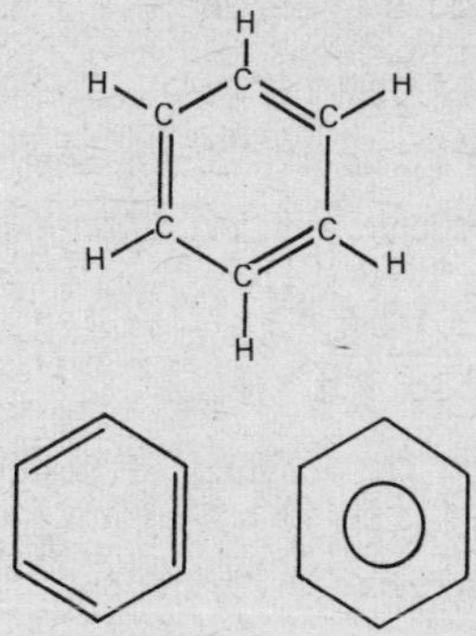

Fig. 2: Kekulé formula of benzene

The modern idea of aromaticity is based not on equilibrium between Kekulé structures but on *resonance between them. The bonds in benzene have characters between double and single bonds: the carbon atoms are held together by six single bonds and the remaining six electrons, from the double bonds, are delocalized over the ring. This is the reason benzene has all its C-C bonds of the same length and undergoes electrophilic substitution reactions. The German chemist Erich Hückel (b. 1896) suggested that for a compound to be aromatic it had to have a planar ring of alternating double and single bonds with $4n + 2$ pi electrons available from the double bonds for delocalization. This is the *Hückel rule*. In benzene, $n = 1$: i.e. there are 6 pi electrons delocalized. The Hückel rule is often used as the definition of aromaticity. *See also* pseudoaromatic.

aromatization. The conversion of an *aliphatic ring compound into an *aromatic compound by dehydrogenation. A typical method is catalytic dehydrogena-

tion using a catalyst such as nickel, platinum, or palladium. Cyclohexane, for example, will yield benzene: $C_6H_{12} = C_6H_6 + 3H_2$.

Arrhenius equation. The equation $v = A \exp (E_A/RT)$, relating the rate constant v of a chemical reaction to the activation energy E_A. A is a constant for a given reaction. [After Svante Arrhenius (1859–1927), Swedish chemist.]

arsenate. Any salt of arsenic acid.

arsenic. Symbol: As. A silvery-grey brittle metalloid element occurring in several minerals, including orpiment, realgar, and arsenopyrite. It is also found in copper and lead ores, the main source being arsenic oxide recovered as a by-product in the smelting process. This is reduced with charcoal to give the metallic element. Two other allotropic forms exist: yellow arsenic and black arsenic. Metallic arsenic is used in some alloys and semiconductors. It is a reactive element, combining directly with the halogens and tarnishing in air. Like phosphorus it exhibits a valency of three (*arsenious* or *arsenous*) and five (*arsenic*). The term "arsenic" is sometimes loosely applied to the oxide, As_2O_3. A.N. 33; A.W. 74.9216; m.pt. 218°C (28 atm); sublimes at 615°C; r.d. 5.72 (metallic).

arsenic acid. A white solid tribasic acid, $H_3AsO_4 \cdot \frac{1}{2}H_2O$, made by the action of hot concentrated nitric acid on arsenic III oxide.

arsenic oxide. Either of two oxides of arsenic. *Arsenic (III) oxide* (arsenic trioxide) is a white odourless tasteless poisonous powder, As_2O_3, manufactured by roasting arsenopyrite ore. It is an amphoteric compound, the anhydride of arsenious acid, and is used in the preparation of other arsenic compounds. It is also used in insecticides, weedkillers, rat poison, and the manufacture of glass. Arsenic (III) oxide is often loosely called *arsenic.* Sublimes at 193°C; r.d. 3.87. *Arsenic (V) oxide* (arsenic pentoxide) is a white deliquescent amorphous solid, As_2O_5, prepared by oxidizing the lower oxide. It is used in dyeing, printing, and insecticides and is the anhydride of arsenic acid. M.pt. $315°C$; r.d. 4.09.

arsenic pentoxide. *See* arsenic oxide.

arsenic trioxide. *See* arsenic oxide.

arsenious. *See* arsenic.

arsenious acid. An acid, H_3AsO_3, obtained in aqueous solutions of arsenic III oxide.

arsenite. Any salt of arsenious acid.

arsenopyrite (mispickel). A white or grey mineral consisting of a mixed sulphide and arsenide of iron, FeS_2.-$FeAs_2$. It is an ore of arsenic and also contains small amounts of gold.

arsenous. *See* arsenic.

arsine. A colourless poisonous gas, AsH_3, with an odour of garlic. It is an unstable compound, decomposing readily to arsenic and hydrogen. M.pt. $-113.5°C$; b.pt. $-55°C$.

artificial silk. *See* rayon.

aryl. An organometallic compound in which a metal atom is directly bound to an aryl group.

arylation. The introduction of an aryl group into an organic compound in a chemical reaction.

aryl group. An organic group derived by removing a hydrogen atom from an *aromatic compound, as in the phenyl group, C_6H_5-.

Ascorbic acid

ascorbic acid (vitamin C). A white crystalline vitamin, $C_6H_8O_6$, present in many fruits and vegetables. It is also synthesized commercially from D-glucose. Vitamin C is an essential component of the diet of man for the prevention of scurvy. M.pt. $192°C$; r.d. 1.65.

asdic. *See* sonar.

ash. The nonvolatile inorganic residue left after combustion of a substance.

aspartic acid. A colourless crystalline optically active nonessential amino acid, $HOOCCH_2CH(NH_2)COOH$, occurring naturally in sugar cane and sugar beet. It is used in making detergents and fungicides. R.d. 1.66.

aspect. The longitudinal positions of two celestial objects relative to one another. *See* conjunction, opposition, quadrature.

aspirin (acetylsalicylic acid). A white crystalline substance, $CH_3COOC_6H_4C$-OOH, used to relieve headaches, neuralgia, etc. It is prepared by the acetylation of salicylic acid. M.pt. $132-6°C$; b.pt. $140°C$; r.d. 1.35.

asphalt. A solid or semi-solid mixture of bitumen with fine mineral matter. Asphalt deposits occur naturally in many parts of the world (notably Trinidad) and asphalt is also manufactured by mixing bitumen and minerals. It is used in paints, varnishes, roof coverings, and road surfaces.

associative. Denoting an operation that is not affected by the way in which the terms operated on are grouped. Multiplication and addition of integers are associative since $(n + m) + l = n + (m + l)$, etc.

astatine. Symbol: At. A radioactive element first prepared by bombarding bismuth with alpha particles. Minute amounts of astatine are also found naturally. The most stable isotope, ^{210}At, has a half-life of 8.3 hours. Little is known about the chemistry of astatine. It belongs to the *halogens and seems to be more metallic than iodine in its behaviour. A.N. 85; m.pt. 302°C; b.pt. 337°C.

asteroid (minor planet, planetoid). Any of a large number of small celestial bodies orbiting the sun, mainly between Mars and Jupiter. The largest of these rocky objects is Ceres with a diameter of 685 kilometres but very few asteroids have diameters greater than 100 km. The orbits of some are very eccentric, such as those of Adonis and Icarus, which approach the path of Mercury. It is estimated that there are about 100 000 asteroids. They may be the remains of a planet that disintegrated or could not be formed because of Jupiter's massive gravitational influence. Such a planet is allowed for by *Bode's law.

astigmatism. An *aberration of lenses and mirrors, in which the curvature is different in two mutually perpendicular planes, so that a single point off the axis of the system is focused as two different images in the form of short lines. The eye can suffer from astigmatism, a defect remedied by the use of toroidal spectacle lenses.

astrolabe. An ancient astronomical instrument used for plotting angular distances on the *celestial sphere, consisting of a graduated disc with an index arm (the alidade) that could be moved and sighted. The angular distance between two objects could be measured by lining up on each in turn and measuring the angle formed at the observer's position.

astronomical unit. Symbol: au. A unit of length equal to the mean distance of the earth from the sun (1.495 00 × 10^{11} metres). It is usually used for measurements of distance within the solar system.

astronomy. The science of the stars, planets, and other celestial bodies of the universe.

astrophysics. The branch of astronomy concerned with the properties and structures of stars and other celestial objects.

asymmetric carbon atom. See optical isomerism.

asymptote. A line to which a curve approaches as its limit as the value of either the function or its *arguments tend to infinity or to some other value for which the function is not defined. More generally, one curve may be an asymptote of another if the difference between the functions tends to zero.

atactic. See polymer.

athermancy. The property of being opaque to infrared radiation. Compare diathermancy.

atmolysis. The process of separating gases by their different rates of diffusion. If two gases are allowed to diffuse through a porous barrier, the one with lower molecular weight will pass through faster.

atmosphere. 1. The region of gas surrounding a planet or other heavenly body. The earth's atmosphere is divided into layers depending on temperature gradient or electrical properties (see illustration). The composition of dry air, by volume, is given in the table.

Atmosphere

composition of air by volume

nitrogen	78·08 %
oxygen	20·94 %
argon	0·933 %
carbon dioxide	0·03 %
neon	0·0018 %
helium	0·0005 %
krypton	0·0001 %
xenon	0·000009 %

In addition, air contains a variable amount of water vapour, dust particles, and usually small quantities of hydrocarbons, sulphur compounds, and other pollutants.
2. Symbol: atm. A unit of pressure equal to 101 325 pascals. Before 1954 the atmosphere was taken to be 760 millimetres of mercury.

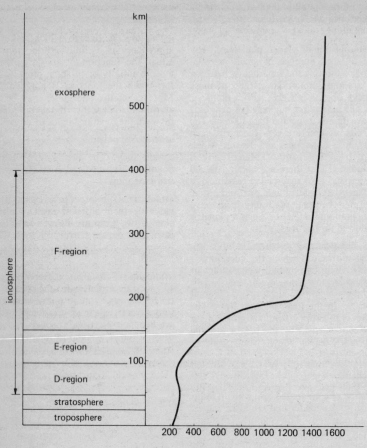

Atmospheric layers   temperature (K)

atom. One of the large number of particles that together make up a given element. An atom is the smallest amount of a chemical element that can take part in a chemical reaction. All atoms of the same element are identical. Atoms were originally thought of as small rigid indivisible particles (*see* Dalton's atomic theory). They are now known to be made up of simpler particles: protons, neutrons, and electrons. The atomic radius is of the order of 10^{-10} metre but almost all its mass is concentrated in a small central region—the *nucleus—with a radius of 10^{-15} metre. The nucleus has a positive charge and is composed of protons and, usually, neutrons. Negatively charged electrons move around this central region, the number of electrons being equal to the number of protons so that the atom is electrically neutral. An atom of a given element has a characteristic number of protons (or electrons)—this is the *atomic number*. Thus, hydrogen atoms have one proton, helium atoms have two, lithium three, etc. The element may also have different *isotopes, containing different numbers of neutrons in their nuclei.

In early models of the atom the electrons were thought of as moving in fixed circular or elliptical paths around the nucleus (see Bohr atom). The modern idea, resulting from wave mechanics, is that no precise path can be given. Instead, our knowledge of the location of electrons is based on probability and each electron is assigned to a region in space (see atomic orbital).

Whichever model is used, the orbital electrons in atoms occupy certain energy levels characterized by quantum numbers. First, the electrons are grouped into *shells denoted by letters K, L, M, N, etc. These are characterized by the *principal quantum number* (n), which may have values 1, 2, 3, 4, etc. Each shell can contain a maximum of $2n^2$ electrons: thus the K shell is full when it contains 2 electrons, the L shell has 8, the M shell 18, etc. The shells have *sub-shells characterized by an *azimuthal quantum number* (l). This gives the orbital angular momentum of the electron and has values 0, 1, 2, ... (n - 1). The first four values are denoted by letters s, p, d, and f respectively. The K shell has only one sub-shell (i.e. the sub-shell is the same as the shell). This is an s sub-shell. The L shell has two sub-shells: an s sub-shell and a p sub-shell. Within any given type of sub-shell, there may be more than one energy level (or atomic orbital). These are characterized by a *magnetic quantum number* (m) with values $-l$, $-(l - 1)$, ... 0 ...($l - 1$), l. If l = 0 (an s-orbital), m = 0. If l = 1 (a p-orbital), m = -1, 0, and 1: i.e. there are three p-orbitals within this sub-shell. Usually these are of identical energy but their energies differ in the presence of an external magnetic field. Similarly there are five d-orbitals and seven f-orbitals. Finally, an electron in an atom is characterized by a fourth quantum number describing its *spin: this can have values of either $+\frac{1}{2}$ or $-\frac{1}{2}$. Each level (or orbital) can

contain a maximum of two electrons with opposing spins.

The pattern of energy levels given by the four quantum numbers determines the *configuration of the atom of a particular element and explains the groupings of elements in the *periodic table.

atom bomb. See nuclear weapon.

atomic energy. Energy obtained by nuclear fission or fusion.

atomic heat. The heat capacity of one mole of a substance. See also Dulong and Petit's law.

atomicity. The number of atoms in a molecule of a given element. Carbon dioxide, for example, has an atomicity of three.

atomic mass unit (dalton). Symbol: amu. A unit of mass equal to one twelfth of the mass of a neutral carbon-12 atom. It is equivalent to 1.660 33 $\times$ 10^{-27} kilogram. This is the unit on the international scale. Formerly, atomic mass units were defined in terms of the mass of the oxygen atom. The physical atomic mass unit was one sixteenth of the mass of the neutral oxygen-16 atom. The chemical atomic mass unit was one sixteenth of the weighed average of the masses of the three naturally occurring isotopes of oxygen. These units were slightly different: 1 amu (international) is equal to 1.000 31 amu (physical) or to 1.000 04 amu (chemical). The atomic mass unit is not to be confused with the atomic unit of mass (see hartree unit). See also atomic weight.

atomic number (proton number). Symbol: Z. The number of protons in the nucleus of a given atom. The atomic number is also the number of electrons in the atom, thus determining the chemical properties of the element.

atomic orbital. A region around the nucleus of an atom in which an electron moves. According to wave mechanics, the electron's location is not described by a fixed orbit but by a probability

Shapes of atomic orbitals

distribution in space, given by the *wave function. In fact, there is a finite probability of finding the electron at any point: the orbitals are regions within which the electron may be found with a high degree of probability.

Each orbital has a fixed energy and contains a maximum of two electrons. It is characterized by three quantum numbers n, l, and m (see atom). The shape depends on the value of l (see illustration).

atomic weight. Symbol: A. The ratio of the average mass per atom of an element to one twelfth of the mass of an atom of the isotope carbon-12. An atomic weight is a ratio and has no units—it is also called the *relative*

atomic mass. It is numerically equal to the atom's mass expressed in *atomic mass units and was formerly defined in terms of the mass of the hydrogen atom and the oxygen atom.

atropine. A poisonous white crystalline alkaloid, $C_{17}H_{23}NO_3$, obtained from deadly nightshade and similar plants. It is used in medicine. M.pt. 114-6°C.

attenuation. Decrease in intensity of a signal, electromagnetic wave, sound wave, beam of particles, etc.

atto-. Symbol: a. Prefix indicating 10^{-18}: 1 attometre (am) = 10^{-18} metre.

audio frequency. A frequency in the range 20 to 20 000 hertz, this being the range in which the human ear can detect sound.

Auger effect. The phenomenon in which an excited ion decays to the ground state and ejects an electron, producing a doubly charged ion. An excited ion has a vacancy in one of its inner electron shells and this can be filled by transition of an electron from an outer shell. The resulting release of energy can be taken up by an emitted photon, as in X-ray fluorescence. The Auger effect is an alternative process in which the energy is used to ionize the singly charged ion. [After Pierre Auger (b. 1899), French physicist.]

auric. See gold.

aurora. A display of luminous arcs, curtains, and streamers of light, ever-changing in shape and intensity and varying in colour from whitish-green to deep red, seen in the sky in the regions of the poles. Apparently linked to *solar flares, the aurorae are caused by charged particles reaching the earth where they are trapped by the earth's magnetic field. They spiral towards the poles where they cause ionization of molecules in the atmosphere, thus producing light. In the north the effect is called the *aurora Borealis,* while in the south it is referred to as the *aurora Australis.*

aurous. See gold.

austenite. A solid solution of carbon or iron carbide in *gamma iron. It is produced in some steels by quenching from a high temperature.

autocatalysis. A type of *catalysis in which the catalyst is a product of the reaction being catalysed. The reaction starts slowly, then speeds up more and more as the amount of catalyst builds up.

autoclave. A thick-walled vessel, usually of steel, designed for carrying out chemical reactions at high temperatures and pressures.

autumnal equinox. See equinox.

avalanche. The process in which a large number of ions are produced by a single ionizing collision, as in a *Geiger counter.

Avogadro constant. Symbol: L or N_A. The number of molecules in one mole of any substance. It has the value 6.022 52 $\times$ 10^{23}. [After Amedeo Avogadro (1776–1856), Italian physicist.]

Avogadro's hypothesis. The principle that equal volumes of all gases at the same temperature and pressure contain the same number of molecules. See also ideal gas.

avoirdupois weight. The system of weights based on the pound. See Appendix.

axiom. A proposition assumed to be true or considered to be self-evident, used as a basis for logical deduction.

azeotropic mixture (azeotrope). A mixture of two liquids in such proportions that it boils without change in composition: i.e. the composition of the vapour is the same as that of the liquid. In general, when a mixture is distilled the more volatile constituent vaporizes first. When the composition reaches that of the azeotropic (or *constant-boiling*) mixture, the constituents distil together. Ethanol and water, for instance, have an azeotropic mixture containing 4.4% of water: further separation cannot be effected by simple distillation.

azide. 1. A salt of hydrazoic acid, containing the ion N_3^-. The azides of heavy metals are often explosive and are used as detonators.
2. An organic compound with the formula RN_3, where R is a hydrocarbon group.

azimuth. The angular distance of a vertical circle passing through a celestial object and the observer's zenith from the south point of the observer's horizon. The azimuth is measured westward from the south point, which is taken as 0°.

azimuthal quantum number. *See* atom.

azine. Any heterocyclic compound having a six-membered ring containing carbon and nitrogen atoms bound by alternate double bonds. Pyridine, C_5NH_5, is the simplest example and is sometimes simply called *azine*. Compounds containing two nitrogen atoms in the ring are *diazines, those with three nitrogen atoms are *triazines, etc.

azobenzene. A yellow or orange crystalline *azo compound, $C_6H_5N_2C_6H_5$, made by reducing nitrobenzene with iron and sodium hydroxide solution. Its main use is in the manufacture of dyes. M.pt. 68°C; b.pt. 297°C; r.d. 1.2.

azo compound. Any of a class of compounds containing the group -N:N- (the *azo group*) linking two aromatic groups. They are usually made by reacting a *diazonium salt with an aromatic phenol or amine. Azo compounds are usually brightly coloured. They form the basis of a class of dyes—*azo dyes*—which usually have a sulphonic acid group attached to one of the rings and are used in the form of their sodium salts.

azulene. An intensely blue crystalline terpene, $C_{16}H_{26}O$; the blue colouring matter of camomile and wormwood. It is used as an anti-inflammatory drug. M.pt. 98°C; b.pt. 170°C; r.d. 0.99.

B

Babbitt metal. Any of various alloys of tin (80-90%), antimony (7-11%), and copper (4-8%), used in bearings. [After Isaac Babbitt (1799-1862), U.S. inventor.]

Babo's law. The principle that the vapour pressure of a liquid is lowered when a solid is dissolved in it, by an amount proportional to the concentration of solid dissolved. *See also* Raoult's law. [After Clemens Heinrich von Babo (1818-99), German chemist.]

back e.m.f. An electromotive force opposing the flow of a current in a circuit. In cells a back e.m.f., reducing the cell's natural e.m.f., can be produced by *polarization. Back e.m.f.s can also arise by induction. In a circuit in which the current is changing, the back e.m.f. is caused by self-induction. Similarly, in a conductor moving in a magnetic field, as in an electric motor, there is a back e.m.f. induced by the changing flux. Induced e.m.f.s are produced according to *Lenz's law.

background. A residual signal in a counter or other detector of radiation observed in the absence of the radiation that is measured. The background signal is caused by cosmic rays, small amounts of radioactive material in the surroundings, etc.

baddeleyite. A black, brown, or yellowish naturally occurring oxide of zirconium, ZrO_2.

Bailey's beads. The phenomenon characterized as a necklace effect surrounding the dark body of the moon just before and after totality during a total solar eclipse. It is caused by sunlight catching the irregular lunar surface features and shining into valleys on the moon's limb. [After Francis Bailey (1774-1844), English astronomer.]

Bakelite. ⓡ Any of various synthetic *phenol-formaldehyde resins. [After Leo Hendrick Baekeland (1863-1944), Belgian chemist.]

baking powder. Any mixture used in baking as a substitute for yeast in generating carbon dioxide and thus causing the dough to rise. The commonest type is a mixture of sodium bicarbonate and either tartaric acid or cream of tartar.

baking soda. *See* sodium bicarbonate.

balance. An instrument for measuring weight, consisting of a rigid beam balanced on a fulcrum at its midpoint with a pan suspended from each end. The object to be weighed is placed in one pan and standard weights added to the other until the beam is horizontal. The accuracy of sensitive balances is increased by use of a *rider*—a small weight moved along the top of the balance beam along a calibrated scale. Laboratory balances have agate knife edges to reduce friction and are enclosed in a case to reduce errors from air currents and temperature changes. The beam has a vertical pointer attached to its midpoint to enable the balance position to be detected. Such instruments are capable of measuring to 0.001 g. Other instruments for weighing use elastic deformation, as in the *spring balance.

balanced reaction. *See* equation.

ballistic galvanometer. A galvanometer with a relatively long period of oscillation, for use in measuring an electric charge. The charge is allowed to flow quickly through the galvanometer. If the natural period of oscillation is long compared with the time for which the current flows, the current passes before the coil or magnet begins to move. In the case of a moving-coil instrument the initial deflection (or "throw") is proportional to the charge.

ballistic pendulum. A heavy pendulum used to measure the velocity of a fast projectile such as a bullet. It consists of a heavy freely suspended mass which the projectile hits—the momentum, and thus the velocity, being obtained from the subsequent deflection of the pendulum.

ballistics. The study of the propulsion and motion of projectiles.

Balmer series. A *spectral series in the visible emission of hydrogen with wavelengths given by $1/\lambda = R(1/2^2 - 1/m^2)$, where R is the *Rydberg constant and m is 3, 4, 5, etc. [After Johann J. Balmer (1825–98), Swiss physicist.]

balsam (oleoresin). Any of various substances consisting of a mixture of a natural *resin and an *essential oil,

exuded from certain trees and shrubs. Balsams are semi-solids or viscous liquids, usually with pleasant aromatic odours. They are used in medicine, perfumery, varnishes, and lacquers.

band spectrum. An emission or absorption *spectrum consisting of a series of bands, each of which is formed of a number of closely spaced lines. Band spectra are produced by molecules. Each band corresponds to a transition between electronic energy levels and the lines within bands result from different vibrational energy levels of the molecule.

band theory (of solids). The application of quantum mechanics to the energies of electrons in crystalline solids. In isolated atoms the orbiting electrons can have only certain fixed energies: i.e. they occupy discrete energy levels. In solids, these levels become bands of allowed energy made up of closely-spaced energy levels, each capable of containing a pair of electrons. The allowed energy bands are separated by bands of forbidden energy.

The band theory of solids explains differences in the electrical conductivity of different materials. When a solid conducts electricity, electrons move through the lattice under the influence of an applied electric field. According to quantum mechanics, they can only gain energy from the field in discrete amounts, corresponding to a difference between two energy levels. Thus in accelerating they make transitions from one energy level to a higher level. This can only occur if the energy band is not completely fitted by electrons: i.e. if there are higher vacant energy levels available, the situation occurring in metals and other good conductors. In electrical insulators the bands are filled and there are no vacant allowed energy levels for the electrons to occupy. *Semiconductors have filled energy bands, in which all the possible energy levels are occupied, but the forbidden band is narrow. In intrinsic semiconductors it is possible for electrons to gain enough thermal energy to move into a higher vacant band. The electrons excited to this band can then conduct electricity and further conduction occurs as a result of *holes left in the lower energy band. The conductivity increases with temperature. Extrinsic semiconductors owe their properties to the presence of energy bands due to impurities, lying in the forbidden gap between the filled band and the higher empty band. Impurities that accept electrons from the filled band create positive holes and produce p-type semiconductivity. Electron donors supply electrons to the empty band and cause n-type semiconductivity.

bandwidth. 1. The range of frequencies over which an amplifier or other electronic device operates. The bandwidth is taken to be the range over which the device performs at some specified fraction of its maximum performance.
2. The frequency range over which most of the power is transmitted in a radio transmission. The bandwidth is defined with reference to a specified fraction of the total power, usually 99%.

bar. A unit of pressure equal to 10^5 pascals. The bar was formerly defined as the pressure resulting from a force of one dyne acting on an area of 1 square centimetre.

Barff process. A process for protecting iron against corrosion by heating the hot metal in steam, thus forming a protective layer of black iron oxide: $2Fe + 3H_2O = Fe_2O_3 + 3H_2$. [After F. S. Barff, 19th-century U.S. engineer.]

barite (barytes, heavy spar). A white or colourless mineral consisting of barium sulphate, $BaSO_4$.

barium. Symbol: Ba. A silvery-white fairly malleable metallic element occurring principally in barite ($BaSO_4$) and witherite ($BaCO_3$). Produced by electrolysis of the fused chloride, it is used in some alloys and as a getter. Barium is one of the *alkaline-earth metals. A.N. 56; A.W. 137.34; m.pt. 850°C; b.pt. 1140°C; r.d. 3.78; valency 2.

barium carbonate. A poisonous heavy white powder, $BaCO_3$, used in rat poison, optical glass, ceramics, paints, and enamels. Barium carbonate occurs in nature as the mineral witherite. Decomposes at 1300°; r.d. 4.27.

barium chloride. A colourless poisonous crystalline salt, $BaCl_2.2H_2O$, used as an additive for lubricating oils. M.pt. 960°C (anhydrous); r.d. 3.10.

barium hydroxide. A white alkaline hydroxide, $Ba(OH)_2$, obtained as a crystalline solid or powder by adding water to barium oxide. The monohydrate, pentahydrate, and octahydrate can also be made. Barium hydroxide is used in sugar refining and in the manufacture of glass. M.pt. 78°C; r.d. 2.13.

barium oxide. A white or yellowish deliquescent basic oxide, BaO, obtained as a powder by fusion of barium sulphate with carbon in an electric furnace. It is used as a dehydrating agent. M.pt. 1920°C; r.d. 5.72.

barium peroxide. A grey-white powder, BaO_2, made by heating barium oxide in a stream of oxygen. The compound is used in the manufacture of hydrogen peroxide. M.pt. 450°C; r.d. 4.96.

barium sulphate. A white heavy insoluble salt, $BaSO_4$, occurring naturally as barite. It is used in paint pigments and in X-ray photography. M.pt. 1580°C; r.d. 4.25–4.5.

Barkhausen effect. The effect observed when a ferromagnetic material is magnetized, in which the magnetization does not increase continuously but in a series of small steps. These are interpreted as the alignment of magnetic domains. The Barkhausen effect can be demonstrated by winding two coils on an iron core. As the current in one is slowly increased, fluctuations, detectable by an earphone or oscilloscope, occur in the other. [After Heinrich

Barkhausen (1881-1956), German physicist.]

barn. Symbol: b. A unit of area equal to 10^{-28} sq. metre. It is used for expressing the cross section of atomic nuclei in scattering experiments.

barograph. A barometer that produces a continuous record of pressure with time: usually an aneroid barometer with the pointer carrying a pen that traces a curve on a moving chart.

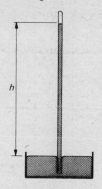

Fig 1: Mercury barometer

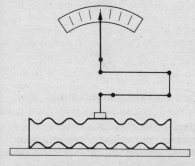

Fig 2: Aneroid barometer

barometer. An instrument for measuring atmospheric pressure. The *mercury barometer* consists of a long glass tube sealed at one end. It is filled with mercury and inverted in an open reservoir of mercury. The atmospheric pressure is the vertical height of mercury that the atmosphere will support (normally 760 mm).

The *aneroid barometer* contains a sealed metal box from which all the air has been removed. It has a corrugated top of very thin metal attached to a pointer by a train of levers. Changes in atmospheric pressure cause movements in the corrugated top, thus causing the pointer to move.

barometric formula. The formula $p = p_0 \exp(-mgx/kT)$, relating the pressure p to the height x above the earth's surface. p_0 is the pressure at the surface ($x = 0$), g the acceleration of free fall, k the Boltzmann constant, m the mass per molecule, and T the thermodynamic temperature, which is assumed to be constant.

barrel distortion. *See* distortion.

barycentre. *See* centre of mass.

baryon. Any of a class of *elementary particles that have half-integral spin and take part in *strong interactions. They generally have a mass larger than that of the *mesons. Baryons are *fermions: they include the proton, the neutron, and the *hyperons. It is possible to assign a quantum number (the *baryon number*, B) to each particle, defined to be 1 for the baryons and -1 for their antiparticles. Particles that are not baryons have a baryon number 0. In any reaction the total baryon number is unchanged. *Compare* lepton.

baryta. *See* barium oxide.

barytes. *See* barite.

base. 1. A substance that will react with an *acid to form a salt and water. Bases turn litmus blue and produce hydroxide ions if dissolved in water. They are usually oxides or hydroxides of metals or solutions of ammonia or other amines. *Strong bases* are substances that are completely dissociated into ions in solution. Sodium hydroxide is an example: it consists of a mixture of Na^+ and OH^- ions. *Weak* bases are partially dissociated. Ammonia, for example, reacts with water to give ammonium ions and hydroxide ions: $NH_3 + H_2O = NH_4^+ + OH^-$. Soluble bases are *alkalis. The idea of a base in chemistry has been extended to include *Lewis bases.

2. The region of a *transistor between the emitter and the collector.

3. The horizontal straight line at the bottom of a geometric figure, or the bottom face of a geometric solid.

4. The number of units grouped together in a positional number system and represented by 1 in the next position. Thus *decimal numbers have a base of ten: *binary numbers have a base of two.

base metal. A common metal of relatively low value, such as iron or lead, as distinguished from noble metals such as gold or other rare or precious metals.

base unit. See unit.

basic. Denoting a compound that is a base or a solution that contains an excess of hydroxide ions.

basic dye. Any of a class of dyes that consist of salts containing large coloured positive ions. Basic dyes can be used on acidic acrylic fibres and as direct dyes for wool and silk. They can also be used on cotton with a tannic acid mordant. *Triphenylmethane derivatives are examples of basic dyes.

basic salt. See salt.

basic slag. A by-product in the manufacture of steel, consisting of a mixture of calcium phosphate, calcium silicate, lime, and ferric oxide. It is used as a fertilizer for its high phosphorus content.

battery. An electric cell or a number of cells connected together. The car battery is a *lead-plate accumulator. The common dry batteries used in torches, radios, etc., are *Leclanché cells.

bauxite. A naturally occurring hydrated aluminium oxide, $Al_2O_3.nH_2O$, containing ferric oxide and silica. It is the most important ore of aluminium.

beats. Regular fluctuations in the intensity of sound heard when two tones of nearly equal frequency are sounded together. The beats result from interference between the tones and have a frequency, the *beat frequency*, given by the difference in the frequencies of the two tones.

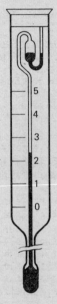

Beckmann thermometer

Beckmann thermometer. A mercury thermometer designed for measuring small changes of temperature. The capillary tube has a scale of about $5°C$ and leads into a reservoir at the top of the thermometer (see illustration). The instrument can be used near any temperature in the range $0-100°C$ by setting the amount of mercury in the bottom bulb using the reservoir. [After Ernst Beckmann (1853–1923), German physical chemist.]

bel. *See* decibel.

bell metal. An alloy of copper (about 80%) and tin, sometimes with small amounts of zinc and lead.

Benedict solution. A blue aqueous solution of sodium carbonate, copper sulphate, and sodium citrate. It is used as a test for reducing agents, such as glucose, in the presence of which it forms a red-to-yellow precipitate. [After Stanley Benedict (1884–1936), U.S. biochemist.]

benzal chloride. *See* benzylidene chloride.

benzaldehyde. A yellowish fragrant volatile oily aldehyde, C_6H_5CHO, occurring naturally in the kernels of bitter almonds. It is made by the oxidation of toluene and used in the manufacture of dyes, perfumes, and flavours. M.pt. $-26\,^\circ$C; b.pt. $178.1\,^\circ$C; r.d. 1.04.

benzanilide. A white or reddish crystalline powder, $C_6H_5CONHC_6H_5$, manufactured from benzoic anhydride and aniline. Benzanilide is used in the synthesis of drugs, dyes, and perfumes. M.pt. $163\,^\circ$C; r.d. 1.32.

benzene. A clear colourless highly flammable liquid, C_6H_6, with a characteristic odour. The simplest *aromatic hydrocarbon, it is obtained by catalytic reforming in the refining of petroleum and from coal tar. Products manufactured using benzene include styrene, detergents, nylon, and insecticides. M.pt. $5.5\,^\circ$C; b.pt. $80.1\,^\circ$C; r.d. 0.88.

benzene problem. *See* aromaticity.

benzenesulphonic acid. A colourless crystalline sulphonic acid, $C_6H_5SO_2OH$, made by sulphonation of benzene and used mainly in organic synthesis. M.pt. $65\,^\circ$C.

benzidine. A white, grey, or reddish crystalline powder, $NH_2(C_6H_4)_2NH_2$, produced by the reduction of nitrobenzene with zinc dust and used in the manufacture of dyes and as an analytical reagent. M.pt. $128\,^\circ$C; b.pt. $400\,^\circ$C.

benzil. A yellow crystalline solid, $C_6H_5COCOC_6H_5$, prepared from benzoin by oxidation with nitric acid and used in organic synthesis. M.pt. $95\,^\circ$C; b.pt. 346–$8\,^\circ$C; r.d. 1.08.

benzine. *See* petroleum ether.

benzoate. Any salt or ester of benzoic acid.

benzoic acid. A white crystalline carboxylic acid, C_6H_5COOH, occurring naturally in many plants and in benzoin gum. It is manufactured by the catalytic decarboxylation of phthalic anhydride and is used in curing tobacco and preserving food. M.pt. $122.4\,^\circ$C; r.d. 1.27.

benzoin. A white or yellowish crystalline optically active ketone, $C_6H_5CH(OH)(COC_6H_5)$, made by the condensation of benzaldehyde in potassium cyanide solution. It is used in organic synthesis. M.pt. $137\,^\circ$C (*dl*); b.pt. $344\,^\circ$C; r.d. 1.31.

benzoquinone. *See* quinone.

benzoyl chloride. A colourless pungent liquid acid chloride, C_6H_5COCl, used as a benzoylating agent. M.pt. $0\,^\circ$C; b.pt. $197.2\,^\circ$C; r.d. 1.21.

benzoyl group. The monovalent group, C_6H_5CO-.

benzoyl peroxide. A white crystalline explosive solid, $(C_6H_5CO)_2O_2$, made by reacting benzoyl chloride with sodium peroxide and used chiefly as a bleaching agent and a catalyst for free-radical reactions. M.pt. 103–$106\,^\circ$C (decomposes); r.d. 1.33.

benzyl alcohol. A colourless liquid, $C_6H_5CH_2OH$, used as a solvent and, in the form of its esters, in perfumery. M.pt. $-15.3\,^\circ$C; b.pt. $205.4\,^\circ$C; r.d. 1.04.

benzyl group. The monovalent group $C_6H_5CH_2$-.

benzylidine chloride (benzal chloride). A colourless liquid, $C_6H_5CHCl_2$, used in making benzaldehyde. M.pt. $-16.4\,^\circ$C; b.pt. $205.2\,^\circ$C; r.d. 1.55.

benzylidine group. The divalent group $C_6H_5CH=$, derived from toluene.

Benzyne

benzyne. A short-lived intermediate, C_6H_4, present in certain reactions.

Bergius process. A process for manufacturing oil from coal by mixing the coal with heavy oil to form a paste that is mixed with a catalyst and heated under hydrogen at high pressure (about 250 atmospheres). A mixture of liquid hydrocarbons is produced. [After Friedrich Bergius (1884–1949), German chemist.]

berkelium. Symbol: Bk. A transuranic *actinide element made, in trace amounts, by neutron bombardment of curium. The most stable isotope, ^{249}Bk, has a half-life of about 300 days. A.N. 97; valency 3 or 4.

berthollide compound. A chemical compound whose atoms are not in simple ratio to each other. Such compounds are said to be nonstoichiometric. An example is titanium oxide, $TiO_{1.8}$, which is deficient in oxygen. [After Claude Louis Berthollet (1748–1822), French chemist.]

beryl. A mineral, $Be_3Al_2(SiO_3)_6$, having several varieties including the gemstones emerald and aquamarine. It is the principal ore of beryllium.

beryllium. Symbol: Be. A light brittle greyish-white metallic element found in several minerals, especially *beryl. The metal is obtained by conversion to the chloride followed by electrolysis or reduction with magnesium. Its main uses are in X-ray windows and in the manufacture of alloys. Beryllium is one of the *alkaline-earth metals. A.N. 4; A.W. 9.0122; m.pt. 1281°C; b.pt. 2450°C; r.d. 1.93; valency 2.

Bessemer process. A process for making *steel by blowing air through molten pig iron to oxidize the carbon. The molten iron is contained in a large cylindrical steel vessel (a *Bessemer converter*) and air is introduced through holes in the bottom. Silicon and manganese impurities are also oxidized and removed as slag. The converter is lined with a basic refractory material to remove phosphorus. In some processes a mixture of air and steam is used to prevent absorption of nitrogen by the steel. When the carbon has been removed, the correct quantity of carbon is added to give the type of steel required. [After Sir Henry Bessemer (1813–98), English inventor.]

beta decay. A type of radioactive decay in which an unstable nucleus ejects either an electron and an antineutrino or a positron and a neutrino. The product has the same mass number but its atomic number has increased by 1. Tritium, for example, with one proton and two neutrons emits electrons. The process is decay of one neutron to give a proton, an electron, and an antineutrino. The product is helium-3, containing two neutrons and one proton. Nitrogen-13 decays by emission of positrons to give carbon-13. See also radioactivity.

beta iron. Iron in the temperature range within which it changes from a ferromagnetic material to a paramagnetic material. See also alpha iron.

beta particle. An electron, especially one ejected by radioactive decay. Streams of beta particles are called *beta rays*.

betatron. A type of particle *accelerator in which electrons are accelerated by a changing magnetic field. The electrons have a fixed orbit in a torus-shaped evacuated vacuum chamber held between the pole-pieces of a ring-shaped magnet. The magnetic field is pulsed, producing a varying magnetic flux linkage with the orbit and inducing a tangential electric field. The betatron thus works by magnetic induction: an accelerating potential is produced in the same way as the voltage that is induced in a coil of wire placed in a varying magnetic field.

The particle's momentum is proportional to the size of the magnetic field and this can be controlled to increase the electron energies and maintain a stable orbit. High energies (up to 300 MeV) are only possible with light particles, i.e. with electrons. The efficiency is limited by practical constraints on the size of the magnet.

BeV. *See* GeV.

bi-. 1. Prefix indicating two. In the names of chemical compounds *di-* is usually used. Thus, carbon bisulphide, CS_2, is more commonly called carbon disulphide.
2. Prefix indicating the presence of hydrogen in an acid salt. The recommended method of naming such compounds is to use *hydrogen* in the name. For example, sodium bisulphate, $NaHSO_4$, is more correctly called sodium hydrogen sulphate.

biaxial crystal. A crystal with two optic axes. *See* double refraction.

bicarbonate. Any acid salt of carbonic acid. They are more correctly called *hydrogen carbonates.*

biconcave. *See* lens.

biconvex. *See* lens.

big-bang theory (superdense theory). The theory that the universe began at some definite point in time by a violent explosion of a superdense collection of high-energy matter—a "primordial fireball". According to the theory, this matter has continued to fly apart, slowing down as it moves outwards and eventually forming galaxies and stars. Thus the observed expansion of the universe is a result of the "big bang" and the age of the universe can be calculated from *red shifts to be 5-50 $\times$ 10^9 years.

Evidence for the big-bang model comes from observations of red shifts at far points in the universe. Since the radiation takes many light-years to reach the earth from distant space, such observations give an idea of conditions existing nearer the time of the universe's origin and they show that the universe used to be more dense and was expanding faster. Background radiation has also been discovered that is thought to be a remnant of the big-bang. The main rival to the big-bang theory, the *steady-state theory, has less experimental support. The velocity at which the universe is expanding is decreasing. In some modifications of the theory it is predicted that the universe will eventually stop expanding and begin to contract under the influence of gravity until once again it is compressed into a superdense state. This can then expand again—the whole cycle repeating itself indefinitely. Such a model is known as the *oscillating universe. See also* expanding universe, black hole.

billion. One thousand million, 10^9; formerly, 10^{12}.

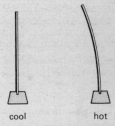

cool   hot

Bimetallic strip

bimetallic strip. A strip consisting of two pieces of different metals welded together so that as the temperature rises the strip bends because of unequal amounts of expansion. Bimetallic strips are used in thermostats and safety devices to break an electric circuit at a preset temperature.

binary. Having two parts or components. A *binary compound,* for example, is any compound of only two elements. A *binary alloy* has two component metals.

binary notation. The system of representing numbers using the base 2, and thus only the digits 0 and 1. The binary number $1101101 = 1 \times 2^6 + 1 \times 2^5 + 0 \times 2^4 + 1 \times 2^3 + 1 \times 2^2 + 0 \times 2^1 + 1 = 64 + 32 + 8 + 4 + 1 = 109$. Similarly, $11.101 = 1 \times 2^1 + 1 + 1 \times 2^{-1} + 0 \times 2^{-2} + 1 \times 2^{-3} = 2 + 1 + 0.5 + 0.125 = 3.625$ in decimal notation. Because the digits 0 and 1 can be represented by the "off" and "on" states of an electric circuit, binary arithmetic is used in digital computers. *Compare* decimal notation.

binary star. A system of two separate but associated stars *(components)* orbiting around a common centre of mass. Three types are observed: *visual binaries* are those where the angular separation of the components is noticeable enough for a telescope to resolve them or even for the naked eye to distinguish them. *Spectroscopic binaries* are those detected by the cyclic variations in the Doppler effect caused by the successive approach and recession of the components. *Eclipsing* or *photometric binaries* are those in which a brighter component is periodically eclipsed by a darker one. Such binaries are also called *Algol variables* after the star Algol in Perseus.

binding energy. The energy that has to be supplied to an atomic nucleus in order to separate it completely into its constituent nucleons. *See also* mass defect.

binocular. Involving the simultaneous use of both eyes. *Binocular vision* is normal vision using two eyes—the two images from slightly different viewpoints enable the observer to gain an impression of depth. *Binoculars* (or *field glasses*) have two *terrestrial telescopes mounted side by side with inverting lenses or prisms to produce an erect image. In prism binoculars (see illustration) the increased path length gives the instrument a longer focal length.

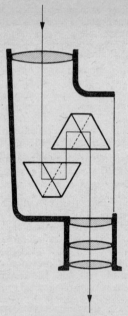

Prism binocular

binomial theorem. The expansion

$$1 + \frac{nx}{1} + \frac{n(n-1)x^2}{2.1} + \frac{n(n-1)(n-2)x^3}{3.2.1}$$

$$+ \frac{n(n-1)(n-2)(n-3)x^4}{4.3.2.1} + - - - -$$

It is valid if n is a positive integer, in which case it has a finite number of terms. Otherwise it is valid only when $-1 < x < 1$. The coefficients of terms in the expansion are the *binomial coefficients*.

biochemistry. The study of compounds and chemical reactions occurring in living matter.

bioluminescence. *Luminescence occurring during biochemical reactions, as observed in fireflies, glow worms, certain bacteria, and many deep-sea animals. It is a particular type of *chemiluminescence in which the compound *luciferin* is oxidized with high efficiency by the enzyme *luciferase*.

biophysics. The application of the theories and techniques of physics to biology.

biprism. A prism made up of two very narrow prisms placed base to base. If positioned in front of a single light source it produces interference fringes.

birefringence. *See* double refraction.

Birkeland-Eyde process. A process for the *fixation of nitrogen by reacting nitrogen and oxygen in an electric arc to give nitric oxide: $N_2 + O_2 = 2NO$. [After Kristian Birkeland (1867-1917) and Samuel Eyde (1866-1940), Norwegian scientists.]

bisector. A line or plane that divides an angle, line, solid, etc., into two equal parts.

bismuth. Symbol: Bi. A pinkish-white brittle metallic element, usually obtained as a by-product in refining other metal ores. It expands when solidified, making it suitable for use in alloys for casting. Its main use is in the production of low-melting alloys, such as Wood's metal. Chemically, it resembles arsenic and antimony, forming covalent bismuth III (*bismuthous*) and bismuth V (*bismuthic*) compounds. Bismuth III compounds can also be ionic, containing the Bi^{3+} ion. Bismuth salts tend to hydrolyse in solution to yield insoluble bismuthyl compounds. A.N. 83; A.W. 208.98; m.pt. 271.3°C; b.pt. 1560°C; r.d. 9.75.

bismuthic. *See* bismuth.

bismuthous. *See* bismuth.

bismuth oxide chloride (bismuth oxychloride). A white lustrous crystalline powder, BiOCl. It is the insoluble product of hydrolysis of bismuth (III) chloride and is used in face powder and as a pigment. R.d. 7.72.

bismuth oxychloride. *See* bismuth oxide chloride.

bit. A unit of information; the minimum amount of information necessary to specify either of two alternative states. In particular, a bit is a binary digit (0 or 1) used in *binary notation. *See also* byte.

bittern. The liquid remaining after sodium chloride has been crystallized from sea water. It is a source of iodine, bromine, and some magnesium salts.

bitumen. Any of various mixtures of hydrocarbons, especially semi-solid mixtures extracted from coal.

bituminous. Containing bitumen or tar.

bivalent. *See* divalent.

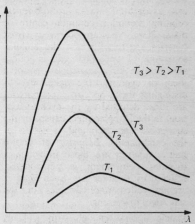

Curves of intensity (I) against wavelength (λ) for black-body radiation

black body. A body that absorbs all the radiation falling on it: i.e. it has an *absorptance of 1. A black body is a hypothetical ideal but a practical approximation can be made by using a small hole in the wall of a constant-temperature enclosure.
The radiation emitted from a black body has a continuous distribution of wavelengths with a characteristic intensity maximum at a particular wavelength (see illustration). The *black-body radiation* lies mainly in the infrared region at

lower temperatures. As the temperature is increased the maximum in the intensity-wavelength curve moves to shorter wavelengths (see Wien's displacement law) until a temperature is reached at which the body is incandescent. A black body is sometimes called a *full radiator*. See also quantum theory.

blackdamp. The air, depleted in oxygen, left in coal mines after an explosion of *firedamp.

black hole. A hypothetical entity produced by the contraction of a star to such an extent that its gravitational field is high enough to prevent light from leaving it. When a star has used up its nuclear fuel it begins to contract under the influence of gravitational forces between its constituent particles. Its eventual fate depends on its mass: in particular, if its mass is large enough it undergoes a supernova explosion and the remaining core of matter forms a *neutron star. This occurs if the core remaining after the supernova has a mass of about 1.4 times that of the sun. If the core has a mass several times that of the sun the gravitational forces are large enough to compress the matter still further, so that its density is even larger than that in a neutron star. Under such conditions a very large mass is concentrated into a very small volume and an immense gravitational field is produced, large enough to prevent any matter, light, or other radiation escaping. The theory of such a phenomenon is treated by the general theory of *relativity but in simple terms a black hole forms when the escape velocity from a star exceeds the velocity of light. The black hole is the region in space (or more correctly in space-time) from which no matter or radiation can escape. The boundary of this region is called the *event horizon*: it is a sphere with a radius given by $2GM/c^2$, where G is the gravitational constant, M is the mass, and c is the velocity of light. This radius is called the *Schwartzschild radius* of the mass.

Although black holes are theoretical concepts there is some evidence that they exist. An isolated black hole would not be detected because it would not emit any form of radiation. However, there is some evidence for the presence of black holes in certain *binary stars that emit X-rays. In the constellation of Cygnus, for example, there is a blue supergiant that revolves around a small invisible companion with a period of 5-6 days. The combination is a variable source of X-rays. The invisible companion must have a mass about 15 times that of the sun and seems to have a diameter of only about 30 miles. Its mass is too large for a stable neutron star and it is thought to be an example of a black hole, which is continually gaining matter by attraction from the supergiant. The matter would not enter the black hole directly but would form a rotating disc around it within which particles of matter would spiral in towards the black hole. In this process the matter is compressed and heated to very high temperatures, at which it emits the observed X-rays.

Within a black hole the behaviour is described by the general theory of relativity and the classical laws of physics are not obeyed. One problem concerns what happens to the collapsing matter within the hole's event horizon. In theory, it should contract indefinitely, finally disappearing at a point at which the curvature of space-time becomes infinite—such a point is called a *singularity*. The singularity that would be formed by a collapsing star is surrounded by an event horizon—a condition known as *cosmic censorship*. Some physicists have postulated the existence of *naked singularities*, in which the point of infinite curvature is not surrounded by an event horizon. Such an entity might "create" matter and light by the reverse process of that occurring in a black hole. The result would be a *white hole*. In particular, a naked singularity might have been the source of the "big bang" postulated in the *big-bang theory of the origin of the universe. However, the behaviour of a system under the bizarre conditions applying at a singularity cannot be treated by present physical theories and the existence of such entities as white holes is still no more than speculation.

black lead. *See* graphite.

blanc fixe. *See* barium sulphate.

blanket. *See* fission reactor.

blast furnace. A furnace for the production of pig iron from iron oxide ore. Typically it consists of a large steel shell lined with refractory bricks. The charge, consisting of alternate layers of ore, coke, and flux, is added to the top and preheated air is blown in through inlet pipes located near the bottom of the furnace. The temperature reaches about 1300°C and the iron oxide is reduced to iron: $2Fe_2O_3 + 3C = 4Fe + 3CO_2$. The flux is often limestone, which at the temperatures used decomposes to lime and carbon dioxide: $CaCO_3 = CaO + CO_2$. The lime reacts with impurities in the ore to form a molten slag of calcium silicate, aluminate, etc. Molten pig iron is tapped from the bottom of the furnace—the process is a continuous one.

blasting gelatin. A gelatinous elastic substance made by mixing nitroglycerin with gun cotton (nitrocellulose), used as a powerful explosive.

bleaching powder (chloride of lime). A white powder made by the reaction of chlorine on slaked lime. It has the formula $CaOCl_2$ and consists of mixed hydrated calcium choride, $CaCl_2$, calcium hydroxide, $Ca(OH)_2$, and calcium hypochlorite, $Ca(OCl)_2$. Bleaching powder acts with dilute acids to give chlorine. It is used as a bleaching agent and disinfectant.

blende. A mineral sulphide of a metal. Zinc blende, for example, is naturally occurring zinc sulphide, ZnS.

blind spot. *See* eye.

blink comparator (blink microscope). A viewing apparatus that can alternately display to the user in rapid succession each of two photographic plates of the same part of the sky, so that minute differences can be observed in the brightness, position, etc., of celestial objects.

block copolymer. *See* polymer.

bloodstone. *See* haematite.

blowpipe. A pipe with a narrow nozzle for directing a jet of air into a flame and producing a more intense heat.

bluestone. *See* copper sulphate.

blue vitriol. *See* copper sulphate.

Board of Trade unit. Symbol: BTU. A practical unit of energy equal to 1 kilowatt-hour. It was formerly used in Britain for the sale of electricity.

Bode's law. An empirical rule relating the mean distances of the planets from the sun, having the form $x = 0.4 + 0.3 \times 2^n$, where x is the distance in astronomical units and $n = -\infty, 0, 1, 2$, etc. The planetary distances may be obtained as given in the table.

The law breaks down badly for Neptune and is completely inoperative for Pluto. The fact that it also allows for a planet between Mars and Jupiter may have a bearing in our understanding of the solar system (*see* asteroid). [After Johann Elert Bode (1747-1826), German astronomer.]

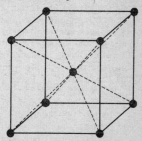

Body-centred cubic unit cell

body-centred. Denoting a crystal structure in which the unit cell has eight atoms at its corners and one at its centre. In a *body-centred cubic* lattice

Bode's Law

planet	n	Bode's law distance	true mean distance
Mercury	− ∞	0·4	0·39
Venus	0	0·7	0·72
Earth	1	1·0	1·00
Mars	2	1·6	1·52
Asteroids	3	2·8	2·65*
Jupiter	4	5·2	5·20
Saturn	5	10·0	9·54
Uranus	6	19·6	19·20
Neptune	7	38·8	30·10
Pluto	8	77·2	39·50

*Average value

(see illustration) the unit cell is a cube. *Compare* face-centred.

Bohr atom. A model of the *atom in which the electrons move around the nucleus in circular orbits with the nucleus at the centre. The electrostatic force of attraction between the negative electrons and the positive nucleus is balanced by an outwards centrifugal force. Thus, the electrons move in orbits in an analogous way to the motion of planets in the solar system. In Bohr's theory the electrons can only move in a number of fixed orbits in which their angular momentum (mvr) is equal to $nh/2\pi$, where h is the Planck constant and n is an integer (1, 2, 3, ...). In these orbits the electron does not radiate electromagnetic radiation. It can be shown that the radii of the Bohr orbits are given by $r = n^2h^2\epsilon_0/Ze^2\pi m$, where ϵ_0 is the electric constant and Z the atomic number. The energies are given by $E = -Ze^4m/8n^2h^2\epsilon_0^2$.
The electron can occupy any of these states: normally it is in the one of lowest energy. Electromagnetic radiation is emitted or absorbed when the electron makes a transition between two orbits: the radiation is in the form of a photon of energy $h\nu$, where ν is the frequency. The integer n is the *principal quantum number.*
The theory gives an explanation of *spectral series and a good value for the *Rydberg constant of hydrogen. Later, the German physicist Arnold Sommer-feld (1868–1951) introduced a refinement in which the electrons move in ellipses precessing around the nucleus. This involved the introduction of an *azimuthal quantum number* and was successful in explaining fine structure in the hydrogen lines. This model is called the *Bohr-Sommerfeld atom.* [After Niels Bohr (1885–1962), Danish physicist.]

Bohr magneton. *See* magneton.

boiling. The process by which a liquid changes into a gas when its *vapour pressure equals the external pressure. When this occurs, small bubbles of vapour can form in the liquid and rise to the surface. The temperature at which this happens is the *boiling point* of the liquid, which depends on the forces between the molecules—a substance such as water in which there are strong intermolecular forces due to hydrogen bonds has a higher boiling point than a substance such as ether, in which intermolecular forces are small. The boiling point also depends on the external pressure—the lower the pressure the lower the boiling point. Boiling points of liquids are usually quoted for standard atmospheric pressure (760 mmHg).

boiling-water reactor (BWR). *See* fission reactor.

bolide. A large and spectacularly bright *meteor that appears like a fireball and may sometimes explode with a loud detonation.

bolometer. An instrument for measuring radiant energy, consisting of a thin strip of platinum blackened on one side and connected into a Wheatstone bridge circuit. The increase of temperature caused by radiant heat falling on the strip is measured by the change of electrical resistance.

Boltzmann constant. Symbol: k. A constant equal to the gas constant R divided by the Avogadro constant L. It is equal to $1.380\,54 \times 10^{-23}$ joules per kelvin. [After Ludwig Boltzmann (1844–1906), Austrian physicist.]

bomb calorimeter. A device for determining heats of combustion, consisting of a sealed steel vessel in which the sample is ignited in an atmosphere of oxygen. Complete combustion occurs and the heat evolved can be calculated from the rise in temperature. The apparatus is extensively used for measuring the calorific value of fuels and of foods. The heat of combustion is measured at constant volume—the pressure rises during the reaction.

bond. A linkage holding together two atoms in a molecule or holding atoms or ions in a solid. The types of chemical bond are the *covalent bond, *electrovalent bond, *coordinate bond, *metallic bond, and *hydrogen bond.

bond angle. The angle between two chemical bonds taken as the angle between lines through the nuclei of each pair of atoms.

bond energy. The energy associated with a chemical bond between two atoms. For a diatomic molecule, the bond energy is simply the heat required to dissociate it into atoms. In a compound such as methane, in which all four C-H bonds are equivalent, the bond energy of the C-H bond is ¼ of the total heat of atomization. In molecules containing more than one type of bond the concept of bond energy is less well defined. Thus the energies of the C-H

bonds in ethane would not necessarily be the same as those in methane. By considering large numbers of similar compounds it is possible to assign average values to types of bonds so that for any molecule the sum of the bond energies approximately equals the heat of atomization. The bond energy differs from the *bond dissociation energy*, which is the energy to break a bond in a compound. Thus, in ethane the bond dissociation energy of the C-C bond is the energy required to break ethane into two methyl groups.

bond length. The distance between the nuclei of two atoms in a compound. It depends on the type of compound. For example the C-H bond in methane is longer than that in benzene.

bone black. *See* animal charcoal.

boracic acid. *See* boric acid.

borate. Any salt or ester of boric acid.

borax (disodium tetraborate). A whitish crystalline hydrated solid, $Na_2B_4O_7.10\text{-}H_2O$, found in some salt lakes and alkaline soils. The pentahydrate and the anhydrous salt can also be prepared. Borax is used in the manufacture of glass and enamels. M.pt. 741°C (anhydrous); r.d. 1.7 (decahydrate).

borax-bead test. A test for the presence of some metals in salts. The sample is mixed with borax and fused in a flame on the end of a piece of wire. A small bead forms with a colour that may be characteristic of the metal present. Cobalt salts, for example, form a blue bead.

borazon. *See* boron nitride.

Bordeaux mixture. A liquid mixture of slaked lime and copper sulphate solution used as a fungicide and insecticide for plants.

boric acid (orthoboric acid, boracic acid). A colourless odourless crystalline solid or powder, H_3BO_3. It is a very weak acid prepared by the addition of hydrochloric acid to borax and is used in ceramics and fire-proofing compositions. M.pt. about 160°C; r.d. 1.44.

boride. Any compound of boron with a more electropositive element. Most metals form borides. They are chiefly hard interstitial materials with metallic conductivity.

boron. Symbol: B. A nonmetallic element found naturally in several minerals, especially borax $(Na_2B_4O_7.10H_2O)$. Usually, it is obtained as a yellow-brown amorphous powder but a yellow crystalline form also exists. It can be made by reducing the oxide with magnesium metal and is used in hardening steels, in control rods for nuclear reactors, and in semiconductors. Boron is a fairly unreactive element. Its most important compounds are boric acid and borax. Many metal *borides exist, most of them prepared by direct combination of the elements at high temperatures. Boron forms covalent compounds with a valency of three, although the boron hydrides are electron deficient. A.N. 5; A.W. 10.811; m.pt. 2300°C; sublimes at 2550°C; r.d. 2.34.

boron nitride. A crystalline solid, BN, made by direct reaction of the elements at high temperatures and pressures. It has two crystalline structures, one resembling that of *diamond and the other that of *graphite. The former dimorph, *borazon*, is one of the hardest substances known. M.pt. 3000°C; r.d. 3.48 (borazon).

boron trifluoride. A colourless gas, BF_3, with a pungent suffocating odour, used as a catalyst in *Friedel-Crafts reactions. Boron trifluoride can act as a *Lewis acid. M.pt. -127°C; b.pt. -100°C.

Bosch process. A process for the manufacture of hydrogen by reacting carbon monoxide with steam over a hot catalyst: $CO + H_2O = CO_2 + H_2$. The raw material used is *water gas, a mixture of carbon monoxide and hydrogen. [After Carl Bosch (1874-1940), German chemist.]

Bose-Einstein statistics. A type of *statistical mechanics applied to the distribution of particles amongst energy levels when the particles are indistinguishable from one another and any number of particles can occupy each energy state. *See also* boson. [After Satyendra Nath Bose (1894-1974), Indian physicist, and Albert Einstein.]

boson. Any particle that obeys Bose-Einstein statistics. Examples are deuterons and photons. *Compare* fermion.

Boyle's law. The principle that a fixed quantity of gas at a constant temperature has a pressure that is inversely proportional to its volume: i.e. $pV = C$, where C is a constant. The law strictly applies only to perfect gases: real gases tend to obey it at low pressures. *See also* gas laws. [After Robert Boyle (1627-91), Irish scientist.]

Brackett series. A *spectral series in the infrared emission of hydrogen with wavelengths given by $1/\lambda = R(1/4^2 - 1/m^2)$, where R is the *Rydberg constant and m is 5, 6, 7, etc.

Bragg's law. An equation for the diffraction of X-rays by crystal planes giving the condition for maxima in the diffraction pattern: $n\lambda = 2d \sin \theta$, where λ is the wavelength, d is the distance between planes, and θ is the angle between the plane and the incident beam. The equation is derived by considering that reflection occurs from atomic planes and interference takes place between X-rays reflected from different planes. *See also* X-ray crystallography, X-ray diffraction. [After Sir Lawrence Bragg (1890-1971), English physicist.]

branched chain. A chain of atoms, usually carbon atoms, in a molecule, with side chains attached. Isobutane, for example, is a branched-chain hydrocarbon whereas butane has a *straight chain.

brass. Any of a large number of alloys containing copper and zinc (up to 40%), sometimes with small amounts of aluminium, tin, manganese, or other metals.

brazing. *See* solder.

breeder reactor. *See* fission reactor.

bremsstrahlung. Electromagnetic radiation produced by the deceleration of charged particles. An electron approaching an atomic nucleus loses energy and the energy lost is taken up by an emitted photon. The effect is observed for electrons of high energy. Bremsstrahlung, in the form of *X-rays having a continuous distribution of wavelengths, is obtained in X-ray tubes.

Brewster's law. The principle that light reflected from a solid surface is plane polarized with a maximum polarization occurring when the angle of incidence (α) is $\tan^{-1}(n)$, where n is the refractive index (i.e. $n = \tan \alpha$). The angle α is the *Brewster angle*. It is the angle at which the reflected ray makes an angle of 90° with the refracted ray. [After Sir David Brewster (1781-1868), English physicist.]

bridge. Any network of four resistors, inductors, capacitors, or other circuit elements in the form of a quadrilateral with an input connected across two opposite corners and the output across the other two corners. Bridges are often used as measuring circuits, as in the *Wheatstone bridge.

Brinell test. A test determining the hardness of metals by forcing a small hard steel ball into the surface. The hardness is characterized by the *Brinell number*—the ratio of the load in kilograms to the area of the depression made. [After Johann August Brinell (1849-1925), Swedish engineer.]

Britannia metal. An alloy of tin (80-90%), antimony (5-15%), and copper (0-3%), used in bearings.

British thermal unit. Symbol: Btu. A unit of energy equal to 1055.055 853 joules. Formerly, the British thermal unit was defined as the amount of heat required to raise the temperature of one pound of water through one degree Fahrenheit.

bromate. Any salt of bromic acid.

bromeosin. *See* eosin.

bromic acid. A colourless unstable acid, $HBrO_3$, made in dilute solution by the action of sulphuric acid on a bromate salt. It is an oxidizing agent.

bromide. Any of certain compounds containing bromine. *See* halide.

bromination. A chemical reaction in which one or more bromine atoms are introduced into an organic compound. The term is usually used to describe reactions involving elemental bromine. An example is the bromination of ethane, under the influence of light, to give bromoethane: $C_2H_6 + Br_2 = C_2H_5Br + HBr$.

bromine. Symbol: Br. A heavy volatile reddish-brown liquid element producing a pungent irritating red vapour. It occurs in natural bromide salts and is manufactured from sea water by displacement with chlorine. Bromine is used in the manufacture of organic chemicals. It is a reactive element belonging to the *halogen group. A.N. 35; A.W. 79.904; m.pt. -7.2°C; b.pt. 58.78°C; r.d. 3.12 (liquid); valency 1, 3, 5, or 7.

bromoethane (ethyl bromide). A colourless flammable liquid alkyl halide, C_2H_5Br, prepared from ethene and hydrogen bromide. It is used in organic synthesis and as a refrigerant. M.pt. -119°C; b.pt. 38.4°C; r.d. 1.43.

bromoform (tribromomethane). A colourless liquid, $CHBr_3$. M.pt. 8°C; b.pt. 150°C; r.d. 2.9.

bromomethane (methyl bromide). A colourless volatile nonflammable liquid alkyl halide, CH_3Br, used as a fumigating agent and solvent. M.pt. -93°C; b.pt. 4.5°C; r.d. 1.73.

bronze. Any of various alloys of copper. The term was originally used for alloys of copper and tin (up to 30%), sometimes with smaller amounts of other elements. *Phosphor bronze*, for example contains small amounts (up to 1%) of phosphorus. The term *bronze* has been extended to certain copper alloys that contain no tin, as in *aluminium bronze*.

Brownian movement. Small random motions of colloidal particles suspended in a liquid or gas. It is caused by random bombardment of the particles by molecules of the liquid or gas. [After Robert Brown (1773-1858), Scottish botanist.]

Brunswick green. A green pigment made by mixing chrome yellow and Prussian blue. It is used in paints.

brush. An electrical contact on an *electric motor or *generator.

brush discharge. A type of gas discharge occurring near sharp points at high voltages, the voltage being lower than that producing a spark or arc. The discharge is seen as short luminous streamers at the tip of the point.

bubble chamber. A device for detecting and investigating elementary particles by the tracks they make in a liquid. Usually liquid hydrogen is used, kept under pressure at a temperature slightly above its boiling point. The operating technique is to reduce the pressure for a few milliseconds, during which a photograph is taken, and reimpose the pressure before the liquid boils. Particles passing through the chamber when the photograph is taken cause a track of small bubbles along their path. Decays and interactions of particles can be investigated. Usually large numbers of photographs are taken from different positions and known electric and magnetic fields are applied, so that the particles can be identified from the curvature of their tracks. *See also* cloud chamber.

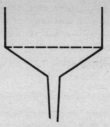

Buchner funnel

Buchner funnel. A type of funnel with a flat perforated tray on which is placed a circular piece of filter paper. Buchner funnels are usually made of porcelain. They are used for filtering liquids by suction. [After Eduard Buchner (1860-1917), German chemist.]

buffer solution. A solution with a definite pH that does not change on dilution or addition of (small) amounts of acid or alkali. Buffer solutions can be made by mixing a solution of a salt of a weak acid or base with the free acid or base in calculated amounts to produce the desired pH. An example of a buffer solution is one containing acetic acid and sodium acetate. The acid partially dissociates to produce hydrogen ions: $CH_3COOH = CH_3COO^- + H^+$. The sodium acetate is a source of more acetate ions. If acid is added, the hydrogen ions combine with the acetate ions to give undissociated acetic acid. The presence of the sodium acetate "buffers" the solution against increases in acidity. Buffer solutions are useful in experiments in which a contant pH is required. Buffers are also important in biological systems, which often only function under neutral or nearly neutral conditions.

bulk modulus. *See* modulus.

bumping. Violent boiling of a liquid caused when the liquid has been heated to a temperature above its normal boiling point, so that when the bubbles do form they have a higher pressure than atmospheric pressure.

Bunsen burner. A gas burner having a vertical metal tube with an air inlet at the bottom. The mixture of gas and air gives a flame with two regions. The cooler inner cone is pale blue and contains incompletely burnt gas, producing a reducing atmosphere. The hotter outer flame is almost invisible. Here all the gas is fully oxidized. The tip of this outer flame is the hottest part, at a temperature of about 1500°C. [After Robert Bunsen (1811–99), German chemist.]

buoyancy. The upthrust on a body that is partially or totally immersed in a fluid. See Archimedes' principle.

burette. A long graduated glass tube with a tap at one end used to deliver known volumes of liquid in titrations, etc.

butadiene. A colourless flammable gas, $H_2C:CHHC:CH_2$, manufactured by the catalytic dehydrogenation of butanes and butenes. It polymerizes readily and is used in making synthetic rubbers. Butadiene is a conjugated diene and can undergo the *Diels-Alder reaction. M.pt. -109°C; b.pt. -4.5°C; r.d. 0.62 (liquid).

butanal (butyraldehyde). A colourless flamable liquid, $CH_3(CH_2)_2CHO$. M.pt -99°C; b.pt. 75.7°C; r.d. 08.

butane. A colourless flammable gaseous *alkane, C_4H_{10}, present in crude petroleum. It is used as a fuel, especially when liquefied under pressure in cylinders, and as raw material for the manufacture of synthetic rubber. Butane is sometimes called *n-butane* to distinguish it from isobutane. M.pt. -138.4°C; b.pt. -0.5°C; r.d. 0.60 (0°C).

butanedioic acid. See succinic acid.

butanoic acid. See butyric acid.

butanol. Either of two isomeric straight-chain alcohols with the formula C_4H_9OH. *Butan-1-ol* (*n*-butyl alcohol) is a colourless flammable liquid, $CH_3(CH_2)_2CH_2OH$, manufactured from butanal and used as a solvent. M.pt.

-89.5°C; b.pt. 117.3°C; r.d. 0.81. *Butan-2-ol* (*sec*-butyl alcohol) is also a colourless flammable liquid, $CH_3CH_2CHOHCH_3$. See also butyl alcohol.

butanone (methyl ethyl ketone). A colourless flammable liquid ketone, $CH_3COC_2H_5$, manufactured by the oxidation of butane and used in lacquers and cleaning fluids. M.pt. -86.4°C; b.pt. 77.6°C; r.d. 0.8.

butenedioic acid. Either of two geometrical isomers with formula HCOO-CH:CHOOCH, *maleic acid and *fumaric acid.

butter of antimony. See antimony chloride.

butter of zinc. See zinc chloride.

butyl alcohol. Any of four isomers of the formula C_4H_9OH: *n*-butyl alcohol $(CH_3(CH_2)_2CH_2OH)$, *sec*-butyl alcohol $(CH_3CH_2CHOHCH_3)$, isobutyl alcohol $((CH_3)_2CHOHCH_3)$, and *tert*-butyl alcohol $((CH_3)_3COH)$. See also butanol.

butyl group. A monovalent group C_4H_9-, derived from butane by removing one atom of hydrogen. The term is often used for the group with a straight chain or carbon atoms: i.e. for the *n-butyl group*. There are three other isomers: the *isobutyl group*, $(CH_3)_2CHCH_2-$; the *sec-butyl group*, $CH_3CH_2CH(CH_3)-$; and the *tert-butyl group*, $(CH_3)_3C-$.

butyl rubber. A type of synthetic rubber made by copolymerization of isobutylene and small amounts of isoprene. Butyl rubber can be vulcanized with sulphur. It has a much lower permeability to gases than natural rubbers and is used in tyre inner tubes.

butyraldehyde. See butanal.

butyric acid (butanoic acid). A colourless liquid fatty acid, CH_3CH_2COOH, with a rancid odour, present, in the form of its esters, in butter. Its main use is in making esters for perfumes and flavourings. B.pt. 163°C; r.d. 0.96.

BWR. A boiling-water reactor. *See* fission reactor.

by-product. A substance obtained at the same time as the main product of a manufacturing process.

byte. A unit of information equal to eight bits. It is used to specify the capacity of a computer store.

C

cadmium. Symbol: Cd. A soft bluish-white metallic element found associated with zinc ores, from which it is obtained as a by-product. It is used in bearing alloys and solder and in electroplated coatings for metal parts. Like zinc, it reacts with oxygen and other nonmetals and dissolves in nonoxidizing acids. Its compounds are similar to those of zinc. A.N. 48; A.W. 112.4; m.pt. 320.9°C; b.pt. 765°C; r.d. 8.65; valency 2.

cadmium cell. *See* Weston cell.

cadmium sulphide. An insoluble yellow or orange solid, CdS, occurring naturally as *greenockite*. It is an important pigment (*cadmium yellow*) and is also used in transistors, fluorescent screens, and small batteries. M.pt. 1759°C (100 atm); r.d. 4.82.

cadmium yellow. *See* cadmium sulphide.

caesium. Symbol: Cs. A soft silvery-white metallic element found in some rare minerals. It can be made by electrolysis of the fused cyanide. Caesium is one of the most electropositive elements (second only to francium) and is used in photoelectric cells and similar devices. It is an *alkali metal. A.N. 55; A.W. 132.905; m.pt. 28.5°C; b.pt. 690°C; r.d. 1.90; valency 1.

caesium clock. An apparatus used for obtaining the steady frequency used in the definition of the *second. The device depends on the fact that in a magnetic field the caesium atom can have two discrete energies due to the spin of its nucleus. A beam of atoms is passed into a region of strong nonuniform magnetic field in which the atoms are separated according to their energy state (*see* Stern-Gerlach experiment). Atoms with the lower energy are allowed to pass through a slit into a region in which there is a uniform magnetic field and a source of radiofrequency radiation with a frequency of 9 192 631 770 hertz. This excites some of the atoms to the higher energy state (*see* nuclear magnetic resonance). The beam is then passed into another nonuniform field in which atoms in the lower state are deflected onto a detector. The signal from this detector is fed to the radiofrequency supply and used to stabilize it on the particular frequency required for the transition. The device is stable to one part in 10^{11}.

calamine. 1. A mixture of zinc oxide with small amounts (0.5%) of iron III oxide, used to counteract inflammation of the skin.
2. Either of two zinc minerals: the carbonate, $ZnCO_3$, or a hydrated silicate, $2ZnO.SiO_2.H_2O$.

calcination. The gradual deposition of calcium carbonate from hard water in pipes, kettles, etc.

calcite (Iceland spar). A colourless crystalline natural form of calcium carbonate, $CaCO_3$. Crystals of calcite show double refraction.

calcium. Symbol: Ca. A soft white metallic element occurring extensively in nature, especially as forms of calcium carbonate (limestone, marble, chalk, etc.). Calcium is made by electrolysis of fused calcium chloride and is used as a reducing agent and getter. It is an *alkaline-earth metal. A.N. 20; A.W. 40.08; m.pt. 850°C; b.pt. 1440°C; r.d. 1.58; valency 2.

calcium carbide (carbide). A grey solid, CaC_2, made by heating limestone with coke. It reacts with water to produce acetylene. R.d. 2.22.

calcium carbonate. A white powder or colourless crystalline solid, $CaCO_3$, occurring naturally as aragonite, calcite, chalk, limestone, and marble. In a precipitated form it is used as a gastric antacid and in toothpowders, flour, and white polishes. R.d. 2.71.

calcium chloride. A white crystalline or granular deliquescent solid, $CaCl_2$, used as a preservative and a drying agent. It can also be obtained as the monohydrate, dihydrate, and hexahydrate. M.pt. $772°C$; r.d. 2.2 (anhydrous).

calcium cyanamide (cyanamide). A grey-black powder, $CaCN_2$, manufactured by heating calcium carbide in nitrogen. The reaction is a method of nitrogen fixation: cyanamide is hydrolysed to ammonia and acetylene and is used as a fertilizer. M.pt. $1300°C$; r.d. 1.08.

calcium cyanide. A highly poisonous colourless or grey solid, $Ca(CN)_2$, used as a plant insecticide and rat poison. Decomposes at $350°C$.

calcium cyclamate. A white powder, $(C_6H_{11}NHSO_3)_2Ca.2H_2O$, with a sweet taste, formerly used as a sweetener in soft drinks.

calcium fluoride. A white powder, CaF_2, occurring naturally as fluorspar. It is the main source of fluorine and its compounds. M.pt. $1360°C$; b.pt. $2500°C$; r.d. 3.2.

calcium hydrate. See calcium hydroxide.

calcium hydroxide (lime, slaked lime, calcium hydrate). A soft white powder, $Ca(OH)_2$, prepared by the action of water on calcium oxide. It is used in cement and in neutralizing acid soil. R.d. 2.24.

calcium nitrate. A deliquescent crystalline solid, $Ca(NO_3)_2$, used in fertilizers and explosives. M.pt. $561°C$; r.d. 2.5.

calcium oxide (lime, quick lime, calx). A white or grey granular solid, CaO, made by roasting natural forms of calcium carbonate. It is widely used, chiefly in

making calcium hydroxide, as a flux in steel manufacture, in making paper, and in water softening. M.pt. $2580°C$; b.pt. $2850°C$; r.d. 3.40.

calcium phosphate. Any calcium salt of a phosphoric acid. Calcium phosphates are found in some rocks and are also present in bone. They are used as fertilizers (see superphosphate). The term is often used for calcium orthophosphate, $Ca_3(PO_4)_2$, which is a white insoluble powder. M.pt. $1670°C$; r.d. 3.14.

calcium sulphate. A white powder, $CaSO_4$ or $CaSO_4.2H_2O$, occurring naturally as anhydrite or gypsum. Calcium sulphate hemihydrate, $CaSO_4.0.5-H_2O$, is plaster of Paris. M.pt. $1450°C$; r.d. 2.96.

calcium sulphide. A colourless crystalline solid, CaS, made by reducing calcium sulphate with carbon at high temperature. It is a luminescent compound. R.d. 2.5.

calculus. The branch of mathematics concerned with *integration and *differentiation of functions. It is divided into the integral calculus and the differential calculus.

calibration. The process of using an instrument or apparatus to take readings of standard or known quantities so that the arbitrary readings given by the instrument can be converted into absolute values.

caliche. See sodium nitrate.

californium. Symbol: Cf. A transuranic *actinide element made, in trace amounts, by neutron bombardment of berkelium. The most stable isotope, ^{249}Cf, has a half-life of 470 years. A.N. 98.

calomel. See mercury chloride.

calomel electrode. A half cell consisting of a mercury electrode covered with a paste of mercury I chloride and mercury in a solution of potassium chloride saturated with mercury I chloride. The system is used as a standard electrode: when the potassium chloride solution is saturated its potential is -0.2415 volts at 25°C with respect to a *hydrogen electrode.

caloric theory. A theory concerning the nature of heat, in which heat is considered to be a weightless fluid (*caloric*). The theory was abandoned in the mid-nineteenth century when experiments on the mechanical equivalent of heat showed that heat is a form of energy.

calorie (international table calorie). A unit of heat energy equal to 4.1868 joules. Formerly, the calorie was defined as the amount of heat required to raise the temperature of one gram of water through one degree centigrade. This value depends on the temperature of the water, leading to considerable confusion in the past about the use of the calorie. The *fifteen-degree calorie* or *small calorie* is the heat required to raise one gram of water from 14.5°C to 15.5°C at standard pressure. It is equivalent to 0.999 690 international table calorie.

calorific value. A measure of the quality of a fuel, equal to the amount of heat produced by complete combustion of unit mass of the fuel.

calorimeter. Any apparatus for making thermal measurements in calorimetry.

calorimetry. The measurement of heat as used in the determination of heat capacities, latent heats, calorific values, heats of combustion, and similar quantities.

calx. 1. A metal oxide formed by roasting an ore.
2. *See* calcium oxide.

camera. 1. An apparatus for producing a photographic image. It consists of a light-tight box with a lens to focus a real image onto photographic film. 2. (**television camera**). A device for converting an optical image into electrical signals. Essentially it consists of a lens system that focuses the image onto a screen

consisting of a *mosaic of light-sensitive particles. An electrical signal is obtained, proportional to the light falling on each element of the mosaic, by using photo-conduction (*see* vidicon) or photoemission (*see* image orthicon).

Camphor

camphor. A colourless crystalline ketone, $C_{10}H_{16}O$, obtained from the wood of the camphor tree. It is used as a plasticizer for celluloid and as an insect repellent. M.pt. 179°C; b.pt. 204°C; r.d. 0.99.

Canada balsam. A transparent resin obtained from certain coniferous trees, used as a cement in optical systems because its refractive index is similar to that of glass.

canal rays. Streams of positive ions obtained from a discharge tube by boring small holes in the cathode. Some of the ions moving to the cathode pass through forming positive rays. They were first obtained by Sir J. J. Thomson in his experiments on charge-to-mass ratios.

candela. Symbol: cd. The base *SI unit of luminous intensity, equal to the luminous intensity of 1⁄600 000 sq. metre of the surface of a black body at a temperature of 2040 kelvins and a pressure of 101 325 pascals. The temperature is the freezing point of platinum. *See also* luminance, candle.

candle. A former unit of luminous intensity. The *international candle* was defined in terms of the luminous intensity of the *Harcourt pentane lamp. The unit was replaced by the candela in 1948.

candlepower. *Luminous intensity as measured in candles.

canonical. Denoting or involving a general mathematical rule or formula.

canonical equations. A set of equations used in classical mechanics to describe the motion of a set of particles. They have the form $dp_i/dt = \partial H/\partial q_i$ and $dq_i/dt = H/\partial p_i$. H is the *Hamiltonian function, q the position coordinate, and p the momentum.

capacitance. Symbol: C. The ability of a system to store electric charge or a measure of this for a particular system, equal to the charge (Q) stored to raise the system's potential (V) by one unit (i.e. $C = Q/V$). The unit of capacitance is the farad.

capacitor (condenser). A device consisting of conductors and insulators with the property of storing electric charge. In the *parallel-plate capacitor* two parallel metal plates are separated by air or some other insulator (the dielectric). If a potential difference V is applied across the plates, each plate acquires a charge Q. The capacitance is Q/V, and is equal to $\epsilon A/d$, where A is the area of the plates, d the distance between them, and ϵ the permittivity of the material between the plates.
Capacitors are used to introduce *reactance into alternating-current circuits. The usual fixed type is constructed of layers of metal foil separated by a thin layer of waxed paper. An *electrolytic capacitor* has two aluminium electrodes separated by paper soaked in electrolyte or by liquid electrolyte. The dielectric is an insulating oxide layer on the anode. In this type one electrode must always have a positive potential with respect to the other. Variable capacitors have a set of moving plates and a set of fixed plates, so that the effective area can be changed.

capillarity. *See* capillary action.

capillary. Any long narrow hole or pore. A *capillary tube* is a tube with a narrow bore, especially a glass tube with thick walls, of the type used in mercury thermometers.

capillary action (capillarity). The movement of liquid through capillary tubes, small pores, etc., as caused by *surface tension.

caproic acid. *See* hexanoic acid.

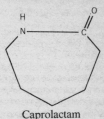

Caprolactam

caprolactam. A white crystalline powder, $C_6H_{11}NO$, used in the manufacture of nylon. M.pt. 70°C; b.pt. 150°C; r.d. 1.02.

capture. Any process in which an additional particle is acquired by an atom, ion, molecule, or atomic nucleus. The term is more usually applied to nuclear processes, as in the capture of a neutron by a nucleus, often followed by emission of a gamma ray (*radiative capture*). Another nuclear process is *K-capture* in which the nucleus acquires an orbiting electron from its K-shell. In the nucleus, a proton is changed into a neutron with associated emission of a neutrino. The atom is left with a vacant state in its K-shell which is filled by an electron from an outer shell with associated emission of an X-ray photon.

carat. 1. A unit of mass used for precious stones, equal to 2×10^{-4} kilogram. Formerly the carat was an Imperial unit of troy measure equal to 4 grains.
2. A measure of the quality of gold, equal to the number of parts of gold (by weight) in 24 parts of a gold alloy. Thus 24 carat gold is pure gold, 22 carat gold is 22 parts gold and 2 parts other metal, etc.

carbamide. See urea.

carbanion. A negatively charged organic ion in which the negative charge occurs on a carbon atom. An example is the ion $(C_6H_5)_3C^-$, which is found in the ionic compound triphenylmethyl sodium. The cyclopentadienyl ion, $C_5H_5^-$, is another example of a stable carbanion: in both these examples the ion is stabilized by resonance. Many carbanions exist only as transient intermediates in chemical reactions. See also carbonium ion.

carbene. A divalent species with the general formula $RR'C$:, where R and R' are atoms or groups. The simplest example is methylene, H_2C:, which can be produced by the photolysis of ketene or by heating diazomethane. Carbenes can exist in two states: one in which the two electrons are paired in the same atomic orbital and the other in which the electrons occupy different orbitals. The latter species is a diradical. All carbenes are extremely reactive, having a transient existence as intermediates in certain chemical reactions. They add to double bonds to form a three-membered cyclopropane ring. They are also able to insert themselves into C–H bonds: methylene for example produces C–CH_3.

carbide. Any compound of carbon with a more electropositive group or element. Carbides are often made by combination of the elements at high temperatures ($2000°C$). Three main types can be distinguished.
Interstitial carbides are formed by iron, cobalt, nickel, and a number of other transition metals. The carbon atoms occupy holes in the metal lattice and the substance has properties similar to those of the metal.
Covalent carbides are formed by silicon and boron. Both compounds are hard high-melting solids in which covalent bonds hold the atoms in the lattice.
Ionic carbides are salts. Two types are known. The methanides appear to contain C^{4-} ions and yield methane on hydrolysis. An example is aluminium carbide Al_4C_3. The acetylides contain the ion $^-C \vdots C^-$, derived from acetylene. They give acetylene on hydrolysis. Calcium carbide, CaC_2, is a common example and the term carbide is often used to refer to this compound.

carbinol. A former name for *methanol. The term was also used to name other alcohols as substituted derivatives of methanol. Ethanol, for example, was methylcarbinol.

carbocyclic. Denoting a chemical compound whose molecules contain a ring of carbon atoms.

carbodiimide. See cyanamide.

carbohydrate. Any of a class of compounds with the formula $C_nH_{2m}O_m$. The carbohydrates include the *sugars and high-molecular-weight compounds such as *starch and *cellulose.

carbolic acid. See phenol.

carbon. Symbol: C. A nonmetallic element present in all living matter, atmospheric carbon dioxide, and many mineral carbonates. There are two allotropic crystalline forms, *graphite and *diamond, and several amorphous forms, including *gas carbon, *charcoal, and *lamp black. The element is widely used in electrodes, colloidal conductive coatings, crucibles, rubber, inks, and carbon fibres. It reacts directly with other nonmetals at high temperatures. Metal *carbides can also be made. Carbon is unique in its ability to form molecules containing long chains of atoms and over a million *organic compounds are known. A.N. 6; A.W. 12.011; m.pt. above $3500°C$; r.d. 3.51 (diamond), 2.25 (graphite); valency 4.

carbon-14. The radioactive isotope of carbon with mass number 14, present in small amounts in nature (see radiocarbon dating). It is made by irradiation of a nitrate in a reactor and extensively used in *tracer studies. Half-life 5720 years.

carbon-14 dating. *See* radiocarbon dating.

carbonate. Any salt of carbonic acid.

carbonation. The introduction of carbon dioxide into a liquid under pressure. Carbon dioxide is only slightly soluble in water forming a weakly acidic solution of carbonic acid. More CO_2 can be dissolved in this solution under pressure, and when the pressure is released the gas comes out of solution in tiny bubbles.

carbon black. A fine powdered form of carbon made by incomplete combustion of hydrocarbons. It is used as a pigment in inks and as a filler for rubber.

$$^1_1H + {}^{12}_{6}C \longrightarrow {}^{13}_{7}N$$
$$^{13}_{7}N \longrightarrow {}^{13}_{6}C + {}^{0}_{1}e$$
$$^1_1H + {}^{13}_{6}C \longrightarrow {}^{14}_{7}N$$
$$^1_1H + {}^{14}_{7}N \longrightarrow {}^{15}_{8}O$$
$$^{15}_{8}O \longrightarrow {}^{15}_{7}N + {}^{0}_{1}e$$
$$^1_1H + {}^{15}_{7}N \longrightarrow {}^{12}_{6}C + {}^{4}_{2}He$$

Carbon cycle

carbon cycle. A cycle of *thermonuclear reactions producing the energy in some stars. It results in the formation of helium nuclei from hydrogen nuclei (see diagram).

carbon dioxide. A colourless odourless noncombustible gas, CO_2, produced by combustion of carbon compounds. It is also a product of respiration and is present in the atmosphere. Carbon dioxide is used in carbonated beverages, fire extinguishers, and for providing an inert atmosphere in welding. Solid carbon dioxide (*dry ice*) is used as a refrigerant. M.pt. -56.6 °C (5.2 atm); sublimes at -78.5 °C; r.d. 1.10 (-37 °C). *See also* carbonic acid.

carbon disulphide. A colourless poisonous flammable liquid, CS_2, which is almost odourless when pure but usually has a strong disagreeable odour caused by sulphur impurities. It is made from natural gas and sulphur. Carbon disulphide is used as a solvent and as a raw material for the manufacture of rayon and carbon tetrachloride. M.pt. -110.8 °C; b.pt. 46.3 °C; r.d. 1.26.

carbon fibres. Fibres of carbon made by heating fibres of rayon or other textiles. The material has an orientated crystal structure giving it a high tensile strength. It is used to reinforce resins, ceramics, or metals to produce strong light temperature-resistant materials.

carbonic acid. The weak disbasic acid, H_2CO_3, produced by dissolving carbon dioxide in water. It is only stable in aqueous solution and cannot be isolated. Its salts are the carbonates.

carbonium ion. A positively charged organic ion in which the positive charge occurs on a carbon atom. An example is the positive triphenylmethyl cation, $(C_6H_5)_3C^+$, which is stabilized by resonance. Carbonium ions exist as short-lived intermediates in many chemical reactions. *See also* carbanion.

carbonize. 1. To treat with carbon.
2. (**carburize**). To change from an organic compound into elemental carbon, or to coat something with carbon by such a process. Normally, organic compounds are carbonized by incomplete oxidation in a restricted supply of air. Complete oxidation would lead to carbon dioxide and water.

carbon monoxide. A colourless odourless flammable highly toxic gas, CO, produced by incomplete combustion of carbon compounds and manufactured by partial oxidation of coke or natural gas. Carbon monoxide is a neutral oxide. It burns with a blue flame and is used as a fuel (*see* water gas) and as a reducing agent in metallurgy. M.pt. -205 °C; b.pt. -190 °C; r.d. 0.97.

carbon tetrachloride (tetrachloromethane). A colourless nonflammable heavy liquid, CCl_4, with a characteristic sweet odour. It is an unreactive compound, made by the chlorination of carbon disulphide or methane and used as a fire extinguishant, refrigerant, and solvent. M.pt. -23 °C; b.pt. 76.5 °C; r.d. 1.59.

carbonyl. Any of a class of transition-metal compounds containing a metal atom coordinated to one or more carbon monoxide molecules. A common example is nickel carbonyl, $Ni(CO)_4$, which was used in the *Mond process.

carbonylation. Any chemical reaction that results in the production of a carbonyl group in a compound. An example is the dehydrogenation of propan-2-ol to acetone using a copper catalyst at $250°C$: $CH_3CHOHCH_3 = CH_3COCH_3 + H_2$.

carbonyl chloride. *See* phosgene.

carbonyl group. The divalent group $=CO$, as present in aldehydes, ketones, carboxylic acids, and metal carbonyls.

Carborundum. ® Silicon carbide used as an abrasive.

carboxylation. Any chemical reaction that results in the production of a *carboxyl group in a compound. The term is usually applied to the direct oxidation of an alkyl group to a carboxyl group, as in the reaction $ArCH_2CH_3 = ArCOOH$, where Ar is an aryl group.

carboxyl group. The organic group $-COOH$, containing a carbonyl group linked to a hydroxyl group. It is the functional group in carboxylic acids.

carboxylic acid. Any of a class of organic acids with the general formula R.CO.OH, where R is usually a hydrocarbon group. The strength of the acid depends on the nature of R: when it is a simple hydrocarbon group, as in acetic acid, the compound is usually a weak acid. Carboxylic acids can be made by oxidation of the corresponding alcohol or *aldehyde. They can be dehydrated to yield *anhydrides and react with halogenating agents to form *acid halides. They also show the usual reactions of acids, forming salts with alkalis and *esters with alcohols. Carboxylic acids are often called *fatty acids*, owing to the fact that esters of many higher

carboxylic acids are found in *fats and oils.

carburize. *See* carbonize.

carbylamine. An isocyanide.

carbylamine reaction. A test for primary amines in which the sample is heated with an alcoholic solution of potassium hydroxide and chloroform. The amine is converted into a carbylamine (isocyanide) which is easily recognized by its unpleasant smell.

Carius method. A technique for the quantitative analysis of sulphur and halogens in organic compounds. The sample is heated in a sealed tube with concentrated nitric acid and silver nitrate. Precipitated silver halides and silver sulphide are separated and weighed.

carnallite. A mineral consisting of a mixed chloride of potassium and magnesium, $KCl.MgCl_2.6H_2O$, used as a source of potassium fertilizers.

carnotite. A naturally occurring vanadate of uranium and potassium, $K_2(UO_2)_2(VO_4)_2.3H_2O$. It is an important ore of uranium and vanadium.

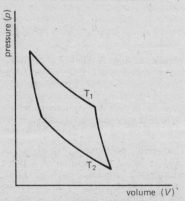

Carnot's cycle

Carnot's cycle. A reversible cycle of four operations applied to the gas in a heat engine. The successive stages are adiabatic compression, isothermal expansion, adiabatic expansion, and isothermal compression, thus restoring the original pressure, volume, and temperature. The Carnot cycle is the cycle of operations occurring in a perfect heat engine. [After Sadi Nicolas Léonard Carnot (1796-1832), French scientist.]

Carnot's theorem. The principle that the efficiency of a reversible heat engine depends only on its temperature range, not on the working substance.

Caro's acid. *See* peroxosulphuric acid.

carrier. 1. The entity, either an electron or positive hole, that carries the electric charge in a conductor or semiconductor. **2.** The normal form of a chemical compound to which the radioactive form is added in order to introduce it into a system for *tracer studies.

carrier gas. The gas used to carry the sample through the column in *gas-liquid chromatography.

carrier wave. The radio wave of constant amplitude and frequency that is modulated (*see* modulation) by the signal in *radio transmission.

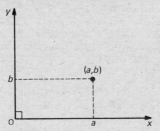

Rectangular Cartesian coordinates

Cartesian coordinates. *Coordinates used to locate the position of a point with respect to intersecting lines (*axes*), each coordinate being the distance from one axis measured parallel to the others. In a plane, a horizontal x axis and a vertical y axis is used, the point (a,b) being a distance a (the *abscissa*) along the direction of the x axis and b (the *ordinate*) along the direction of the y axis. Usually the axes are perpendicular (*rectangular coordinates*). In three dimensions a z axis is also used. [After René Descartes (1596-1650), French philosopher and mathematician.]

cascade. A process arranged in a series of consecutive stages so that the product of one stage is used as the starting material for the next. In the separation of *isotopes by diffusion, the enriched material from one stage is further enriched in the next, and so on.

cascade liquefier. A system for liquefying a gas by cooling it below its critical temperature and applying a pressure. A gas with a high critical temperature is first liquefied by pressure, then allowed to evaporate under reduced pressure thus lowering its temperature. This gas cools a second gas below its critical temperature, so that it can also be liquefied and evaporated to reduce the temperature still further. In this way the temperature is reduced in stages. If air is used liquid oxygen can be produced. The process cannot be used for hydrogen or helium because their critical temperatures are too low.

cascade shower. *See* shower.

case hardening. A method of hardening the surface of metals, particularly steel, by chemical reaction to produce a hard layer of metal carbide. A typical method involves dipping the hot iron into molten sodium cyanide and then quenching it.

casein. A white amorphous tasteless powder, the principal protein in milk. It is extracted from milk curds and is used in casein plastics, paper coatings, and glues.

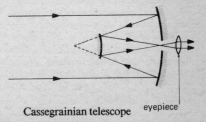

Cassegrainian telescope   eyepiece

Cassegrainian telescope. A type of astronomical *telescope in which the light is reflected by a large concave mirror onto a smaller convex mirror, and from there through a hole in the main mirror into the eyepiece (see illustration). [After N. Cassegrain, 17th-century French astronomer.]

cassiopeium. *See* lutetium.

cassiterite. A black, brown, or white mineral consisting of tin IV oxide, SnO_2. It is the principal ore of tin.

cast iron. A brittle iron-carbon alloy containing between 2% and 4.5% of carbon and small amounts of other elements such as sulphur, phosphorus, and silicon. The carbon is present as cementite or graphite.

castor oil. A pale yellow viscous oil extracted from the seeds of the castor plant. It is used as a laxative and is also a raw material for the manufacture of resins, plastics, and lubricants.

catalysis. The process by which the rate of a chemical reaction is increased by the presence of another substance, the *catalyst*, that does not appear in the stoichiometric equation for the reaction. The catalyst only affects the rate of attainment of equilibrium, not the position of equilibrium, and it follows that the rates of the forward and reverse reactions must be affected equally. Thus dehydrogenation catalysts, such as platinum and nickel, are also good hydrogenation catalysts.
A distinction is generally made between *homogeneous catalysis*, in which the entire reaction occurs in a single phase, and *heterogeneous catalysis*, in which the catalyst and the reactants are in different phases. In the latter case the catalyst is usually a solid and the reactants are gases.

cataphoresis. *See* electrophoresis.

catechol (pyrocatechol, 1,2-dihydroxy-benzene). A colourless crystalline substance, $C_6H_4(OH)_2$, used as a developer in photography. M.pt. 105°C; b.pt. 240°C; r.d. 1.4.

catenary. The curve in which a uniform rope would hang under gravity when suspended at each end. It has the equation $y = a(e^{x/a} + e^{-x/a})/2$.

catenation. The formation of chains of atoms in molecules.

cathetometer. A device for measuring heights and lengths at a distance, consisting of a small telescope fitted with cross-wires and mounted so that it can move along a graduated vertical scale.

cathode. An electrode with a negative electric charge, as in a cell, thermionic valve, rectifier, etc. *Compare* anode.

cathode glow. *See* glow discharge.

cathode-ray oscilloscope (CRO). An instrument for displaying changing electrical signals on a *cathode-ray tube. The signal to be investigated is applied, via an amplifier, to the vertical electrostatic deflecting plates of the tube. The spot on the screen is deflected by an amount proportional to the strength of the signal. The beam is also deflected horizontally by the time base, which moves the spot horizontally across the screen at a uniform rate, returning it quickly to the starting point and repeating the sweep. The trace on the screen is a graph of signal against time.

cathode rays. Streams of electrons emitted from the cathode of a gas-discharge tube, vacuum tube, etc.

cathode-ray tube (CRT). An electronic tube for producing a visual display from electrical signals, as used in television sets, radar screens, and cathode-ray oscilloscopes. It consists of an electron gun focusing a beam of electrons onto a fluorescent screen. The beam can be deflected, from side to side or up and down, by electrostatic or magnetic fields. Electrostatic deflection is effected by potentials applied to two pairs of plates mounted inside the tube on either side of the axis. In electromagnetic deflection wire coils are mounted outside the tube and the deflecting field is produced by a current in these coils.

cation (positive ion). An ion with a positive charge. Such ions are attracted to the cathode during electrolysis. *Compare* anion.

caustic. 1. A curve or surface that is the envelope of light rays reflected or refracted at a curved surface. Such rays are not focused to one point. They are all tangential to the cusp-shaped caustic curve (*see* spherical aberration).
2. Denoting a corrosive alkaline substance.

caustic potash. *See* potassium hydroxide.

caustic soda. *See* sodium hydroxide.

Cavendish's experiment. An experiment for determining the gravitational constant and demonstrating the inverse-square law of *gravitation. A long thin beam with a small lead sphere at each end was suspended by a wire. Large lead spheres were placed close to the smaller spheres, first to one side of each sphere and then to the other. Using the system as a torsion balance, it was possible to calculate *G* from the change in angular deflection. [Performed in 1798 by Henry Cavendish (1731–1810), English scientist.]

cavitation. The formation of small vapour-filled bubbles in a liquid in regions in which it has a high velocity. Cavitation occurs around propellers, in pumps, and in similar devices. The cavities formed in the low pressure region are carried into a region of higher pressure where they collapse, causing pitting of the surface.

celestial equator. The great circle on the *celestial sphere parallel to the earth's equator and perpendicular to the great circle passing through the celestial poles.

celestial mechanics. The branch of astronomy concerned with the relative motion of celestial objects under the influence of gravitational forces.

celestial poles. The northern and southern points in the sky where the earth's axis would intersect the *celestial sphere.

celestial sphere. The sky thought of as the inner surface of a sphere of infinite radius, centred on the position of the observer or the centre of the earth, upon which celestial objects appear to be projected. The fixed position of the stars are specified with respect to the celestial equator and the celestial poles; the positions of celestial objects as they change relative to the observer are given with reference to the celestial poles, the horizon, and the zenith. Other fixed coordinate systems are based on the ecliptic and the plane of the Galaxy.

celestine. A mineral, white, red, or blue in colour, consisting of strontium sulphate, $SrSO_4$. It is a source of strontium salts.

cell. 1. A device for converting chemical energy directly into electrical energy, consisting of two electrodes dipping into an electrolyte. Ions are produced or discharged at each electrode so that one acquires a positive potential and the other a negative potential. When the electrodes are connected through a circuit, a current flows. A *primary cell* is one in which the electrical energy is produced by the reaction occurring in the cell. A *secondary cell* (or *accumulator) is one used for storing electricity.
2. Any of various similar devices having electrodes and an electrolyte used for electrolysis, study of electrochemical processes, etc.

Cellophane. Ⓡ A thin transparent cellulose sheeting made by the *viscose process, used mainly in wrapping and packaging.

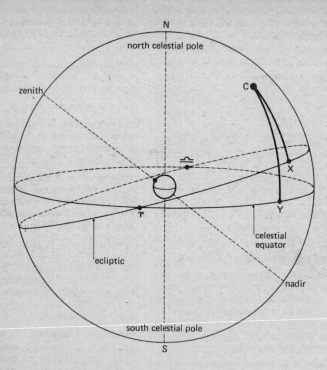

♈ first point of Aries; vernal equinox

♎ first point of Libra; autumnal equinox

C celestial object

♈ Y right ascension of C (in hours anticlockwise from ♈)

♈ X celestial longitude of C (in degrees anticlockwise from ♈)

YC declination of C

XC celestial latitude of C

Celestial coordinates

celluloid. A highly flammable thermoplastic substance made from cellulose nitrate with a camphor plasticizer. It is used in film.

cellulose. A white solid *carbohydrate, $(C_6H_{10}O_5)_n$, found in all plants as the main constituent of the cell wall. Cellulose can be hydrolysed to glucose by acids. The usual source is wood pulp: it is used in making *rayon and cellulose acetate plastics.

cellulose acetate. A synthetic material made by treating cellulose (usually wood pulp) with acetic anhydride or similar acetylating agents. It is used with plasticizers in making many moulded and extruded articles and in making acetate film. Cellulose acetate fibre is used in acetate *rayon.

cellulose nitrate. *See* nitrocellulose.

Celsius scale. The official name for the *centigrade scale. Temperatures are measured in *degrees Celsius* (˚C), which are identical to degrees centigrade. The term has been recommended since 1948 but *centigrade* remains in widespread use. [After Anders Celsius, 1701–44, Swedish astronomer.]

cement. A powdered mixture of calcium silicates and aluminates made by heating limestone with clay. When mixed with water it hardens to a solid mass.

cementite. Iron carbide, Fe_3C, present in cast iron and steel.

centi-. Symbol: c. Prefix indicating one hundredth (10^{-2}: e.g. 1 centimeter (cm) $= 10^{-2}$ m).

centigrade scale. A *temperature scale in which the melting point of ice is $0˚$ and the boiling point of water is $100˚$. Temperature is measured in *degrees centigrade* (˚C). Since 1948 the recommended name has been the *Celsius scale*.

centimetric waves. *Microwaves with wavelengths in the range 1–10 centimetres.

central force. A force on a moving body that is always directed towards a fixed point. A planet, for example, experiences a central force directed towards the sun.

central processing unit (CPU). The part of a digital *computer in which operations are performed on the data.

centre of curvature. *See* lens, mirror.

centre of gravity. A fixed point through which the weight of a body can be considered to act, irrespective of the body's position.

centre of mass. A fixed point at which the mass of a body can be considered to act. In a uniform gravitational field the body's centre of mass coincides with its *centre of gravity.

centrifugal force. A force that appears to act radially outwards on a body that moves in a curved path. It is equal and opposite to the *centripetal force. A body with mass m moving in a circular path of radius r with velocity v suffers a centrifugal force mv^2/r.

centrifugal separation. A method of separating *isotopes by using a large centrifuge. The technique can be applied to both liquid and gaseous samples. Under the action of the centrifugal force, radial concentration gradients are set up in which the lighter isotope tends to concentrate nearer the axis.

centrifuge. An apparatus for separating suspended particles in a liquid by centrifugal force. It consists of a rapidly rotating bar or wheel with tubes attached containing the suspension, free to pivot outwards at high speed. *See also* ultracentrifuge.

centripetal force. A force acting radially inwards constraining a body to move in a curved path. It is equal and opposite to the *centrifugal force.

centroid. The point associated with a geometric figure or solid at which the centre of mass would occur if the geometric object were made of homogeneous material. The centroid of a plane figure, for example, is the centre of mass of a thin sheet of material with the same shape and dimensions as the figure.

ceramic. Any of various hard high-melting nonmetallic inorganic materials, including pottery, porcelain, enamels, and refractory substances. Many ceramics are metal silicates and aluminosilicates, obtained by fusing clay. Other examples include metal oxides, borides, carbides, and nitrides and such compounds as ferrites and titanates. *See also* cermet.

Cerenkov radiation. Bluish light emitted when charged particles move through a medium at a velocity greater than the velocity of light in that medium. The effect is used as the basis for a particle counter (*Cerenkov counter*) by detecting the light with a photomultiplier. [After Pavel Alekseyevic Cerenkov (b. 1904), Soviet physicist.]

ceric. *See* cerium.

cerium. Symbol: Ce. A grey malleable ductile metallic element belonging to the *lanthanide series. It forms salts with a valency of 3 (*ceric*) or 4 (*cerous*). A.N. 55; A.W. 132.9055; m.pt. 28.4°C; b.pt. 678.4°C; r.d. 1.87.

cermet. Any of a class of materials made by sintering or bonding together powdered ceramic and metal. Cermets are hard and resistant to corrosion and high temperatures.

cerous. *See* cerium.

cetane rating. A number measuring the ignition qualities of diesel fuel. It is the volume percentage of *cetane* (hexadecane: $C_{16}H_{34}$) in a mixture of cetane and α-methylnaphthalene ($C_{10}H_7CH_3$) with the same ignition qualities as the fuel tested. A similar measure, the *octane rating, is used for petrol.

CGS units. A system of *units based on the centimetre, the gram, and the second. The system includes the dyne as

the unit of force and the erg as the unit of energy. The calorie is a unit of heat energy.

Two types of electrical unit are used in the CGS system: *electrostatic units in which the electric constant, ϵ_0, is unity and *electromagnetic units in which the magnetic constant, μ_0, is unity. The constants are related by $1/(\epsilon_0\mu_0)^{1/2} = c$, where c is the velocity of light.

In practice, *Gaussian units* have usually been used. In these the units of electrical quantities (*Q, V, I, E,* etc.) are the same as the electrostatic units and the units of magnetic quantities (*H, B, ϕ,* etc.) are the same as the electromagnetic units. Equations including quantities from both groups contain a factor c.

chain reaction. Any process in which a product of one reaction initiates a further reaction, giving more product, which in turn causes further reaction, and so on.

In a *nuclear chain reaction* a series of *fission reactions occur. A uranium-235 nucleus, for example, may capture a neutron and undergo *fission. The fragments include two or three further neutrons. Under favourable conditions these may cause further fissions, leading to even more neutrons. The reaction becomes self-sustaining with the number of free neutrons increasing—it is said to be *supercritical.* On the other hand, the neutrons may not induce sufficient fissions to cause a self-sustaining process—the reaction is *subcritical.*

In a *chemical chain reaction* the reaction is maintained by the product of one stage taking part in the next. Chain reactions often occur when free radicals are produced: a free radical may attack a molecule, removing a hydrogen atom and leaving another radical, which in turn attacks another molecule, etc.

chair configuration. *See* configuration.

chalcogens. The elements oxygen, sulphur, selenium, tellurium, and polonium, belonging to group VIA of the periodic table.

chalcopyrite (copper pyrites). A yellow lustrous naturally occurring mixed sulphide of iron and copper, $(Cu,Fe)S_2$, important as the main ore of copper.

chalk. A soft porous natural form of calcium carbonate, $CaCO_3$, produced by shells of tiny marine organisms.

change of state. The change of a substance from one *state of matter to another. Changes of state include the processes of melting, boiling, freezing, condensation, and sublimation.

characteristic. See logarithm.

charcoal. An amorphous form of carbon made by the destructive distillation of wood or other organic matter. It is used as an absorbent (see activate).

charge. 1. Symbol Q. The property of certain elementary particles that enables them to attract or repel one another by *electromagnetic interation. Charge is conventionally designated as positive or negative, such that like charges repel one another and unlike charges attract. The electron has the smallest natural unit of negative charge with the proton having an equal amount of positive charge. A body or system has a negative charge if it contains more electrons than protons and a positive charge if it contains more protons than electrons. Quantitatively, charge is taken to be the product of an electric current and the time for which it flows. If the current is not constant, the time integral is used. Charge is measured in coulombs. See also electrostatic units, electromagnetic units.
2. (of a capacitor) The electric charge on the positive plate of a parallel-plate capacitor.
3. The total amount of electricity stored in an accumulator. It is usually measured in ampere-hours.

charge-transfer complex. A type of complex formed by partial electron transfer between two different molecules. Iodine forms such complexes with certain aromatic compounds.

charm. A postulated property of certain elementary particles arising from the idea that a multiplet of particles with SU3 symmetry (see unitary symmetry) may be a subgroup of a larger group of particles with symmetry SU4. The octet of mesons, for example, may be a subgroup of a multiplet of fifteen particles. The grouping of particles using the SU3 symmetry group implies that elementary particles may be composed of three types of *quark. If SU4 symmetry groups do occur an extra quark is required having a charge 2/3, zero strangeness, and the extra property of charm. The additional seven mesons would consist of combinations of a charmed quark with the other types of quark. One of these, containing the charmed quark and its antiquark, would have zero charm but the others would have a charm of +1 or -1. Similarly, charmed baryons may be expected to occur.

So far no charmed particles have been detected but the concept helps to explain certain particle-scattering processes and may also account for the existence and properties of the *psi particle.

Charles' law. The principle that a fixed quantity of gas at constant pressure has a volume that is inversely proportional to its temperature. If the thermodynamic temperature is used the law has the form $V = AT$, where A is a constant for all gases. The law strictly applies only to ideal gases: real gases tend to obey it at low pressures.

If the volume of a gas is known at $0°C$, the volume increases or decreases by 1/273 of this value for each degree rise or fall in temperature. This is sometimes given as a statement of Charles' law. The principle is also called *Gay-Lussac's law.* See also gas laws. [After Jaques Charles (1746–1823), French physicist.]

chelate. An inorganic metal complex in which there is a closed ring of atoms, caused by attachment of a *ligand to a metal atom at two points. An example

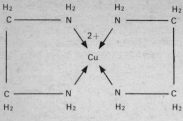

Copper chelate

is the complex ion formed between ethylene diamine and cupric ions, [Cu-$(NH_2CH_2CH_2NH_2)_2]^{2+}$.

chelating agent. A molecule, ion, or radical that can form chelates. Chelating agents are capable of coordinating to a metal atom at more than one point. If the *ligand coordinates at two points it is said to be *bidentate*: ligands that coordinate at three points (forming two rings) are *tridentate*, etc. Chelating agents are often used for removing metal ions from solution or neutralizing their chemical activity. See sequestration.

chemical combination, laws of. Three laws describing the way in which elements combine together to form compounds. They are the laws of *constant composition, *multiple proportions, and *reciprocal proportions. The third is derivable from the first two. See also Dalton's atomic theory.

chemical equilibrium. The state of a reversible chemical reaction when its forward and reverse reactions occur at the same rates, so that there is no apparent change with time in the amounts of products and reactants present. The relative amounts of products and reactants are then given by the *equilibrium constant.

chemical equivalent. See equivalent weight.

chemical reaction. A process in which one or more chemical substances change into other substances by breaking or forming chemical bonds.

chemiluminescence. *Luminescence occurring as a result of chemical reaction. Part of the energy of the reaction is taken up by exciting the electrons so that molecules of one of the products appear in excited states, from which they decay by emission of light. *Luminol*, for example, is a crystalline compound that emits blue light when oxidized with hydrogen peroxide. See also bioluminescence.

chemisorption. *Adsorption in which the adsorbed substance is held by chemical bonds.

chemistry. The branch of science concerned with the elements and their compounds. Chemistry involves the study of effects produced by the orbiting electrons of atoms, rather than by atomic nuclei. It is divided into *inorganic, *organic, and *physical chemistry.

Chile saltpetre. See sodium nitrate.

Chinese white. See zinc oxide.

chip. See integrated circuit.

Chladni's figures. Regular patterns obtained from fine powder scattered on a vibrating plate. The usual method of obtaining such patterns is to mount a horizontal plate on a central pillar and vibrate it by an electromagnet or mechanical vibrator. The powder collects along nodal lines of the plate where the vibration is least. [After Ernst Chladni (1756–1827), German physicist.]

chloral (trichlorethanal). A colourless mobile oily liquid, CCl_3CHO, made by the chlorination of acetaldehyde and used in the manufacture of DDT. M.pt. -57.5°C; b.pt. 97.8°C; r.d. 1.51.

chloral hydrate. A colourless aromatic poisonous solid *diol, $CCl_3CH(OH)_2$, prepared by the hydrolysis of chloral and used as a sedative and in the manufacture of DDT. M.pt. 57°C; b.pt. 96.3°C; r.d. 1.91.

chloramine. A colourless explosive liquid, NH_2Cl, produced by the action of sodium hypochlorite on ammonia. M.pt. $-66°C$.

chlorate. Any salt of chloric acid.

chloric acid. A strong oxidizing acid, $HClO_3$, stable only in aqueous solution. It is formed along with chlorous acid when chlorine dioxide is dissolved in water. When concentrated, it decomposes into chlorine and perchloric acid. Its salts are the chlorates.

chloride. Any of certain compounds containing chlorine. *See* halide.

chloride of lime. *See* bleaching powder.

chlorination. 1. Treatment with chlorine, as in the disinfection of water.
2. A chemical reaction in which one or more chlorine atoms are introduced into an organic compound.

chlorine. Symbol: Cl. A greenish-yellow poisonous nonflammable gas with a pungent odour, occurring in many minerals, especially in common salt. It is obtained by electrolysis of solutions of sodium chloride. The element is used in the manufacture of paper and many organic chemicals. It is also employed in water treatment (chlorination) and in bleaching textiles. It is a reactive element belonging to the *halogen group. A.N. 17; A.W. 35.43; m.pt. $-100.98°C$; b.pt. $-34.6°C$; r.d. 2.5 (air = 1); valency 1, 3, 5, or 7.

chlorine oxide. Any of four oxides of chlorine, all unstable explosive compounds. The lowest is *chlorine monoxide*, Cl_2O, which is a brownish-yellow gas formed by the action of chlorine on mercury II oxide. It is the acid anhydride of hypochlorous acid. M.pt. $-116°C$; b.pt. $2°C$. *Chlorine dioxide*, ClO_2, is a reddish-yellow gas made by the action of concentrated sulphuric acid on potassium chlorate at low temperature. It is a mixed anhydride, yielding a mixture of chlorous and chloric acids in water. M.pt. $-59°C$; b.pt. $11.0°C$. The two higher oxides are both unstable liquids. *Chlorine trioxide*, Cl_2O_6, is made by reaction between chlorine and ozone. M.pt. $3.5°C$. *Chlorine heptoxide*, Cl_2O_7, is produced by dehydrating perchloric acid. M.pt. $-91.5°C$; b.pt. $82°C$.

chlorite. Any salt of chlorous acid.

chloroacetic acid. Any of three acids that can be produced by reacting chlorine with acetic acid in the presence of red phosphorous. *Monochloroacetic acid*, $CH_2ClCOOH$, is a colourless crystalline solid. M.pt. $55.6°C$ (β form); b.pt. $189°C$; r.d. 1.37. *Dichloroacetic acid*, $CHCl_2COOH$, is a colourless liquid. M.pt. $-4°C$; b.pt. $193°C$; r.d. 1.57. *Trichloroacetic acid*, CCl_3COOH, is a white crystalline solid. M.pt. $57.5°C$; b.pt. $197.5°C$; r.d. 1.63. All three acids are used as intermediates. The presence of chlorine atoms in the methyl group makes the chloroacetic acids stronger than acetic acid itself, trichloroacetic acid being the strongest.

chlorobenzene (monochlorobenzene). A colourless mobile flammable liquid, C_6H_5Cl, with an almond-like odour. It is manufactured by catalytic chlorination of benzene and is used as a raw material in making phenol, aniline, and DDT and as a solvent for paints. M.pt. $-45°C$; b.pt. $132°C$; r.d. 1.11.

chloroethane (ethyl chloride). A flammable gaseous alkyl halide, C_2H_5Cl, with an ethereal odour and a burning taste. It is made by the addition of hydrogen chloride to ethene and used as a refrigerant, local anaesthetic, and alkylating agent. M.pt. $-137°C$; b.pt. $12.3°C$; r.d. 0.9.

chloroethene (chloroethylene, vinyl chloride). An easily liquefied gas, CH_2:-$CHCl$, with an ethereal odour, manufactured by chlorination of ethene. It is readily polymerized, hence its use in the manufacture of *polyvinyl chloride. M.pt. $-160°C$; b.pt. $-14°C$; r.d. 0.91.

chloroform (trichloromethane). A colourless volatile liquid, $CHCl_3$, with a characteristic sweet odour. It is made by reacting chlorinated lime with acetone, acetaldehyde, or ethanol. Chloroform is now principally used in making fluorocarbon compounds. It is also a solvent and anaesthetic. M.pt. $-63.5°C$; b.pt. $61°C$; r.d. 1.48.

chloromethane (methyl chloride). A colourless flammable gaseous alkyl halide, CH_3Cl, made by the chlorination of methane. Chloromethane is used as a refrigerant and a local anaesthetic. M.pt. $-97°C$; b.pt. $-24°C$; r.d. 0.92.

chloroplatinic acid. See platinum chloride.

chloroprene. A colourless liquid, $H_2C:CHCCl:CH_2$, made from vinyl acetylene and hydrochloric acid. It is the monomer for the manufacture of neoprene rubbers. B.pt. $59°C$; r.d. 0.96.

chlorosulphonic acid. A colourless or light yellow fuming highly corrosive liquid, $ClSO_2OH$, used in the manufacture of detergents and pharmaceuticals. M.pt. $-80°C$; b.pt. $158°C$; r.d. 1.77.

chlorous acid. An unstable acid, $HClO_2$, formed together with chloric acid when chlorine dioxide is dissolved in water. Its salts are the chlorites.

choke. A coil of wire used as an *inductor to impede the flow of an alternating current in a circuit.

chord. A straight line joining two points on a curve.

choroid. See eye.

chromate. Any salt of chromic acid.

chromatic aberration. An *aberration of lenses caused by the fact that the refractive index of a material depends on the wavelength of the light, so that different colours are focused at different points. Images formed in this way are surrounded by fringes of spectral colours. Chromatic aberration can be avoided by using *achromatic lenses.

chromatography. Any of several related techniques for separating and analysing mixtures by selective adsorption or absorption in a flow system.

Column chromatography is used for separating fairly large quantities of mixtures in solution. The apparatus is a vertical glass column filled with powdered activated alumina or similar material. The sample is introduced into the top and washed slowly through the column with the solvent. This is known as *elution* of the sample. The rate at which a component of the mixture travels through the alumina depends on the extent to which it is adsorbed: a strongly adsorbed substance would have more affinity for the solid (the *stationary phase*) than for the solvent (the *moving phase*) and would move down the column slowly. A weakly adsorbed component would be washed through quickly. The mixture thus separates as it moves through the column: if the components are coloured they are visible as bands. The usual technique is to collect the solution (the *eluate*) as it drips from the column and collect the separate fractions. Column chromatography is extensively used in organic laboratory preparations. An extension of the technique—*ion-exchange chromatography*—can be used for separating substances that exist as ions in solution by adsorption on an *ion-exchange resin. It can be used for separating inorganic compounds with a high degree of efficiency. For example, it can be employed to separate the closely similar compounds of different lanthanides.

Gas chromatography is a similar technique in which the sample is a mixture of gases or vapours and the column is a long thin tube packed with absorbent. The sample is carried through by a continuous stream of a *carrier gas*, such as hydrogen. Gas chromatography can be employed for separation of mixtures but its main use is in analysis. The stream of gas emerging from the column is monitored electronically to detect the presence of components, which are identified by the time they take to pass through the column. In gas chromatography the stationary phase is often a thin layer of grease or other liquid deposited on solid particles. The method is called *gas-liquid chromatography*.

Paper chromatography is another analytic technique for separating mixtures in solution. The stationary phase is a strip of porous paper and the sample, in the form of a spot near one edge, is eluted by solvent permeating through the paper. The components separate into spots which are carried along by the moving solvent. If the compounds are colourless they can be made visible by spraying with dye or by their fluorescence in ultraviolet light. They are identified by the rate at which they move. A similar technique is *thin-layer chromatography* in which the stationary phase is made by spreading a thin even layer of alumina slurry on a glass plate and drying the plate in an oven.

chrome alum. *See* chromium potassium sulphate.

chrome yellow. A yellow pigment consisting of lead chromate, $PbCrO_4$.

chromic. *See* chromium.

chromic acid. A hypothetical acid, H_2CrO_4, known in the form of chromate salts. The name is also sometimes used for chromium trioxide.

chromite. A black or brownish-black mixed oxide of iron and chromium, $FeO.Cr_2O_3$, important as the only commercial chromium ore.

chromium. Symbol: Cr. A hard grey lustrous *transition element, which can be given a high silvery polish, found in chromite ($FeO.Cr_2O_3$), from which it is extracted by the *thermite process. The metal is used in alloys, particularly stainless steels, and in shiny electroplated coatings on steel parts. It is not attacked by oxygen and is generally resistant to corrosion, although it does react with oxygen, sulphur, the halogens, and other nonmetals at high temperatures. Although it dissolves in

nonoxidizing acids nitric acid makes the chromium passive to attack. Chromium has a variable valency, forming chromium II (*chromous*) and chromium III ionic compounds. Many complexes exist. The element also forms compounds in a +6 oxidation state (*chromic*), as in chromic oxide, CrO_3, and chromic acid, H_2CrO_4. A.N. 24; A.W. 51.996; m.pt. 1890°C; b.pt. 2482°C; r.d. 7.19; valency 2, 3, or 6.

chromium oxide. Any of several oxides of chromium. The lowest oxide is *chromium II oxide* (chromous oxide), CrO. *Chromium III oxide* (chromic oxide), Cr_2O_3, is an amphoteric green solid used as a pigment. M.pt. 2435°C; b.pt. 4000°C; r.d. 5.21. *Chromium VI oxide* (chromium trioxide), CrO_3, is a red deliquescent solid used as an oxidizing agent and in the manufacture of many chemicals. It is the acid anhydride of chromic acid and is sometimes called chromic acid itself. M.pt. 196°C; r.d. 2.7-2.8. *Chromium IV oxide* (chromium dioxide), CrO_2, is a black solid.

chromium potassium sulphate (chrome alum). A violet-red efflorescent crystalline *alum, $K_2SO_4.Cr_2(SO_4)_3.24H_2O$, used as a mordant. M.pt. 89°C; r.d. 1.81.

chromophore. The group responsible for the colour of a dye or other chemical substance. In azo dyes, for example, the chromophore is the azo group, $-N=N-$.

chromosphere. The gaseous atmosphere of the *sun.

chromous. *See* chromium.

chromyl chloride. A dark red liquid, CrO_2Cl_2, used as an oxidizing agent in organic synthesis. M.pt. 96.5°C; b.pt. 116°C; r.d. 1.91.

chronon. The time taken for a photon to travel a distance equal to the diameter of the electron: i.e. about 10^{-24} second.

cinnabar. A bright red naturally occurring form of mercury II sulphide, HgS, important as the principal ore of mercury.

cinnamic acid (3-phenylpropenoic acid). A white crystalline carboxylic acid, $C_6H_5CH:CHCOOH$, occurring naturally in certain balsams and essential oils. It exists in two isomeric forms, the *trans* isomer being the common one. Esters of cinnamic acid are used in perfumery and medicine. M.pt. 133°C; b.pt. 300°C; r.d. 1.28 (*trans*).

circle. The locus of a point that moves so that it is always a fixed distance from a fixed point. The area is πr^2 and the circumference $2\pi r$, where r is the radius. A circle is an ellipse with an eccentricity of zero.

circuit. An arrangement of conductors, resistors, capacitors, etc., through which a current can flow.

circularly polarized light. *See* polarization.

circular measure. Measure of angles in *radians.

circular mil. A unit of area equal to the area of a circle whose diameter is 0.001 inch.

circumcentre. The centre of the circle that passes through all three vertices of a triangle. Since this point is equidistant from the vertices, it is the point of intersection of the perpendicular bisectors of the three sides of the triangle.

circumference. The line enclosing any simple closed curve, especially a circle; the length of this line. In the case of a circle of radius r, the circumference is $2\pi r$.

circumpolar Describing stars or constellations that always remain above the horizon when observed from a given geographical latitude.

cis. Indicating the position of a group or atom in a molecule adjacent to the position of a specified group or atom: i.e. on the same side of a double bond or central atom. *See* geometrical isomerism.

cis-trans isomerism. *See* geometrical isomerism.

citrate. Any salt or ester of citric acid.

citric acid. A white crystalline solid, $HO_2CCH_2C(OH)(CO_2H)CH_2CO_2H$ with an acid taste, occurring widely in both plant and animal cells. It is extracted by industrial fermentation of molasses or citrus fruits. M.pt. 153°C; r.d. 1.67.

cladding. The process of coating one metal by another, as to prevent corrosion.

Clark cell. A type of primary cell in which the positive electrode is mercury coated with a paste of mercury sulphate, the negative electrode is zinc, and the electrolyte is saturated zinc sulphate solution. It was formerly used as a standard of e.m.f., the value being 1.4345 volts at 15°C. [After William Mansfield Clark (1884-1964), U.S. chemist.]

Clark process. *See* hard water. [After Thomas Clark (1801-67), Scottish chemist.]

classical. Relating to physical theories that do not use either the *quantum theory or the theory of *relativity.

clathrate. A compound formed by physical trapping of molecules of one substance in holes in the crystal lattice of another. Clathrates are produced by crystallizing a mixture of the two substances. Quinol, for example, has holes in its lattice that can trap oxygen, methane, methanol, and other compounds with small molecules. The ratio of quinol:compound can vary up to a maximum value of 3:1. Ice can form clathrate compounds with argon, krypton, and xenon. In such compounds no chemical bond is formed between the host compound and the trapped molecule. *See also* zeolite.

cleavage. Splitting of a crystal to form smooth surfaces. Cleavage occurs along planes of atoms in the crystal.

Clinical thermometer

clinical thermometer. A mercury thermometer designed for taking the temperature of the body. It has a constriction in the capillary just above the bulb so that the thread of mercury remains in position after reaching its maximum value. It is returned to the bulb by shaking. The instrument is usually calibrated in the range 95–115°F, normal body temperature being 98.4°F.

cloud chamber. An apparatus for investigating elementary particles by the tracks they produce on passing through a supersaturated vapour. In the simplest type, the vapour is subjected to a sudden adiabatic expansion, which cools the vapour and produces supersaturation. Particles passing through the chamber leave a stream of ions along their path, which serve as nuclei for the formation of minute drops of liquid, thus producing a visible track. This apparatus is often called the *Wilson cloud chamber* after its inventor.

Clusius column. An apparatus for separating isotopes, consisting of a long vertical column many metres high with a heated wire running down its axis. The material to be separated must be gaseous. A radial temperature gradient builds up in the tube between the hot wire and the wall. The lighter molecules tend to diffuse towards the wall and heavier molecules collect around the wire as a result of *thermal diffusion. Convection currents are set up and the lighter isotope collects at the top of the tube.

cluster. Any of a number of collections of stars that move together in the same direction. Two main types exist: *open* or *galactic clusters* consist of relatively small numbers of widely spaced young stars and lie chiefly in the spiral arms of the Galaxy. *Globular clusters* are found in the vast spherical halo that encompasses the central nucleus and consist of enormous numbers of closely packed older stars. Open clusters are near neighbours whereas globular clusters are much more remote.

coagulation. The formation of macroscopic agglomerations in *colloids leading, in some cases, to their eventual solidification.

coal gas. A fuel gas produced by the destructive distillation of coal, consisting of a mixture of hydrogen (about 50%), methane (35%), carbon monoxide (8%), and other hydrocarbons and smaller amounts of nitrogen, oxygen, and carbon dioxide.

coal tar. A black tar produced by the destructive distillation of coal, consisting of a mixture of aromatic hydrocarbons, including benzene, toluene, xylene, and naphthalene together with phenol and some free carbon.

cobalt. Symbol: Co. A grey hard *transition element found in several ores, including smaltite ($(Co,Ni)As_2$) and cobaltite (CoAsS). It is usually obtained by roasting, to give the oxide, followed by reduction with aluminium. Cobalt is a ferromagnetic element and is used in making ferromagnetic alloys as well as in high-speed and stainless steels. It forms two series of compounds: cobalt II (*cobaltous*) and the less stable cobalt III (*cobaltic*) ionic salts as well as many complexes. Cobalt also forms a few compounds in other oxidation states. A.N. 27; A.W. 58.933; m.pt. 1495°C; b.pt. 2870°C; r.d. 8.9; valency 2 or 3.

cobalt chloride. A dark red crystalline solid, $CoCl_2.6H_2O$. The anhydrous salt is blue. Cobalt II chloride (cobaltous chloride) is the only stable chloride of cobalt. M.pt. 86.7°C; r.d. 1.9.

cobaltic. *See* cobalt.

cobaltous. *See* cobalt.

cobalt oxide. Any of three oxides of cobalt, all used in pigments and ceramics. *Cobalt II oxide* (cobaltous oxide) is a green-grey powder, CoO, prepared by heating cobalt II carbonate in nitrogen. M.pt. 1935°C; r.d. 6.5. *Tricobalt tetroxide* (cobalto-cobaltic oxide) is a steel-grey powder, Co_3O_4, obtained by heating other cobalt oxides. It changes to cobalt II oxide when heated above 1000°C. R.d. 6.1. Cobalt III oxide (cobaltic oxide, cobalt black) is a steel-grey or black powder, Co_2O_3, made by heating cobalt III hydroxide. Decomposes above 895°C; r.d. 4.8-5.6.

cocaine. A colourless crystalline solid alkaloid, $C_{17}H_{21}NO_4$, extracted from the leaves of the coca tree. Cocaine and its derivatives are used as local anaesthetics. When taken they can be addictive. M.pt. 98°C.

codeine. A white poisonous crystalline alkaloid, $C_{18}H_{21}NO_3$, extracted from opium or made by methylating morphine. Codeine and its derivatives are narcotic and analgesic drugs.

coefficient. 1. A constant multiplier of a variable in an algebraic expression. In $5x^3 + 2x^2$, 5 and 2 are coefficients.
2. A constant measuring some physical property, as in the coefficient of *friction, the coefficient of *expansion, the coefficient of *viscosity, or the coefficient of *restitution.

coefficient of expansion. *See* expansion.

coercive force. The magnetic field strength required to remove the residual magnetism of a ferromagnetic material. If the substance is initially saturated, the coercive force is known as the *coercivity. See* hysteresis.

coherent. Relating to electromagnetic radiation that has a single phase. Light from normal sources is not coherent. It consists of a large number of waves all out of phase with one another. *Laser radiation, on the other hand, is coherent. Two sources of electromagnetic radiation are said to be coherent if they produce waves that are in phase.

coherent units. A system of *units in which a derived unit is produced simply

by multiplying or dividing base units without introducing a numerical factor. The *SI units form a coherent system.

coinage metals. The group of metals copper, silver, and gold.

colatitude. *See* polar coordinates.

colcothar. Red iron oxide, Fe_2O_3.

cold cathode. A cathode that emits electrons by *field emission rather than by thermionic emission.

colligative property. A property that depends on the number of particles of substance present in a substance, rather than on the nature of the particles. Examples are osmotic pressure (*see* osmosis) and the *elevation of boiling point, *depression of freezing point, or change in vapour pressure (*see* Raoult's law) of solvents caused by dissolved substances.

collimator. 1. A system of slits and (sometimes) lenses for producing a parallel beam of light, other electromagnetic radiation, or particles.
2. *See* finder.

collodion. A thin film of nitrocellulose obtained by coating a surface with nitrocellulose dissolved in a volatile solvent and allowing the solvent to evaporate.

colloid. A substance made up of small particles larger than atoms or molecules but too small to be seen by a normal microscope. Usually the particles are dispersed in a medium, giving a *colloidal solution* or *colloidal suspension*. Colloid particles have diameters in the range $10^{-9} - 10^{-5}$ metre. They can be detected by the *ultramicroscope.
Colloid systems have a variety of properties depending on the nature of the components. They are broadly classified into *reversible* and *irreversible colloids* depending on whether the components can be separated and the dispersion regenerated by mixing. Another classification is into *sols, *gels, and *emulsions.

cologarithm. The *logarithm of the inverse of a number expressed with a positive mantissa. Thus if $\log_{10}(25) = 1.3979$, then $\log_{10}(1/25) = -1.3979$ or $-2 + 0.6021$. This is usually written $\overline{2}.6021$ and is the cologarithm of 25. It is used to avoid subtracting mantissae in a calculation.

colorimeter. An instrument used in colorimetric analysis.

colorimetric analysis. A method of quantitative analysis in which the concentration of a coloured solute in a solution is estimated by the intensity of the colour, as by comparing it with standard solutions.

colour. The visual sensation caused by the wavelength of the light falling on the *eye. Colour is detected by the cones in the retina. Colour is also a property of the light itself and of the object reflecting or transmitting the light.
Light of a single wavelength has one of the *spectral colours*: red, orange, yellow, green, blue, indigo, and violet. In general, a colour will be determined by a distribution of wavelengths, but one will predominate: this is its *hue*. Colours with no hue, such as white and black, are *achromatic colours*. *Chromatic colours* can be regarded as a combination of a spectral colour with a certain proportion of white. The *saturation* is the extent to which it is free from white. A third parameter used to describe colours is the brightness or vividness of the colour. In the case of a coloured light this is its *luminosity*: in the case of an object or pigment it is called *lightness*.
A difference between coloured lights and coloured pigments also occurs in mixing colours. The former mix directly giving a different colour sensation by an *additive process. Pigments mix by a *subtractive process. *See also* primary colours, complementary colours.

colourant. Any substance used to impart colour. Technically, colourants are classified into *dyes, which are applied

from solution, and *pigments, which are insoluble substances.

colourtron. A type of *cathode-ray tube used in colour television, having three electron guns, one for each *primary colour. The screen is covered with many triangular groups of three coloured phosphor dots, one for each colour.

colour vision. *See* eye.

columbite-tantalite. A black or brown naturally occurring mixed oxide, $(Fe,-Mn)(Nb,Ta)_2O_6$, used as a source of niobium and tantalum.

columbium. *See* niobium.

column chromatography. *See* chromatography.

coma. 1. An *aberration of lenses in which a point object off the axis of the system gives a comet-shaped image. The effect is similar to spherical aberration.
2. The diffuse luminous region surrounding the head of a *comet.

combination. A selection of a number of entities from a set of entities irrespective of the order of selection. The number of combinations of r elements from a set of n is denoted ^nC_r and is given by $n!/r!(n - r)!$. *See also* permutation.

combustion. The combination of a substance with oxygen to produce heat, light, and sometimes flame. In a burning gas the heat of reaction is sufficient to maintain the process and light is produced by luminescence of excited molecules or ions or by incandescence of small particles of solid produced in the reaction. In most cases solids and liquids burn when the temperature is high enough for the release of flammable vapours to occur, either by simple vaporization or by decomposition. Usually combustion involves a complex series of reactions in which carbon is converted to carbon dioxide and hydrogen is converted to water. The term combustion is also used to describe any similar process in which heat and light are generated by chemical reaction, as in the burning of sodium in chlorine. Slow combustion is a process in which a small amount of heat and little or no light is produced by reaction with oxygen.

comet. Any of a considerable number of celestial objects having a bright head and long conspicuous tail. Revolving about the sun in very eccentric orbits, comets apparently consist of a nucleus composed mostly of ice but also containing dust and other material. When the comet approaches the sun, the surface of the nucleus begins to vaporize, giving rise to the *coma. Under the influence of solar radiation or the stream of ionized gas ejected by the sun, the gas and dust in the coma are pushed out behind (away from the sun) to form the tail, which may be several million kilometres in length.

common denominator. A number that is a multiple of the *denominators of two or more fractions.

common difference. See arithmetic progression.

common fraction. See fraction.

common logarithm. See logarithm.

common ratio. See geometric progression.

common salt. See sodium chloride.

commutative. Denoting an operation for which the order of the terms does not affect the result. For example, since $x + y = y + x$ and $xy = yx$ for all real numbers, addition and multiplication are said to be commutative operations.

compass. An instrument for indicating the directions north, south, etc. A magnetic compass is simply a magnetized needle pivoted at its centre and free to move in a horizontal plane. It lines up along the earth's lines of force, thus pointing in the direction of magnetic north.

The gyro-compass contains a *gyroscope rotated by an electric motor and mounted so that it is free to swing in

any plane. The axis of rotation is aligned along a north-south direction and maintains this direction irrespective of the movements of the mounting. Unlike the magnetic compass, the instrument is unaffected by local irregularities in the direction of the earth's magnetic field.

compiler. A computer program for converting information written in a programming language into machine code.

complement. The angle that, together with a given angle, makes a right angle. Thus 30° is the complement of 60°

complementary colours. A pair of colours that give white light when combined together. Orange, for instance, is the complementary colour of blue. Two pigments of complementary colours mix to give black.

complex. A compound in which atoms or groups are bound to metal atoms or ions by *coordinate bonds. An example of a complex is the bright blue copper ammine formed when copper sulphate is dissolved in ammonia solution. It contains the ion $[Cu(NH_3)_4]^{2+}$, in which the four ammonia molecules each donate an electron pair in forming bonds with the Cu^{2+} ion. The species produced is a complex ion. Negative complex ions are also formed, as in the ferrocyanide ion $[Fe(CN)_6]^{4-}$ and the ferricyanide ion $[Fe(CN)_6]^{3-}$. The ferrocyanide ion, for example, can be regarded as produced by coordination of six cyanide ions (CN^-) to a central iron II ion (Fe^{2+}). The resulting complex ion has an overall charge of 4-. Neutral complexes are also known: the $Pt(NH_3)_2Cl_2$ molecule can be formed by coordination of two (neutral) ammonia molecules and two (negative) chloride ions to a Pt^{2+} ion. Nickel carbonyl and ferrocene are other examples.

The complexes formed by transition metals do not, in general, follow the rule that the metal ion should have the configuration of an inert gas. Often they are paramagnetic because they contain unpaired electrons and many of them

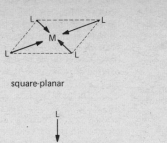

square-planar

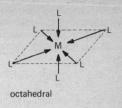

octahedral

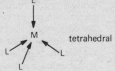

tetrahedral

Simple shapes of complexes

are coloured as a result of the coordinated groups. Hydrated copper sulphate, for instance, owes its blue colour to the complex ion $[Cu(H_2O)_4]^{2+}$ (*see* water of crystallization).

The groups bound to the central metal atom or ion are known as *ligands*. They are molecules, such as the ammonia or water molecule, that can donate electron pairs or they are ions, such as the cyanide and chloride ions, that are also capable of providing pairs of electrons. *Chelating agents are ligands that can form more than one coordinate bond to the metal. The number of ligands surrounding a metal atom or ion in a complex is the atom's *coordination number*. In $[Cu(NH_3)_4]^{2+}$, the copper has a coordination number of 4. Complexes can exhibit a variety of coordination numbers and shapes. Some of the more common ones are shown in the illustration.

complex fraction. *See* fraction.

complex number. A number expressed in the form $a + ib$, where a and b are real numbers and i is $\sqrt{-1}$. *See also* Argand diagram.

component. 1. Any of two or more vectors that add together to produce a given vector.
2. The number of distinct independently variable chemical constituents defining the composition of each phase of a system. In a system of ice and water there is only one component. *See* phase rule.

composite number. An integer that is not prime.

compound. A substance composed of two or more elements chemically combined in fixed proportions. Compounds are produced by the formation of chemical bonds between different atoms and they usually have properties that are different from those of their constituent elements. *See also* molecule.

compound microscope. *See* microscope.

compound pendulum. *See* pendulum.

compressibility. The reciprocal of the bulk *modulus of a substance.

Compton effect. The scattering of photons by free electrons, resulting in an increase in the wavelength of the photon (i.e. a decrease in photon energy) and an increase in the velocity of the electron. [After Arthur Holly Compton (1892-1962), U.S. physicist.]

computer. Any of various devices for processing data according to predetermined instructions. By far the most widely used computers are electronic *digital computers which store and process information in digital form. The other main type of computer, the *analog computer, accepts a continuous input. *Hybrid computers have some of the features of both types.

concave. *See* lens.

concavo-convex. *See* lens.

concentrated. Denoting a solution that contains a high concentration of solute.

concentration. The quantity of solute present in a given quantity of solvent. *See also* solubility.

concentric. Denoting circles that have the same centre and lie in the same plane.

conchoidal fracture. A type of fracture in which the break has a curved face with a series of concentric rings. Conchoidal fracture occurs in amorphous solids.

condensation polymerization. *See* polymerization.

condenser. 1. A piece of apparatus used for cooling, and thus condensing, a vapour. A common type is the *Liebig condenser.
2. *See* capacitor.

conductance. Symbol: *G*. **1.** A measure of the ability of a circuit to pass a steady current, equal to the reciprocal of the resistance. It is measured in siemens.
2. A similar quantity used for alternating-current circuits, equal to the real part of the *admittance. It is equivalent to $R/(R^2 + X^2)$, where R is the resistance and X is the reactance.

conductiometric. Denoting an experimental technique that depends on measurements of electrical conductivity. A *conductiometric titration* is one in which the end point is determined by monitoring the conductivity of the mixture as one reactant is added.

conduction. 1. (of heat). Transfer of heat through a material from one point to another point at lower temperature without transfer of mass occurring. In general, conduction occurs as a result of collisions between atoms or molecules—atoms in the hotter region of the material have higher kinetic energies and communicate this energy to neighbouring atoms. Thus the energy is passed on and not carried by movement of atoms as it is in *convection. Conduction of heat by this mechanism occurs in most liquids, gases, and non-metallic solids: these materials are usually poor thermal conductors. In metals, which are usually good thermal conductors, the mechanism is different. The energy is carried by electrons moving through the lattice and colliding with atoms: this is the reason that good conductors of heat are also good conductors of electricity—in both cases the *conductivity depends on the motion of electrons.
2. (of electricity). Transfer of electric charge through a material as a result of an applied electric field. In all cases electrical conduction depends on the presence of charge carriers that can move under the influence of an applied electric field. In liquid conductors (*electrolytes) the charge is carried by positive and negative ions. In gases the majority of the charge is carried by positive ions and electrons (*see* discharge). In most solids the carriers are electrons although positive holes transport the charge in some semiconductors (*see* band theory).

conductivity. 1. Symbol: σ. A measure of the ability of a material to allow the flow of electric current, equal to the reciprocal of the *resistivity: $\sigma = 1/\rho$. Conductivity is measured in siemens per metre.
2. *See* thermal conductivity.

cone. A surface or solid having a plane base and a vertex above the base, bounded by lines (*generators*) joining the vertex to the base. If the base is a circle the solid is a circular cone. A right circular cone has its vertex above the centre of the circle. Such a cone has a lateral surface area $\pi r h$, where h is the slant height, and a volume of $(1/3)\pi r^2 s$, where s is the altitude.

cones. *See* eye.

configuration. The arrangement of electrons orbiting the nucleus of a given atom. Thus the electron configuration of oxygen is 2,6; these numbers of electrons are present in the first two shells. The configuration can also be written $1s^2 2s^2 2p^4$—the $1s$, $2s$, and $2p$ denote the subshell and the superscript denotes the number of electrons present.

confocal. Denoting *conic sections that have the same foci.

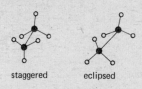

staggered   eclipsed

ethane

chair   boat

cyclohexane

Conformations

conformation. One of the possible shapes that a molecule may have resulting from rotation of groups about single bonds. The illustration shows two possible conformations for the ethane and cyclohexane molecules.

congruent. Denoting geometric figures that are identical in all respects.

conic. A type of curve that is the locus of a point that moves so that its distance from a fixed point (the *focus*) is a constant fraction of its distance from a fixed line (the *directrix*). The constant fraction is the *eccentricity*, e, of the conic. Types of conic include the *ellipse ($e < 1$), the *parabola ($e = 1$), and the *hyperbola ($e > 1$). The circle is a special case of the ellipse with $e = 0$. Conics are also called *conic sections*

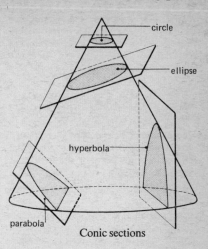

Conic sections

because they can be obtained by taking various sections of a cone (see illustration).

conjugate. A complex number whose real part is equal to the real part of the given complex number and whose imaginary part is equal and opposite in sign to that of the given complex number. Thus $a + ib$ is the conjugate of $a - ib$.

conjugate solutions. Two distinct solutions in equilibrium, each consisting of two liquids, one liquid being the solvent in one of the solutions and the solute in the other. Conjugate solutions are formed under certain conditions from mixtures of partially miscible liquids. Aniline and water, for example, can form two layers. One is a solution of water in aniline and the other a solution of aniline in water.

conjugation. Interaction occurring in molecules that contain alternate double and single bonds, characterized by delocalization of electrons in the double bonds. In butadiene, for example, the two double bonds are separated by one single bond and are said to be *conjugated*. Molecules of this type have extra stability in addition to the stability expected in a molecule containing two double bonds and one single bond. This results from the fact that electrons are not localized in the double bonds, but can move into the region of the single bond. *Benzene is another example of a conjugated molecule.

conjunction. The coincidence in celestial *longitude between two celestial objects. The planets Mercury and Venus are in conjunction twice during their orbits. *Inferior conjunction* occurs when the two planets lie between the earth and the sun; *superior conjunction* occurs when the sun lies between the earth and the planets.

conservation law. Any principle involving the idea that some property of a system is conserved during changes occurring in the system. Mass-energy, charge, baryon number, lepton number, parity, and similar properties are conserved in certain types of reaction.

conservation of energy. *See* energy.

conservation of mass. *See* mass.

conservation of mass-energy. *See* mass-energy.

conservation of momentum. *See* momentum.

consistent. Denoting a set of assumptions or propositions that involve no contradiction.

constant composition, law of. The principle that a given compound always contains the same elements in the same proportions. Water, for example, must always contain hydrogen and oxygen in the ratio 2:16 by weight (i.e. H_2O). The law is one of three laws of *chemical combination. It is sometimes called the *law of definite proportions*.

constellation. A grouping of stars, usually an apparent one caused by the illusion of perspective. The ancients named these shapes after animals and heroes of mythology; Ptolemy listed 48 northern and equatorial constellations in his *Almagest*. With the discovery of the stars of the southern hemisphere, the number has been raised to 88.

contact potential. A small potential difference existing between two different solids in contact. The value is characteristic of the materials and varies with temperature (*see* thermocouple).

contact process. An industrial process for the manufacture of sulphuric acid by the catalytic oxidation of sulphur dioxide. A mixture of sulphur dioxide and air is passed over a heated vanadium pentoxide or platinum catalyst. The sulphur trioxide produced is absorbed in concentrated sulphuric acid to give fuming sulphuric acid, which is then diluted.

continuous phase. See dispersion.

continuous spectrum. A *spectrum that has a continuous distribution of emitted or absorbed wavelengths. Hot solids have continuous emission spectra in the visible or infrared regions. The radiation is produced by changes in vibrational energy of the atoms of the solid.

continuum. 1. *See* space-time continuum.
2. A continuous region of a spectrum.

control grid. The grid that is used to control the current in a *thermionic valve, distinguished from the screen grid and suppressor grid.

control rod. See fission reactor.

convection. The process in which heat is transferred through a fluid by motion of the fluid. In *natural convection* the motion occurs as a result of a vertical temperature difference. Warm fluid expands and its density decreases, causing it to rise through the surrounding cooler fluid. Its place is taken by cool fluid and cyclic *convection currents* are produced. Convection is the main process of heat transfer in liquids and gases. *Forced convection* is the transfer of heat by an imposed flow of fluid, such as a current of air produced by a fan.

conventional current. An electric current considered as flowing from a point at high potential to one at lower potential. In fact the charge is usually carried by electrons, which flow in the opposite direction.

convergent. 1. Designating an infinite *sequence, the terms of which tend to some limiting value as the sequence progresses. For example, the sequence 1, 1/2, 1/3, 1/4, ... 1/n .... converges to zero as n becomes very large.
2. Designating an infinite *series that has a finite sum. A series is strictly defined as convergent if the sequence of its partial sums converges. The series 1/2 + 1/4 + 1/8 + ... 1/2^n + ... is convergent: its partial sums are 1/2, (1/2 + 1/4), (1/2 + 1/4 + 1/8), etc., forming a sequence that converges to 1, the *sum* of the series.

converging lens. *See* lens.

conversion electron. An electron ejected as a result of *internal conversion.

convex. *See* lens.

convexo-concave. *See* lens.

coolant. A fluid used to transfer heat, as in a fission reactor.

coordinate bond (dative bond). A type of *covalent bond in which one of the atoms supplies both of the electrons. It can be thought of as a combination of transfer and sharing of electrons. Coordinate bonds are also called *semipolar bonds.*

coordinate geometry. *See* analytic geometry.

coordinates. Numbers representing the position of a point with respect to reference lines or points. In a plane two numbers are necessary to specify the position: in three-dimensional space three are required. Commonly used coordinate systems are *Cartesian coordinates, *polar coordinates, and *cylindrical coordinates. *See also* analytic geometry.

coordination number. *See* complex.

Copernican system. The system of celestial mechanics formulated by the Polish monk Nicolaus Copernicus (1473-1543). It superseded the *Ptolemaic system in respect of the crucial ideas that the earth underwent daily rotation and the sun occupied the central place around which orbited the earth, moon, planets, and stars. The theory was of critical importance in laying the foundations of modern astronomy.

copolymer. *See* polymer.

copolymerization. *See* polymerization.

copper. Symbol: Cu. A reddish lustrous malleable and ductile *transition element found in malachite ($CuCO_3.Cu(OH)_2$), cuprite (Cu_2O), and chalcopyrite (($Cu,Fe)S_2$). The metal is usually obtained from the sulphide by smelting and electrolytic refining. A certain amount of copper is found native. Copper has a high thermal and electrical conductivity, second only to silver, and is used in electric circuitry and cables. It is also used in alloys such as brass and bronze. The element reacts with oxygen, sulphur, and the halogens and dissolves in concentrated nitric and other oxidizing acids. There are two series of ionic salts: copper I (*cuprous*) and the more important copper II (*cupric*) compounds. The metal also forms many complexes in the I and II oxidation states. A.N. 29; A.W. 63.54; m.pt. 1083°C; b.pt. 2595°C; r.d. 8.96; valency 1 or 2.

copperas. *See* iron sulphate.

copper carbonate. *See* malachite.

copper chloride. Either of two chlorides of copper. *Copper I chloride* (cuprous chloride) is a white crystalline solid, CuCl or Cu_2Cl_2, prepared from copper and copper II chloride solution. It is used as a catalyst, preservative, and fungicide. M.pt. 430°C; b.pt. 1490°C; r.d. 4.14. *Copper II chloride* (cupric chloride) is a brown powder, $CuCl_2$, or green crystalline hydrate, $CuCl_2.2H_2O$, prepared by reacting chlorine with copper. The more stable chloride of copper, it is used as a mordant and a catalyst. M.pt. 620°C; r.d. 3.4.

copper nitrate. A blue crystalline solid, $CuNO_3.3H_2O$, prepared by treating copper or copper II oxide with nitric acid. It is used in dyes and insecticides. Copper II nitrate (cupric nitrate) is the only stable nitrate of copper. M.pt. 114.5°C; r.d. 2.3.

copper oxide. Either of two basic oxides of copper. *Copper I oxide* (cuprous oxide) is a red-brown powder, Cu_2O, prepared by oxidation of finely divided copper. It is used in ceramics and paints. M.pt. 1235°C; b.pt. 1800°C; r.d. 5.7. *Copper II oxide* (cupric oxide) is a brown-black powder, CuO, made by heating copper II carbonate. It is used in coloured glazes, insecticides, and as a catalyst. M.pt. 1326°C; r.d. 6.32.

copper sulphate. A blue crystalline solid (*blue vitriol*), $CuSO_4.5H_2O$, prepared by addition of dilute sulphuric acid to copper II oxide. It is widely used in pesticides, mordants, and electroplating. The white anhydrous salt is obtained by heating. Copper II sulphate is the only stable sulphate of copper and is the most important copper salt. R.d. 2.3 (pentahydrate). *See also* hydration.

core. 1. The part of a *fusion reactor in which the reaction occurs.
2. The part of a transformer, electric motor, or similar device that constitutes the magnetic circuit. The core is made of iron or other ferromagnetic material and is often laminated to reduce power losses caused by eddy currents.
3. A small semiconductor or a ring of ferrite used to store information in a computer *memory.

Coriolis force. A force considered to act to account for the motion of a body as it appears to an observer in a rotating frame of reference. For example, if an object is projected from the centre of a rotating disc along a radius, an observer outside the disc would say that the object was travelling in a straight line above the disc's surface and that the disc was rotating underneath. To an observer on the disc, moving with constant angular velocity, the object would appear to be moving in a curved path away from the radius on which he was situated. In fact it would appear to have a tangential acceleration given by $-2(dr/dt)(d\theta/dt)$, where dr/dt is the velocity of the object in a radial direction and $d\theta/dt$ is the angular velocity of the observer. This acceleration is the result of an apparent force.

The Coriolis force, like the centrifugal force, is a "ficticious" effect used to simplify calculations for observations made with respect to rotating systems. It is used to explain certain phenomena caused by the rotation of the earth, such as the motion of air along the isobars of a depression or anticyclone or the motion of water flowing through the plug hole of a bath. [After Gaspard Gustave de Coriolis (1792–1843), French physicist.]

cornea. *See* eye.

corollary. A result following without further detailed proof from a result that has already been proved.

corona. The gaseous outer region lying above the chromosphere of the *sun.

corona discharge. A luminous electrical *discharge produced in air around a conductor at a high critical potential. The discharge, which is often accompanied by a hissing sound, is produced in high electric fields and its formation depends on the diameter and roughness of the conductor. Corona discharges are responsible for power loss in transmission lines. They are also sometimes seen as balls of light around points on church towers, aircraft, and ship masts, where the phenomenon is called *St. Elmo's fire* or a *corposant*.

corposant. *See* corona discharge.

corpuscular theory. *See* light.

corrosion. Chemical reaction of a solid, usually a metal, with some substance in the environment leading to eventual destruction or denaturing of the surface. The most commonly observed corroding process is the *rusting of iron exposed to air and moisture. Most corrosive reactions are between a solid and a liquid or gas, but solid-solid corrosion also occurs; if two metals of widely different *electronegativities are joined together the electrochemical reaction which slowly takes place leads to the destruction of the metallic structure near the junction.

corrosive sublimate. *See* mercury chloride.

cosecant. *See* trigonometric function.

cosine. *See* trigonometric function.

cosmic background. *See* microwave background.

cosmogony. The branch of *cosmology that deals with the origin and evolutionary development of the *universe.

cosmology. The study of the universe in its entirety, incorporating its origins and development (cosmogony) and a description of it in physical terms as it is observed at present.

containment. The process of confining the plasma to a region away from the walls of a *fusion reactor.

cotangent. *See* trigonometric function.

coulomb. Symbol: C. The derived SI unit of electric charge, equal to the charge transported by a current of one ampere in one second. [After Charles Augustin Coulomb (1736–1806), French physicist.]

coulombmeter. *See* voltameter.

Coulomb's law. The principle that the force between two point charges is proportional to the magnitude of the charges and inversely proportional to the square of the distance between them. Thus, two charges q_1 and q_2

placed a distance d apart each experience a force Cq_1q_2/d^2. The constant C depends on the system of units used. In SI units the force is in newtons, the distance in metres, and the charge in coulombs and the force in free space is given by $q_1q_2/4\pi\epsilon_0d^2$, where ϵ_0 is the *electric constant. Sometimes a similar equation is used for the force between charges in a medium: $F = q_1q_2/4\pi\epsilon_r\epsilon_0d^2$, where ϵ_r is the relative *permittivity. However, this extension of Coulomb's law is not strictly correct. The law was used as the basis of the definition of the *electrostatic unit of charge (esu). In these units, $F = q_1q_2/Kd^2$, where F is expressed in dynes, d in centimetres, and q_1 and q_2 in esu. The constant K, is the *permittivity and it has the value unity for free space.

Coumarin

coumarin (1,2-benzopyrone, cumarin). A colourless crystalline solid, $C_9H_6O_2$, with a fragrant vanilla-like odour and a burning taste. Coumarin is a deodorizing and odour-enhancing agent.

counter. Any of various instruments for detecting and determining numbers of particles and photons. A counter consists of a detector which produces an electric pulse as a result of the impact of an individual particle. Associated with this is an electronic circuit for counting the pulses formed. Counters are extensively used in nuclear physics, cosmic-ray studies, etc. *See* Geiger counter, Cerenkov radiation, crystal counter, scintillation counter, ionization chamber.

counter-glow. *See* gegenschein.

couple. Two parallel forces acting together in opposite directions at different points, thus producing a turning effect. *See* moment, torque.

coupling. Interaction between different effects in a system.

covalent bond. A type of chemical bond in which atoms are held together by shared pairs of electrons. Hydrogen atoms, for example, have a single electron and the carbon atom has an electron configuration of 2, 4. One carbon atom can form four covalent bonds to four hydrogen atoms by donating one electron to each bond, the other electron being supplied by the hydrogen atom: the electrons in each bond have opposite *spins. As a result of this process eight electrons are moving in the field of the carbon nucleus, giving it the stable electron configuration of the inert gas neon. Similarly, each hydrogen atom has two electrons, giving it the electron configuration of helium.
Covalent compounds of this type are composed of individual molecules. In the solid state they form molecular *crystals in which the molecules are held together by fairly weak forces. For this reason they are low-melting and volatile and tend to be insoluble in water, dissolving instead in *nonpolar solvents. Each shared pair of electrons constitutes a *single bond*. It is also possible for two atoms to bond by sharing two (*double bond*) or three (*triple bond*) pairs of electrons. Theories of covalent bonding are based on the use of *molecular orbitals.

covalent crystal. *See* crystal.

CPU. *See* central processing unit.

Crab nebula. The expanding gaseous nebula M1 in the constellation Taurus. It is identified as the remnant of a supernova observed by Chinese astronomers in 1054. Growing at a rate of 1100 km s^{-1}, it lies at a distance of some 4000 light years from the sun. The diffuse central parts of the nebula give rise to a continuous spectrum, while a bright line emission spectrum is produced by the luminous filaments surrounding the nucleus. The nebula contains a *pulsar.

creep. Slow permanent deformation of a metal caused by continuous stress.

creosote. Either of two mixtures distilled from tars. The creosote used for preserving wood is *coal-tar creosote*, a yellow or brown mixture of aromatic hydrocarbons and some phenols (including cresols and naphthols). *Wood-tar creosote* is used in pharmacy for the treatment of bronchitis. It is obtained by distilling wood tar, especially from beechwood, and is largely a mixture of phenols.

cresol. Any of three isomers, CH_3-C_6H_4OH, obtained from coal tar. A mixture of cresols is used as a germicide.

critical angle. The smallest angle of incidence at which electromagnetic radiation suffers *total internal reflection.

critical damping. *See* damping.

critical mass. The minimum mass of fissile material for which a *chain-reaction is self-sustaining.

CRO. *See* cathode-ray oscilloscope.

Crookes dark space. *See* glow discharge.

Crookes radiometer. An instrument for detecting infrared radiation. It consists of an evacuated bulb containing four vertical vanes of thin metal mounted on a freely-rotating horizontal wire cross piece. Each vane is blackened on one side and polished on the other. The black faces absorb more radiation than the polished faces and, at low pressure, molecules rebounding from these faces have more momentum than those from the bright faces. The arrangement is such that the vanes rotate in the presence of a source of heat radiation.

[After Sir William Crookes (1832–1919), English chemist.]

cross-linkage. A chain of atoms linking two longer chains in a *polymer.

cross product. *See* vector product.

cross section. Symbol: σ. A measure of the probability of some specified process occurring when a beam of particles or photons interacts with nuclei, atoms, or molecules, expressed as an effective area. For example, if gas-phase atoms are ionized by a beam of electrons then only a certain proportion of the total number of collisions leads to ionization. The ionization cross section is the area that each atom would present to the beam if every collision led to ionization. The cross section is not the physical size of the atom: it depends on the process and on the energy of particles. It is often used in particle physics to describe the interaction of particles with atomic nuclei.

crotonaldehyde. A colourless flammable liquid, $CH_3CH:CHCHO$, used as an accelerator in producing rubbers and as a lachrymatory war gas. M.pt. –69°C; b.pt. 102°C; r.d. 0.85.

crotonic acid (but-2-enoic acid). A white crystalline carboxylic acid, $CH_3CH:CHCOOH$, made by the oxidation of crotonaldehyde and used in synthesizing resins, polymers, plasticizers, and drugs. M.pt. 72°C; b.pt. 185°C; r.d. 0.97.

crown glass. *See* optical crown.

CRT. *See* cathode-ray tube.

crucible. A dish-shaped vessel suitable for heating chemicals to high temperatures.

cryogenics. The branch of physics concerned with the production of very low temperatures and the study of phenomena occurring at these temperatures.

cryohydrate. A crystalline solid containing water and a salt in definite proportions, obtained by freezing solutions of certain salts. Cryohydrates are *eutectic mixtures.

cryolite. A colourless or white naturally occurring form of sodium aluminium fluoride, Na_3AlF_6, used in the production of aluminium.

cryometer. A thermometer suitable for use at low temperatures.

cryoscope. Any instrument or apparatus for determining a freezing point.

cryoscopy. The determination of freezing points, especially in the determination of molecular weights by freezing-point depression.

cryostat. A vessel or enclosure that can be maintained at a constant low temperature.

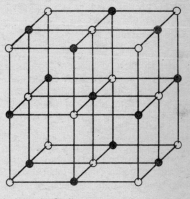

Sodium chloride

crystal. 1. A piece of a solid substance with a regular geometrical shape (the *crystal habit) caused by a regular arrangement of the atoms, ions, or molecules making up the solid. The particular type of arrangement is the *crystal structure* and any solid that has such a regular array is said to be *crystalline*, otherwise it is *amorphous. If the regular pattern is continued with the same orientation throughout the solid, the piece of material is a *single crystal*: many crystalline substances are polycrystalline. *Ionic crystals* are composed of a three-dimensional array of positive and negative ions: an example is sodium chloride (see illustration). *Covalent crystals* have covalent bonds extending throughout the whole solid and may be thought of as very large molecules: an example is *diamond. In *molecular crystals* individual molecules (which are covalent) are held together by *van der Waals forces. Most organic compounds crystallize in this way. They are soft and low-melting as a result of the weakness of the forces. **2.** A *piezoelectric crystal used in electronic equipment, such as the crystal pick-up and crystal oscillator.

crystal counter. A type of *counter in which a crystal of certain materials is connected across a high potential difference. Each particle of ionizing radiation hitting the crystal makes it momentarily conducting and produces a pulse of current.

crystal form. An arrangement of crystal planes in three dimensions, having a particular symmetry. Real crystals are characterized by having either one particular form or a combination of forms. The form of a crystal is distinct from its *crystal habit.

crystal habit. The external appearance of a crystal. The habit depends on the internal pattern of atoms and also on the way in which the crystal has grown.

crystalline. Having a *crystal structure rather than an *amorphous structure.

crystalline lens. *See* eye.

crystallite. One of the small grains in a polycrystalline solid.

crystallography. The study of crystals and crystal structures. *See* X-ray crystallography.

crystalloid. A chemical compound that is crystalline and soluble in water. The term is now obsolete, being formerly used to distinguish compounds that undergo dialysis from *colloids* (which do not dialyse).

crystal microphone. A type of microphone in which vibrations of a diaphragm are used to vibrate a *piezoelectric crystal, thus producing a varying e.m.f.

crystal oscillator. A type of oscillator circuit in which a fixed frequency is produced by the vibrations of a quartz crystal. The frequency of such a circuit can be constant to one part in 2×10^7. Crystal oscillators are used as standards of frequency and in the *quartz clock.

crystal pick-up. A pick-up used on record players in which vibrations produced by the record groove are converted into a varying e.m.f. by a piezoelectric crystal.

crystal plane. A plane of atoms, ions, or molecules in the lattice of a crystal.

crystal spectrometer. A type of spectrometer for use with X-rays, in which the radiation is dispersed by diffraction from the face of a single crystal. *See* X-ray diffraction.

The seven crystal systems

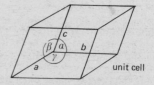

unit cell

cubic	$a=b=c$	$\alpha=\beta=\gamma = 90°$
tetragonal	$a=b\neq c$	$\alpha=\beta=\gamma = 90°$
orthorhombic	$a\neq b\neq c$	$\alpha=\beta=\gamma = 90$
hexagonal	$a=b\neq c$	$\alpha=\beta = 90°; \gamma = 120°$
trigonal	$a=b\neq c$	$\alpha=\beta=\gamma \neq 90°$
monoclinic	$a\neq b\neq c$	$\alpha=\gamma = 90° \neq\beta$
triclinic	$a\neq b\neq c$	$\alpha\neq\beta\neq\gamma$

crystal system. Any of seven classes into which all crystals are classified, depending on the lengths and angles between the edges of the unit cell. The table shows the characteristics of the seven crystal systems. Sometimes the rhombohedral system is included under the hexagonal system.

cube. 1. A regular solid with six equal square faces.
2. The third power of a number: $x^3 = x \times x \times x$.

cubic. *See* crystal system.

cumarin. *See* coumarin.

cumene (isopropyl benzene). A colourless flammable liquid hydrocarbon, $C_6H_5CH(CH_3)_2$, distilled from petroleum and used as a fuel additive and solvent. M.pt. $-96°C$; b.pt. $153°C$; r.d. 0.86.

cupellation. A method of refining precious metals such as gold and silver by melting them in a shallow porous vessel (a *cupel*). Impurities such as lead are oxidized and absorbed into the walls of the cupel.

cupric. *See* copper.

cuprite. A bright red lustrous naturally occurring oxide of copper, Cu_2O, used as an ore.

cuprous. *See* copper.

curie. Symbol: Ci. A unit of radioactive disintegration rate equal, for a specified radioisotope, to the amount of the isotope that produces 3.7×10^{10} disintegrations per second. After Marie Curie (1867–1934), French physicist born in Poland.]

Curie's law. The principle that the *susceptibility of a paramagnetic substance is inversely proportional to its thermodynamic temperature (T): $\chi = C/T$, where C is a constant, the *Curie constant*, for the material. A modification of this law, the *Curie-Weiss law*, has the form $\chi = C/(T - \theta)$, where θ is a fixed temperature. Ferromagnetic materials lose their *ferromagnetism above a fixed temperature (the *Curie point*) and display paramagnetic behaviour. They then follow the Curie-Weiss law, θ being the temperature at the Curie point.

curium. Symbol: Cm. A silvery transuranic *actinide element made by neutron bombardment of americium. It was originally known before americium, being made by bombardment of plutonium with alpha particles. The most stable isotope, ^{248}Cm, has a half-life of 4.7×10^5 years. A.N. 96; r.d. 7; valency 3.

curl. A differential operator; curl A is $\nabla \times A$, where ∇ is the operator *del. Thus, curl $A = i \times \partial A/\partial x + j \times \partial A/\partial y + k \times \partial A/\partial z$.

current. Symbol: I. A flow of electric charge. The magnitude of the current is equal to the charge carried per unit time and is measured in amperes. In metals the charge is transported by electrons: in semiconductors the carriers may be positive holes. By convention the current in a circuit is taken to flow from regions of positive potential to regions of negative potential. This is known as a *conventional current*. In fact, the electrons, which carry the charge, flow in the opposite direction. *See also* direct current, alternating current.

current balance (ampere balance). An instrument for determining an electric current by directly measuring the force produced between conductors. Several types exist: in the commonest a flat horizontal coil is suspended from one arm of a balance so that it lies between two fixed flat horizontal coils. The current to be measured is passed through the three coils in opposite directions and the value can be calculated by the force produced, as determined by the balance. This type of instrument is used in defining the value of the *newton.

current density. Symbol: j. The electric current flowing per unit area at a point in a conductor, beam of charged particles, etc.

cursor. A transparent slider with a vertical hair-line on it, used on slide rules to facilitate reading of the scales.

curvature of field. An *aberration of optical systems in which the images of off-axis points lie on a curved surface rather than on a plane.

cyanamide. 1. (carbodiimide). A colourless deliquescent crystalline solid, HN:C:NH, made by reaction between carbon dioxide and hot sodamide. M.pt. 41°C.
2. See calcium cyanamide.

cyanate. See cyanic acid.

cyanic acid. A volatile liquid, HN:C:O. The compound with this formula is sometimes called *isocyanic acid* and the term *cyanic acid* is used for the isomer HOC:N. This is also known as *fulminic acid.* Covalent derivatives of both acids are known. *Isocyanates* have the formula RNCO and *cyanates* have the formula ROCN. Similarly ionic compounds can be formed from both acids and again there is some confusion in naming the salts. Compounds containing the ⁻NCO ion are called *cyanates* or *isocyanates.* Those containing the ⁻OCN ion are *cyanates* or *fulminates.*

cyanide process. A method of extracting *gold from its ores by dissolving it in potassium cyanide.

cyanogen. A colourless highly poisonous flammable gas, $(CN)_2$, with an odour of almonds. It can be prepared by heating mercury cyanide and has been used as a fumigant, war gas, and rocket fuel. M.pt. -28°C; b.pt. -21°C; r.d. 1.81.

cycle. A sequence of changes occurring in a system and eventually restoring the system to its original state. The term is often used to describe one such set of changes in a system that is periodically varying. In an alternating current or voltage or in a wave, the cycle is the sequence of values that is periodically repeated. In one cycle an alternating current may increase to a maximum, decrease to zero, increase to a maximum in the opposite direction, and decrease to zero again. The time taken for a cycle is its *period.

cycle per second. See hertz.

cyclic. 1. Denoting a chemical compound that has a ring of atoms in its molecular structure. Cyclic compounds can be *aromatic, as with benzene, or *alicyclic, as with cyclohexane. When used to describe a class of compounds, the term is usually taken to mean that the functional group forms part of the ring. For instance, a *cyclic ether* is a compound such as dioxan, in which the -O- group is bound to other atoms in the ring. *Compare* acyclic.
2. Denoting a plane geometric figure that can be inscribed in a circle.

cyclization. The conversion of an organic compound that has an open chain into a compound containing a ring. Many methods exist, most of them involving standard types of reaction in which one end of the chain reacts with the other and *ring closure* occurs.

cyclohexane. A colourless mobile flammable liquid, C_6H_{12}, prepared by the distillation of petroleum or the hydrogenation of benzene. Cyclohexane is used in manufacturing nylon and as a solvent. M.pt. 6.5°C; b.pt. 81°C; r.d. 0.78.

cyclohexanol. A colourless oily cyclic alcohol, $C_6H_{11}OH$, with a camphor-like odour. It is manufactured by the hydrogenation of phenol and used in making nylon and celluloid and as a solvent. M.pt. 23°C; b.pt. 161°C; r.d. 0.94.

cyclohexanone. A colourless or pale yellow oily cyclic ketone, $C_6H_{10}O$, made by the oxidation of cyclohexanol. Cyclohexanone is used in making adipic acid and caprolactam and is a widely used solvent. M.pt. -47°C; b.pt. 157°C; r.d. 0.95.

cycloid. A curve that is the locus of a point on the circumference of a circle as that circle rolls along a straight line. It is a special case of a *trochoid.

cyclopentadiene. A colourless liquid hydrocarbon, C_5H_6, obtained by cracking petroleum. With electropositive metals it forms the cyclopentadienyl anion, $C_5H_5^-$.

Cyclopentadiene

cyclotron. A type of particle *accelerator in which the particles move in spiral paths under the influence of a uniform vertical magnetic field and are accelerated by an electric field of fixed frequency. The magnetic field is produced by a powerful electromagnet. The particles move inside two hollow D-shaped metal electrodes (the *dees*) separated by a small gap. The time taken to describe a semicircle is $\pi m/Be$, m being the mass and e the charge of the particle, and is independent of the path length, thus allowing the particles to be accelerated by an electric field of

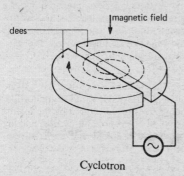

Cyclotron

fixed frequency applied across the gap between the dees. The frequency is such that the particles are accelerated every time they cross the gap. When they reach the edge of the device they are deflected away onto the target. The energy obtained in the cyclotron is limited by relativistic effects. At high particle energies the increase in mass causes the time taken to complete a semicircle to be significantly larger. The motion of the particles then falls out of step with the frequency of the electric supply and a limiting energy is reached. Typical energies obtained are of the order of 25 MeV: higher energies are

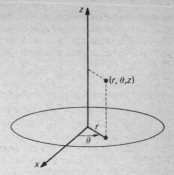

Cylindrical coordinates

obtained in the *synchrocyclotron and the *synchrotron.

cylindrical coordinates. *Coordinates used to locate the position of a point in space by its *polar coordinates in one plane and its perpendicular distance from that reference plane.

cysteine. A colourless crystalline amino acid, $HSCH_2CH(NH_2)COOH$.

cystine. A white crystalline amino acid, $HOOCCH(NH_2)CH_2SSCH_2CH(NH_2)COOH$.

D

Dacron. ⓇA type of *polyester resin used in synthetic fibres.

dalton. *See* atomic mass unit. [After John Dalton (1766–1844), British chemist.]

Dalton's atomic theory. The first modern attempt (1808) at an explanation of the behaviour of chemical elements. Dalton postulated that every element consists of indivisible particles called atoms. All atoms of a given element are identical and atoms can neither be created nor destroyed. Compound elements (compounds) are formed by combination of atoms of different elements in simple ratios to form compound atoms (molecules). The theory was able to interpret the laws of *chemical combination and the law of conservation of *mass.

Dalton's law of partial pressures. The principle that at constant temperature the total pressure exerted by a mixture of (ideal) gases in a definite volume is equal to the sum of the individual pressures each gas would exert if it occupied the same total volume alone: i.e. $p = p_1 + p_2 + p_3$ ..., where the individual pressures, p_1, etc., are the *partial pressures* of the gases.

damping. The progressive decrease in amplitude of oscillations with time. An oscillating system is damped by forces caused by viscosity, friction, magnetic fields, etc. The resisting force is approximately proportional to the velocity. Three types of damping are distinguished according to the way in which the system returns to its equilibrium position when displaced. If it performs a large number of dying vibrations, it is said to be *underdamped*. An *overdamped* system, on the other hand, returns to equilibrium slowly without oscillation. If the system just fails to oscillate it returns to equilibrium in its minimum time: it is then said to be *critically damped*.

Daniell cell. A type of primary cell in which the positive electrode is a copper rod immersed in copper sulphate solution and the negative electrode is an amalgamated zinc rod in dilute sulphuric acid, the two electrolytes being separated by a porous pot. At the positive electrode copper ions gain electrons from the rod and copper is deposited. At the negative electrode zinc atoms dissolve as zinc ions. The e.m.f. is about 1.08 volts. [After J. F. Daniell (1790-1845), British physicist.]

daraf. A unit of elastance: the reciprocal of the farad.

dark-field illumination. Illumination of a sample in a microscope by light from the side rather than underneath. Small specimens are then more easily seen as bright objects on a dark background. See ultramicroscope.

dasymeter. An instrument for determining the density of gases by measuring the upthrust produced on a tube of known volume.

dating. Determination of the age of rocks, archaeological finds, etc., y physical methods. The principal methods include *radiocarbon dating, *rubidium-strontium dating, *potassium-argon dating, *thermoluminescence, and *palaeomagnetism.

dative bond. See coordinate bond.

daughter. A nuclide produced by decay of another nuclide (the *parent).*

Davisson-Germer experiment. An experiment in which a beam of electrons was directed onto the plane face of a single crystal of nickel in a high vacuum and the angular distribution of reflected electrons was investigated. It was found that the electrons were diffracted by the lattice (see Bragg's law) and showed a maximum intensity at one particular angle. The experiment, performed in 1927, demonstrated the existence of the *de Broglie waves predicted three years earlier.

Davy lamp. A safety oil lamp used by miners. The flame burns inside a metal mesh cage or mantle, which cools the hot gases from the flame by conduction, preventing the flame from igniting any inflammable gases outside the lamp. The nature of the flame can indicate the presence of flammable gas. [Invented by Sir Humphrey Davy (1778-1829), English scientist.]

day. The time taken for the earth to make one rotation on its axis. It is measured in various ways (see time).

DDT (dichlorodiphenyltrichloroethane). A colourless crystalline solid, $(ClC_6H_4)_2$CHCCl_3$, with a slight aromatic odour, manufactured by the condensation of chloral with chlorobenzene. It is a powerful insecticide and has been widely used in agriculture, although this has now been restricted.

de Broglie wave. A wave associated with a moving particle, having a wavelength (the *de Broglie wavelength*) given by $\lambda = h/mv$, where m is the particle's mass, v its velocity, and h the Planck constant. Under certain conditions particles of matter display wave-like behaviour—the wave character of the electron was first demonstrated in the *Davisson-Germer experiment. The effect is not confined to elementary particles—diffraction of beams of atoms can also occur from crystal surfaces. *See also* wave mechanics. [After Prince Louis Victor de Broglie (b.1892), French physicist.]

debye. Symbol: D. A unit of electric dipole moment equal to $3.335\ 64 \times 10^{-30}$ coulomb metre. [After Peter Debye (1884-1966), Dutch physicist and physical chemist.]

deca-. Symbol: da. Prefix indicating 10.

decahydrate. A solid compound with ten molecules of *water of crystallization per molecule of compound, as in sodium sulphate decahydrate, $Na_2SO_4.10H_2O$.

decahydronaphthalene. A colourless flammable liquid, $C_{10}H_{18}$, with a menthol-like odour. It is a bicyclic hydrocarbon, made by hydrogenation of naphthalene, and is used as a solvent and fuel. M.pt. -43°C; b.pt. 195°C; r.d. 0.89.

decay (disintegration). Spontaneous disintegration of atomic nuclei: one nuclide (the *parent*) disintegrates yielding another type of nuclide (the *daughter*) and ejecting an electron or alpha particle and sometimes a gamma ray. The term also refers to the fall with time in the activity of a radioactive sample. *See* radioactivity, alpha decay, beta decay.

decay constant. *See* radioactivity.

deci-. Symbol: d. Prefix indicating one tenth (10^{-1}).

decibel. A measure of power level on a logarithmic scale. Although the fundamental unit is the *bel*, the decibel (0.1 bel) is almost always used. A given amount of power P has a power level in decibels of $10 \log_{10}(P/P_0)$, where P_0 is a reference value. The decibel is used particularly for expressing the intensity of sound. *See also* phon.

decimal system. The common system of notation using the base ten and digits 0–9.

declination (variation). The angle made between the direction of the horizontal components of the earth's field at a point and the direction of true north at that point. *See* terrestrial magnetism.

decomposition. The breakdown of a chemical compound into simpler molecules.

decrepitation. The roasting of crystals until they emit a crackling sound or until this sound stops. The sound is produced by abrupt changes in crystal structure, usually due to the evolution of *water of crystallization.

defect. A discontinuity in the regular structure of a crystal. A *point defect* is one caused by a *vacancy or *interstitial. A *line defect* or *dislocation* involves a number of atoms. A simple type occurs when part of one plane of atoms is missing from the lattice.

definite proportions, law of. *See* constant composition.

definition. The sharpness of the image formed by an optical system.

degaussing. The process of eliminating the magnetic field of an object, as by opposing it with an equal field produced by coils of wire.

degenerate. 1. Denoting distinct energy levels in an atom, molecule, etc., that are equal in energy.
2. Denoting matter in stars in which the electrons have been removed from the

atoms so that the material consists of a packed mass of electrons and atomic nuclei.

degradation. A chemical reaction in which a compound is decomposed in several stages.

degree. 1. A unit of plane angles, equal to 1/360 of a complete revolution.
2. An interval in certain scales of measurement, as in a *temperature scale.
3. The sum of the exponents of the variables in a mathematical expression. For example $7x^6$ is of degree six, $5x^2yz^4$ is of degree seven. When an expression has several terms, its degree is that of its highest term; $9x^3 + 2x^2 + 3$ has degree three. It is also possible to speak of a degree in specified variables; x^7yz^3 is of degree seven in x, one in y, and three in z.

degree of freedom. 1. One of a number of modes by which an atom or molecule can take up energy. Any species has three translational degrees of freedom, corresponding to motion in three directions. In addition molecules have rotational and vibrational degrees of freedom. Linear molecules have two rotational degrees corresponding to rotation about two axes perpendicular to the axis (energy cannot be taken up by rotation about the axis). Nonlinear molecules have three degrees of freedom. The vibrational degrees of freedom correspond to vibrations between the atoms in the molecule and depend on the number of atoms.
2. The number of independent variables (temperature, pressure, etc.) that fix the state of a system. *See* phase rule.

dehydration. 1. The removal of water from a mixture or from a compound containing water of crystallization.
2. The removal of hydrogen and oxygen from a compound in the ratio 2:1. Ethanol, for example, can be dehydrated to give ethylene: $C_2H_5OH = C_2H_4 + H_2O$.

Deimos. *See* Mars.

del. Symbol: ∇. A differential operator: $\nabla A = i\, \partial A/\partial x + j\, \partial A/\partial y + k\, \partial A/\partial z$, where i, j, and k are unit vectors. *See also* Laplace's equation.

dekatron. A type of electron tube containing a low pressure of gas and ten small cathodes arranged in a circle about a central anode. The electron arrangement is such that only one cathode can function at a time and a glow discharge is transferred from one cathode to the next by a voltage pulse. Dekatrons are used as visible counters.

deliquescence. Absorption of water by a hygroscopic solid resulting in the formation of a solution.

delocalization. The existence in some molecules of electrons that are not confined to the region of one particular bond or atom but can move over the whole molecule. When this effect occurs the molecule is more stable than it would be if the electrons were localized. *See* aromaticity, resonance.

delta iron. The allotropic form of iron stable above $1403°C$. It has a body-centred cubic crystal structure.

demodulation. The process of separating the signal from a modulated carrier wave.

denature. To render alcohol unfit to drink by the addition of a small amount of some other substance. Ethanol is usually denatured with poisonous methanol (*see* methylated spirits).

dendrite. A crystal with a branching structure, giving it a tree-like appearance.

denominator. *See* fraction.

densitometer. An instrument for measuring the extent to which a material transmits or reflects light, usually using a standard light source and a photoelectric cell. Densitometers are used for measuring the amount of darkening of a photographic film and thus for finding the intensity of the radiation in spectrographic analysis, X-ray diffraction, and similar techniques.

density. 1. Symbol: ρ. The mass per unit volume of a substance or material. Density is measured in $kg\,m^{-3}$, $g\,cm^{-3}$, $lb\,ft^{-3}$, etc. *See also* relative density, vapour density.
2. In general, any measure of the intensity of some property with respect to distance, area, or volume. For example, the current density at a point in a conductor is the electric current per normal unit area at that point. The charge density is the electric charge per unit volume (volume charge density) or per unit area (surface charge density).

dependent variable. *See* variable.

depolarization. The minimization of polarization effects in electrochemical cells by stirring the electrolyte, by increasing the operating temperature, or by use of a chemical reaction, as in the Leclanché cell.

depression of freezing point. A decrease in the freezing point of a solvent when a substance is dissolved in it. The depression is a *colligative property: for pure substances it is proportional to the number of molecules or ions dissolved. Measurement of the elevation is a method of determining molecular weights.

derivative. *See* differentiation.

derived unit. *See* unit.

desiccator. A device for drying or for keeping materials dry. The common form is an air-tight compartment, which may be evacuated, in which the material is placed on a mesh stand, below which is a hygroscopic substance such as phosphorus pentoxide or calcium oxide.

desorption. The opposite process to adsorption: i.e. the removal of molecules, ions, etc., from a surface into the gas phase.

destructive distillation. A process in which a substance or mixture of substances is heated leading to decomposition and formation of volatile products, which can be collected. For example, the destructive distillation of wood produces methyl alcohol, the destructive distillation of coal produces coal gas and coal tar.

determinant. A mathematical expression consisting of a square array of numbers or variables, used in the solution of simultaneous equations.

detector. 1. Any device or instrument showing the presence of or for measuring some effect or physical property, such as streams of particles, electromagnetic radiation, heat, etc.
2. The circuit used to separate the signal from a carrier wave. *See* demodulation.

detergent. Any substance that can be used as a cleansing agent. The term is usually applied to substances that resemble soap in their action but are not made by saponification of fats or oils. Like soap, detergents are surfactants which have a hydrophobic (oil-soluble) and a hydrophilic (water-soluble) part in their molecules. The hydrophobic portion is usually a long hydrocarbon chain. There are three types of detergent depending on the nature of the hydrophilic part. The most important, *anionic detergents,* are salts of organic acids: the negative ion consists of a hydrocarbon chain attached to a negative (hydrophilic group). Commonly anionic detergents are salts of sulphonic acids, containing ions of the type RSO_2O^-. In *cationic detergents* (or *invert soaps*) the active agent is a positive ion. Usually these compounds are quaternary ammonium salts of simple acids. The cation has the form RNH_3^+, in which R is a long hydrocarbon chain (the hydrophobic part) and $-NH_3^+$ is the hydrophilic part: the hydrogen atoms may be replaced by small organic groups. *Nonionic detergents* have neutral molecules in which the hydrophobic part is a polar group, usually containing oxygen atoms.

Synthetic detergents are extensively used as household and industrial cleansing agents and in processes requiring *surfactants. Unlike soap, they are unaffected by hard water. Household detergents contain other substances such as perfume, whitening agents (see fluorescence), etc.

detonation. An explosion, often a small explosion used to initiate a second, larger one.

deuterated. Indicating that a chemical compound contains deuterium atoms in place of hydrogen atoms, as in deuterated methane, CD_4.

deuterium (heavy hydrogen). Symbol: D, ^2H. The hydrogen *isotope with one proton and one neutron in its atomic nucleus. It occurs naturally with the normal isotope (0.0156%) and is prepared by the electrolysis of water (see electrolytic separation). The properties are almost identical to those of normal hydrogen. It is used as a *tracer in chemical reaction studies. A.N. 1; A.W. 2.014.

deuterium oxide. See heavy water.

deuteron. A nucleus of a deuterium atom.

deviation. The difference between a particular value of an observation and either the true value or the average value of a number of observations. The mean deviation is the average of the absolute (positive) values of a set of deviations. See also standard deviation.

devitrification. The destruction of the amorphous structure of a *glass, by crystallization. It results in brittleness and loss of transparency. Repeated heating and cooling of the glass eventually causes devitrification but the most common cause is the presence of foreign material, such as carbon, when the glass is in a molten state.

Dewar flask (vacuum flask). A vessel for maintaining its contents at a temperature above or below that of the surrounding air, having double walls with the space between them evacuated to prevent loss of heat by conduction. Usually the flask is made of thin glass, silvered on the inside surface to prevent loss of heat by radiation. [After Sir James Dewar (1842–1923), Scottish chemist.]

dew point. The temperature at which water vapour begins to condense from humid air as the air is cooled. It is the temperature at which the air is saturated and depends on the *humidity.

dextrorotatory (dextrorotary). Denoting a chemical compound that is capable of imparting a clockwise rotation to the plane of polarization of polarized light, as observed by someone facing the light's direction of motion. See optical activity.

dextrose. See glucose.

diacetyl. A yellow liquid ketone, $CH_3COCOCH_3$, which smells of butter. It is used as a flavouring. M.pt. -4°C; b.pt. 88°C; r.d. 0.99.

dialysis. The separation of mixtures by selective diffusion through a *semipermeable membrane.

diamagnetism. A type of magnetic behaviour in which the material has a low negative *susceptibility that is independent of temperature. A diamagnetic sample will tend to move away from an applied magnetic field. The effect is due to the orbital motion of electrons in the atom or molecule. When a field is applied to the sample, the electrons change their orbit so that they tend to resist the application of the field by producing an opposing field. All substances exhibit diamagnetism but this is often swamped by paramagnetic or ferromagnetic behaviour. See also magnetism.

diaminoethane (ethylene diamine). A colourless viscous liquid amine,

NH$_2$CH$_2$CH$_2$NH$_2$. It is prepared by reacting dichloroethane with ammonia and is a solvent, stabilizer for rubber latex, and corrosion inhibitor. M.pt. 8.5°C; b.pt. 116°C; r.d. 0.89.

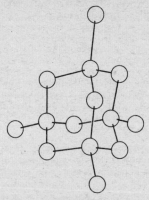

Diamond structure

diamond. A naturally occurring allotrope of carbon. It is an extremely hard crystalline substance. When pure it is transparent and has a high refractive index and dispersive power. Coloured highly impure varieties are used for drills and other cutting tools. The substance owes its hardness to its covalent crystal structure.

diamond-ring effect. The phenomenon occurring just after totality in a total solar eclipse when the dazzling face of the sun (the photosphere) becomes visible from behind the body of the moon and before the corona has had time to fade.

diastereoisomer. *See* optical isomerism.

diathermancy. The property of transmitting infrared radiation. *Compare* athermancy.

diatomic. Having molecules composed of two atoms. Hydrogen, H$_2$, and hydrogen chloride, HCl, are both examples of diatomic gases.

diazine. Any of three isomeric azines, C$_4$N$_2$H$_4$. The isomers are pyrazine, pyrimidine, and pyridazine.

pyridazine   pyrimidine

pyrazine   Diazines

diazo compound. Any of a class of compounds of the formula R:N:N, where R is an alkyl group. The term is also used for azo compounds, which contains the group -N:N-, and for diazonium salts.

diazonium salt. Any of a class of salts with the general formula ArN$_2$$^+X^-$, where Ar is an aryl group. Diazonium salts are made by using an amine dissolved in an acid and adding sodium nitrite to form nitrous acid, which reacts with the amine. They are intermediates in making *azo compounds.

dibromoethane (ethylene dibromide). A colourless oily liquid, BrCH$_2$CH$_2$Br, made by the addition of bromine to ethene. It is used as a solvent. M.pt. 9°C; b.pt. 121°C; r.d. 2.2.

dicarboxylic acid. An acid that has molecules containing two carboxyl groups.

dichlorobenzene. Any of three isomeric compounds, C$_6$H$_4$Cl$_2$, prepared by the direct catalytic chlorination of benzene. The most widely used isomer is *1,4-dichlorobenzene*, a white crystalline solid used as a moth repellent and deodorant. M.pt. 53.1°C; b.pt. 174°C; r.d. 1.28. The other two isomers are colourless volatile liquids: *1,2-dichlorobenzene* is used as an insecticide and an intermediate for making dyes. M.pt. -17°C; b.pt. 180.5°C; r.d. 1.30.

dichlorodifluoromethane. A colourless odourless fairly inert nonflammable gas, CCl_2F_2, manufactured by reacting carbon tetrachloride with hydrogen fluoride and used as a refrigerant, aerosol propellant, and solvent.

dichloroethane (ethylene dichloride). A colourless odourless flammable liquid, $ClCH_2CH_2Cl$, made by the addition of chlorine to ethene. It is used in making vinyl chloride and as a solvent. M.pt. -35°C; b.pt. 83°C; r.d. 1.3 (20°C).

dichloromethane (methylene chloride). A colourless volatile liquid, CH_2Cl_2, prepared by the chlorination of chloromethane. It is used as a solvent and as a local anaesthetic. M.pt. -97°C; b.pt. 40°C; r.d. 1.3.

dichromate. A salt containing the divalent ion $Cr_2O_7{}^{2-}$. Dichromates are salts of the hypothetical *dichromic acid*, $H_2Cr_2O_7$.

dichromic acid. *See* dichromate.

dielectric. A non-conductor of electricity, especially a substance with an electrical *conductivity of less than 10^{-6} siemen.

dielectric constant. *See* relative permittivity.

butadiene   maleic acid   phthalic acid

Diels-Alder reaction

Diels-Alder reaction. A type of chemical reaction in which a conjugated diene reacts with a compound containing a double bond (the *dienophile*) to give a compound containing a benzene ring (see illustration). [After Otto Diels (1876-1954) and Kurt Alder (1902-1958), German chemists.]

diene. An organic compound containing two carbon-carbon double bonds in its molecules.

dienophile. *See* Diels-Alder reaction.

diethylamine. A colourless flammable liquid amine, $(C_2H_5)_2NH$, with an odour resembling that of ammonia. It is prepared by reacting chloroethane with ammonia and is used in the rubber and petroleum industries and in the manufacture of dyes and pharmaceuticals. M.pt. -50°C; b.pt. 55°C; r.d. 0.71.

diethylene glycol. A colourless odourless hygroscopic syrupy liquid, $CH_2OH-CH_2OCH_2CH_2OH$, manufactured by heating ethylene oxide with ethylene glycol. It is used as a solvent, antifreeze, and drying agent. M.pt. -8°C; b.pt. 245°C; r.d. 1.12.

diethyl ether (diethyl oxide, ether). A very light volatile colourless flammable liquid, $(C_2H_5)_2O$, manufactured by the reaction between sulphuric acid and ethanol. It is used as an anaesthetic and solvent. M.pt. -116°C; b.pt. 34.5°C; r.d. 0.71.

diethyl oxide. *See* diethyl ether.

differential calculus. The branch of calculus concerned with *differentiation.

differential equation. An equation containing derivatives. The order of the highest derivative is the order of the equation: for example $d^2y/dx^2 = -ax$ is a second-order differential equation.

differentiation. A mathematical process concerned with finding the way in which a function changes with respect to one of its variables. If y is a function of x, written $y = f(x)$, the *derivative* of y with respect to x is the limit of a small change in y divided by a small change in x as the change in x becomes infinitesimally small: i.e. the limit of $\Delta y/\Delta x$ as Δx tends to zero. In general the derivative, which is written dy/dx, Dy, or y′, will also be a function of x. The derivative of a function at a particular value of x is the gradient of a curve of the function at that point. The differentiation of distance with respect to time gives velocity. Further differentiation can be performed. The second derivative is written d^2y/dx^2, etc. The number of differentiations is the *order* of the derivative. *See also* partial derivative.

diffraction. 1. The formation of light and dark bands (*diffraction patterns*) around the boundary of a shadow cast by an object or aperture. The effect results from *interference of the light from the edges of the object.
2. Interference of waves reflected or scattered from an array of lines, atoms, etc. *See* diffraction grating, X-ray diffraction, electron diffraction.

diffraction grating. A glass plate or metal mirror ruled with large numbers of closely spaced parallel lines. If the line separation is comparable with the wavelength of light the grating reflects or transmits light with *interference so that a series of widely spaced bright lines are produced. The condition for such a maximum is $d(\sin i + \sin \theta) = n\lambda$, where d is the spacing between lines on the grating, i the angle of incidence, θ the angle the reflected or transmitted light makes with the normal, and n is an integer, 0, 1, 2, etc. If $n = 1$, the value of θ for a given value of i depends on the wavelength. Thus polychromatic light is separated into a spectrum by the grating. Higher-order spectra are obtained for higher values of n. Diffraction gratings are used in spectrometers for dispersing light and ultraviolet radiation.

diffuse reflection. Reflection in which the angle of reflection does not equal the angle of incidence. It is distinguished from *specular reflection and occurs when the reflecting surface is uneven, surface irregularities being comparable in size to the wavelength of the reflected radiation.

diffusion. 1. Motion of a substance as a result of the random thermal motion of its atoms or molecules. Two liquids or gases placed in contact will slowly become completely mixed as a result of this process. Gases obey *Graham's law of diffusion: the rate of diffusion is inversely proportional to the density. Thus heavy molecules diffuse more slowly than light molecules, a fact exploited in the separation of isotopes of uranium by diffusion of the volatile hexafluoride through a porous barrier. Solids also undergo diffusion (as can be demonstrated by *tracer studies) but the process is very slow at normal temperatures.
2. Irregular reflection or transmission of light involving scattering.

diffusion pump. A type of vacuum pump in which gas is removed from a system by a stream of mercury or oil vapour. The vapour issues from a nozzle and molecules of gas in the system diffuse into this jet and are carried away. The vapour is condensed inside the pump. Diffusion pumps can only operate at low pressures and are usually backed by a mechanical pump.

digit. Any of the symbols used in a notation for the integers. The number of different digits used in such a system is the *base* of the system. The *binary system has two digits and the decimal system has ten.

digital computer. A type of *computer in which information is stored and handled in the form of discrete units. Modern computer systems are often extremely complex and sophisticated pieces of equipment, but all consist of a few basic units.
The *input system is the part used to feed instructions and data into the device: the instructions are usually written in the form of a programming language. Here the information is converted into the form in which it is stored and handled. In nearly all digital computers some type of *binary notation is used. Thus, numbers, letters, and other

symbols are represented by a string of binary digits (0 and 1), each of which can be represented within the computer by one of two alternative states of a component: i.e. by the presence or absence of a current pulse, voltage, or magnetized region. The information, including the program and the data, is then held in the computer *store.

Operations performed on the data are carried out by electronic circuits that form the *central processing unit. Basically, these contain combinations of switching circuits (gates), which manipulate data drawn from the store in the form of a string of pulses. Thus, combinations of gates are capable of adding, subtracting, and performing simple logical operations such as comparison. The resulting information is again stored until it can be passed to the *output system, where it is converted back into usable form and printed, displayed on a screen, etc. Within the computer operations are governed by the *control*, the electronic circuitry that controls transfer of data from input to store, store to central processing unit, etc.

The characteristic features of digital computers are their ability to store information and their ability to perform operations according to a preset program. They are capable of performing complex calculations in very short times and are also used extensively for information processing: i.e. for sorting vast amounts of information, storing and recovery of data, etc. *See also* analog computer, hybrid computer.

digitalin. A white powder, $C_{36}H_{56}O_{14}$, extracted from the leaves and seeds of digitalis (foxglove) and used as a cardiac stimulant. M.pt. 210°C.

digital voltmeter. A type of *voltmeter in which the reading is given in digital form rather than by a pointer on a scale. Digital voltmeters usually have a solid-state amplifier producing a steady voltage, which is compared with a reference voltage inside the instrument. The reference voltage increases linearly and at the same time pulses are gener-

ated at a constant rate and counted. When the reference voltage reaches the measured voltage the pulses stop, the whole process being repeated every few seconds.

dihydrate. A solid compound with two molecules of *water of crystallization per molecule of compound, as in sodium bromide dihydrate, $NaBr.2H_2O$.

dihydric. Denoting an alcohol or phenol containing two hydroxyl groups per molecule.

dilatancy. A phenomenon shown by some *non-Newtonian fluids in which the viscosity increases as the rate of shear increases. In other words, the fluid becomes more viscous the faster it moves. The effect is observed in some suspensions and pastes: it is caused by friction between the suspended particles. The opposite effect, *thixotropy, is more common.

dilatometer. Any apparatus suitable for measuring changes in volume: typically a glass bulb with a narrow capillary tube attached so that expansion or contraction of a liquid in the apparatus can be measured by the distance moved by the liquid's meniscus in the capillary tube.

dilution. The process of reducing the concentration of a solute by addition of solvent.

dimensions. The set of powers of basic independent physical quantities in terms of which a given physical property is defined. Many quantities can be expressed in terms of mass (M), length (L), and time (T). For example, area has dimensions of L^2, density has dimensions of ML^{-3}, force has dimensions of MLT^{-2}, etc. *Dimensional analysis* is a method of testing the correctness or partially deriving equations by use of dimensions. It depends on the fact that each term of an equation must have the same dimensional formula. Thus, in Einstein's equation, $E = mc^2$, both sides of the equation have dimensions of ML^2T^{-2}.

dimer. A compound whose molecules are formed by addition of two molecules of a simpler compound (the *monomer*). A reaction leading to formation of a dimer is a *dimerization*. Such reactions are often reversible, leading to equilibrium between the monomer and the dimer. An example is the dimerization of nitrogen dioxide: $2NO_2 = N_2O_4$.

dimethylamine (DMA). A colourless gaseous *amine, $(CH_3)_2NH$, prepared by a reaction between ammonia and methanol and used in tanning and the manufacture of soap. M.pt. -96°C; b.pt. 7°C; r.d. 1.05.

dimethyldichlorosilane. A colourless flammable liquid, $(CH_3)_2SiCl_2$, manufactured by reacting silicon with chloromethane. It is used in producing silicone oils, rubbers, and resins. M.pt. -86°C; b.pt. 70°C; r.d. 1.06.

dimethyl ether (methyl ether, wood ether). A colourless flammable gas, CH_3OCH_3, prepared by the dehydration of methanol. It is used as a refrigerant, solvent, and propellant for sprays. M.pt. -141°C; b.pt. 24°C; r.d. 0.66.

dimorphism. The existence of two solid (crystalline) forms of the same chemical substance. Dimorphism in elements is commonly referred to as *allotropy. Some dimorphs exist as modifications in the same crystal system, as in ammonium chloride where both forms are cubic, or they may crystallize in different crystal systems, as in hexagonal and cubic silver iodide. Each dimorphic form is stable within a particular temperature and pressure range and, at a given pressure, the temperature at which one form becomes the other is fixed and called the *transition point*. Existence of more than two forms is *polymorphism*.

dinitrobenzene. Any of three white crystalline isomeric compounds with the formula, $C_6H_4(NO_2)_2$. Direct nitration of benzene gives *1,3-dinitrobenzene*, which is used in the production of dyes and celluloid. M.pt. 90°C; b.pt.

303°C; r.d. 1.58. The 1,2- and 1,4-isomers are prepared by oxidation of the appropriate nitroaniline.

diode. 1. An electronic component consisting of a semiconductor junction with two attached electrodes. The p-n junction acts as a rectifier: if a positive voltage is applied to the p-region a current flows whereas very little current flow occurs in the opposite direction. **2.** A *thermionic valve with two electrodes, an anode and a cathode. Current flows through the valve when the anode has a positive potential with respect to the cathode but no electrons flow when it is negative with respect to the cathode. Diode valves can be used as rectifiers.

diol (glycol). A dihydric *alcohol; an alcohol containing two hydroxyl groups per molecule.

dioptre. A unit of power of a lens, equal to the power of a lens with a focal length of one metre. The power in dioptres is the reciprocal of the focal length in metres. By convention, a converging lens has a positive power and a diverging lens has a negative power.

dip (inclination). The angle that the earth's magnetic field makes with the horizontal at a point on the earth's surface. *See* terrestrial magnetism.

dip circle (inclinometer). An instrument for measuring magnetic dip, consisting of a magnetized needle pivoted to swing in a vertical plane over an angular scale.

diphenylamine (DPA). A colourless crystalline solid amine, $(C_6H_5)_2NH$, prepared by heating aniline with anilinium chloride. Diphenylamine is used in the manufacture of dyes and stabilizers. M.pt. 53°C; b.pt. 302°C; r.d. 1.16.

diphosphate. *See* phosphate.

diphosphoric acid. *See* phosphoric acid.

dipole. The electric charges of equal magnitude and opposite sign a small distance apart. The *dipole moment* is the product of the magnitude of one charge multiplied by the distance between the charges. Molecules can have permanent dipole moments when there is unequal sharing of the electrons in the bond between two atoms. *See also* magnetic dipole.

dipole aerial. *See* aerial.

direct current (d.c.). An electric current that has a fairly constant value and always flows in the same direction.

direct dye. A *dye that is attached directly to its substrate: i.e. without the use of a mordant.

directrix. *See* conic.

disaccharide. A *sugar with molecules containing two monosaccharide units.

discharge. 1. Removal of electric charge, as from a capacitor or cell.
2. Conduction of electricity through a gas or other insulating material. A simple gas discharge can be produced by placing two metal electrodes in a gas with a potential difference between them. Under normal conditions conduction does not occur because the gas molecules are neutral. If an external source of ionizing particles or radiation is available gas molecules can be ionized: the electrons move towards the anode and the ions move towards the cathode, thus producing a flow of electric current. If the electric field between the electrodes is high the electrons may gain enough kinetic energy to ionize other molecules and large numbers of ions and electrons can be formed (*see* avalanche). When this occurs the discharge can become self-sustaining and independent of the external source of ionization. Usually the discharge is initiated spontaneously by cosmic rays or field emission at a high potential difference. The voltage required to maintain the discharge is much lower.
In general, discharges in gases consist of a luminous *plasma containing ions, electrons, free radicals, and excited

atoms and molecules in addition to the normal molecules. The colour of the plasma depends on the nature of the gas. A complex set of processes is occurring, including ionization, recombination of ions and electrons, secondary emission from the electrodes, and excitation of molecules by impact of electrons (*see* fluorescence). The precise behaviour depends on a variety of factors such as the pressure of gas, the potential difference and the current in the circuit, and the sizes, shapes, materials, and separation of the electrodes (*see* glow discharge, Townsend discharge, arc, spark, corona discharge, brush discharge).
Discharges in gases are usually produced in glass or quartz discharge tubes containing gas at low pressures and electrodes sealed through the tube. Discharges can also be produced without electrodes by applying a radiofrequency field. The plasma is formed by acceleration of electrons in the gas. Discharge tubes are extensively used in studies of plasmas and as sources of light and ultraviolet radiation. *See also* fluorescent lamp.

disintegration. *See* decay.

disk (disc). A direct-access storage device used in computers, consisting of a flat circular plate coated with a layer of magnetic iron oxide. The disk rotates on a spindle and the data is stored on a series of concentric tracks access being made by means of a head moved radially over the disk's surface. *See* store.

dislocation. *See* defect.

disodium tetraborate. *See* borax.

disperse phase. *See* dispersion.

dispersion. 1. The separation of a complex wave into component parts according to some characteristic such as frequency or wavelength, as in the separation of visible radiation into colour components by diffraction. The term is particularly used for the way in which a transparent medium can disperse polychromatic radiation as a

result of the dependence of refractive index on wavelength. The *dispersive power* (ω) of a medium is defined by the equation $\omega = (n_1 - n_2)/(n - 1)$, where n_1 and n_2 are the refractive indices of the medium for two specified wavelengths and n is the mean refractive index.
2. A suspension of solid, liquid, or gaseous particles of colloidal size or larger, in a homogeneous medium. The medium forms the *continuous phase* and the suspended material is the *disperse phase. See also* colloid.

dispersion force. A very weak force existing between atoms and molecules. Dispersion forces result from mutual interaction between the electron-nucleus dipoles of two different atoms.

dispersive power. *See* dispersion.

displacement. 1. A chemical reaction in which one atom or group replaces another in the molecule. A typical example is the hydrolysis of an alkyl halide, $H_2O + CH_3Cl = HCl + CH_3OH$, in which the chlorine atom is replaced by a hydroxyl group. *See also* substitution.
2. Symbol: *D*. A measure of the electric flux density in a medium, given by ϵE, where ϵ is the permittivity and *E* is the electric field strength in free space. *See* relative permittivity.

disproportionation. A chemical change involving simultaneous reduction and oxidation of the same compound, as in the reaction $2CuCl = Cu + CuCl_2$.

dissociation. The breakdown of a molecule into radicals, ions, atoms, or simpler molecules. The term is often used to denote reversible decompositions, such as the ionization of acids in aqueous solution.

dissociation constant. The *equilibrium constant of a dissociation reaction. Acid dissociation constants are the equilibrium constants for reactions of the type: $HA + H_2O = H_3O^+ + A^-$. Since the concentration of water is unity, the dissociation constant *K* is equal to $[H_3O^+][A^-]/[HA]$. *See also* pK.

distillation. A process used to purify or separate liquids by evaporating them and recondensing the vapour. The liquid sample is boiled in a flask attached to a *condenser. In this way a pure liquid can be obtained free of dissolved impurities. Distillation is also used for separating mixtures of two or more liquids by exploiting differences in their boiling points. The mixture is heated until it reaches the temperature at which one of the components boils. This fraction is then collected as it distils over at a constant temperature. When this component has been removed, the temperature again begins to rise until it reaches the boiling point of another component, which is collected in a separate receiver. *See also* fractional distillation, vacuum distillation, destructive distillation.

distortion. 1. In general, any undesirable change in the waveform of a signal.
2. A type of *aberration of lenses in which the magnification of the optical system varies with the distance from the axis. Thus a square object may have an image in which the square has sides that bulge outwards (*barrel distortion*) or curve inwards (*pincushion distortion*). Distortion can be reduced by suitable stops.

disulphur dichloride. *See* sulphur chloride.

diterpene. *See* terpene.

div. *See* divergence.

divalent (bivalent). Having a valency of two.

divergence (of a vector). The scalar product $\nabla . F$, where ∇ is the operator del. It is usually written div *F* and is equal to $i. \partial F/ \partial x + j. \partial F/ \partial y + k. \partial F/ \partial z$.

divergent. Designating a sequence or series that is not *convergent.

diverging lens. *See* lens.

divisor. A quantity divided into another quantity. In $a/b = c$, *b* is the divisor.

D-layer. *See* ionosphere.

DMA. *See* dimethylamine.

dodecahedron. A polyhedron with twelve faces. A *regular dodecahedron* has congruent equilateral pentagons as its faces.

dodecanoic acid. *See* lauric acid.

dolomite. A mineral consisting of a mixed carbonate of magnesium and calcium, $CaCO_3.MgCO_3$, used as a refractory material and as a source of magnesium and its compounds.

donor. 1. An atom that donates electrons in an extrinsic *semiconductor, thus increasing the number of electrons in the conduction band.
2. The atom or molecule that supplies the pair of electrons in forming a *coordinate bond. *Compare* acceptor.

doping. The addition of small amounts of impurity to a *semiconductor to modify its electrical properties. Doping can be effected by diffusion or by ion implantation.

Doppler effect. A change in frequency observed for a wave when relative motion occurs between the source and the observer. If the observer is moving towards the source, the frequency has a higher value: if the observer is moving away the frequency is less. The apparent frequency f' is given by $f' = f(c - v_o)/(c - v_s)$, where f is the true frequency, c the velocity of the waves, v_o the observer's velocity, and v_s the source's velocity. The effect occurs for all forms of wave motion, including electromagnetic waves (*see* red shift, Mössbauer effect). [After Christian Johann Doppler (1803–53), Austrian physicist.]

Doppler radar. *See* radar.

dose. A measure of exposure to ionizing radiation. The *exposure dose* is the flux of radiation to which anything is exposed. It is measured in roentgens. The *absorbed dose* measures the amount of absorption of radiation: it is the energy absorbed per unit mass. The unit is the rad. The effect of radiation on human tissue is given by the *equivalent dose*. This is the absorbed dose multiplied by a factor that expresses the biological effect of the radiation used. It is measured in rems.

dosemeter. Any of various instruments or devices used to measure radiation dose. The techniques in common use include measurement of the blackening of a photographic film, determination of the amount of chemical reaction produced in a solution, use of ionization chambers, and measurements of induced *thermoluminescence.

dosimetry. Measurement of radiation dose using a *dosemeter.

dot product. *See* scalar product.

double bond. *See* covalent bond.

double refraction (birefringence). A phenomenon in which certain anisotropic crystals split a ray of incident light into two refracted rays. One ray, the *ordinary ray*, obeys the normal law of refraction. The other ray, the *extraordinary ray*, has a refractive index that depends on direction in the crystal. The two rays are both plane-polarized at right angles to each other (*see* polarization). Double refraction occurs because the crystal is not isotropic and the light propagates at different velocities in different directions: the ordinary ray propagates with spherical wavefronts and the extraordinary ray with ellipsoidal wavefronts. Crystals that show double refraction have either one or two directions along which the light is not doubly refracted. Such a direction is an *optic axis*.

double salt. A mixed salt formed by crystallizing two simple salts in equivalent proportions. The *alums are common examples of double salts.

DPA. *See* diphenylamine.

Döbereiner's triads. Groups of three chemically related elements in which one element has properties intermediate between those of the other two and also has a mass number that is the average of the other two. Examples of such triads are F, Cl, and Br, and Li, Na, and K. They represent an early attempt at classifying the chemical elements. [After Johann Wolfgang Döbereiner (1780–1849), German chemist.]

D-region. *See* ionosphere.

dry cell. A type of primary cell in which the electrolyte is a paste rather than a liquid. The common dry batteries are a form of *Leclanché cell.

drum. A direct-access storage device used in computers, consisting of a cylinder coated with a layer of magnetic iron oxide. The drum rotates on a spindle and the data is stored on a number of tracks around the cylinder's surface, access being made by a set of fixed heads, one for each track. *See* store.

dry ice. *See* carbon dioxide.

drift tube. *See* linear accelerator.

Dulong and Petit's law. The principle that the *atomic heat of all solid elements is approximately equal to $3R$, where R is the gas constant. The value is 25 J K^{-1} mol^{-1}. The law holds for elements with simple crystal structures at normal temperatures. At lower temperatures the atomic heat (molar heat capacity) is proportional to the cube of the thermodynamic temperature. [After Pierre Dulong (1785–1838) and Alexis Petit (1791–1820), French physicists.]

Dumas' method. 1. A technique for measuring the densities of vapours by weighing the sample in a large bulb of known volume and mass.
2. An analytical technique for determining the amount of nitrogen in organic compounds. The sample is mixed with copper oxide and heated to oxidize the nitogen present to nitrogen oxides. These are passed over hot copper, which reduces them to nitrogen gas. This is estimated by a volumetric method. [After Jean Baptise André Dumas (1800–84), French chemist.]

Duralumin. ® An alloy of aluminium and small amounts of copper (about 4%), magnesium, manganese, and silicon.

dwarf. A dwarf star; a star of small diameter and relatively high density. It is usually some ten magnitudes fainter than an ordinary star of the same spectral type. *See also* white dwarf.

dye. A *colourant applied to a fabric, paper, leather, etc., from solution. Dyes are usually organic compounds owing their colour to the presence of conjugated double bonds in their molecules (*see* chromophore). In some types of dye metal atoms are also present. For the colour to be fast the substrate must have an attraction for the dye, a property know as *substantivity*. Dyes can be classified into *substantive dyes*, which are held on the substrate by ionic forces or intermolecular attraction, and *reactive dyes*, which actually form covalent bonds with the substrate. Other classifications are into *mordant dyes, *direct dyes, *acid dyes, *basic dyes, and *vat dyes.

dynamic equilibrium. *See* equilibrium.

dynamics. The branch of mechanics concerned with the motion of bodies and the forces producing motion.

dynamic viscosity. *See* viscosity.

dynamo. *See* generator.

dynamometer. Any of various instruments for measuring power. The power generated by a machine can be determined in a variety of ways including measurement of the heat produced by friction when working under a load, measurement of electrical energy produced by turning a generator, and measurement of the twist produced in a rotating shaft.

dyne. A CGS unit of force equal to the force that, applied to a body with a mass of one gram, produces an acceleration of one centimetre per second per second. It is equivalent to 10^{-5} newton.

dynode. See electron multiplier.

dysprosium. Symbol: Dy. A soft silvery metallic element belonging to the *lanthanide series. A.N. 66; A.W. 162.5; m.pt. 1409°C; b.pt. 2335°C; r.d. 8.54; valency 3.

E

earth. The third planet of the solar system in order of succession from the sun. The earth is not quite spherical, being flattened slightly at the poles because of axial rotation. It has a central core surrounded by a mantle and overlaid by a thinnish outer crust. Some 70% of the surface is covered by water and the whole planet is surrounded by a gaseous *atmosphere.
Equatorial diameter: 12 756.776 km; mass: 5.98×10^{24} kg; relative density: 5.52; period of axial rotation: 23 hours 56 minutes 4 seconds; period of revolution round the sun: 365.256 days; mean distance from the sun: 149 598 000 km; inclination of equator: 23°27′ to the ecliptic.

earth. An electrical connection between a piece of equipment and the earth. Such connections are made for safety reasons so that an appliance cannot become "live" if a fault develops or they are made to produce a fixed reference potential, considered to be zero. In the U.S. the term *ground* is used.

earthshine. Light from the sun reflected from the earth's surface producing a faint illumination of the dark part of the moon.

ebonite. See vulcanite.

ebullition. The bubbling or effervescence of a liquid; boiling.

eccentricity. See conic.

echelon. A type of diffraction grating made by stacking glass plates so that there is a constant offset between each plate to form a series of equal steps with widths of about 1 mm. The plates have to be of equal thickness to within less than one wavelength of light. The echelon can be used as a reflection grating or light can be transmitted through it parallel to the steps. The device has a very high resolution and is used for studies of the fine structure of spectral lines.

echo. 1. A sound repeated as a result of the reflection of sound waves from a surface.
2. The reflected signal in radar.

echolocation. A technique, such as radar or sonar, in which an object is located by the echo reflected from it.

echo sounding. A technique for measuring the depth of the sea by transmitting a pulse of sound and measuring the delay before the reflected pulse is detected. See also sonar.

eclipse. The temporary blocking of the light from the sun (*solar eclipse*) or moon (*lunar eclipse*) as viewed from the earth. Solar eclipses occur when the earth, moon, and sun are directly in line so that the dark body of the moon passes in front of the sun and casts its shadow on the earth's surface. Observers directly in the moon's shadow (*umbra*) see a total eclipse: those on the fringes of it (*penumbra*) see only a partial one. A lunar eclipse takes place when the moon passes into the shadow cast by the earth. Again the eclipse may be total or partial depending on how much of the moon is obscured.

eclipsed conformation. See conformation.

ecliptic. The great circle on the *celestial sphere that marks the earth's orbit around the sun and along which the sun apparently moves relative to the constellations of the zodiac. The circle of the ecliptic is inclined to the plane of the *celestial equator at an angle of 23°27′. Because the orbits of the planets in the solar system are roughly in the same plane these objects also tend to lie in the region of the ecliptic.

Solar eclipse (not to scale)

eddy currents. Alternating electric currents induced in the cores of transformers and other equipment by the changing magnetic fields associated with the equipment. The current causes a power loss in the device. Eddy currents can be reduced by using laminated iron, made up of varnished strips, or by using a *ferrite core.

EDTA (ethylenediaminetetracetic acid). A white powder, $(HOOCCH_2)_2N-(CH_2)_2N(CH_2COOH)_2$, used as a *chelating agent.

effective value. See root mean square.

effervescence. Emission of bubbles of gas from a liquid, especially when the gas is formed *in situ* by a chemical reaction.

efficiency. See machine.

efflorescence. The production of a powdery solid: either from crystals, by the spontaneous loss of water of crystallization, or from a solution by evaporation. Sometimes the term is used to refer to the powdery deposit itself rather than the process.

effort. See machine.

effusion. The flow of a gas from a vessel through a small orifice, with no pressure gradient, due to thermal motion of the gas molecules. The volume of gas effusing in unit time is inversely proportional to its molecular weight. Thus effusion can achieve partial separation of gases of different molecular weights.

EHF. See extremely high frequency.

eigenfunction. One of a set of allowed wave functions of a particle in a given system as determined by *wave mechanics.

eigenvalue. One of a set of allowed energies of a particle in a given system as determined by *wave mechanics.

einstein. A unit of energy of electromagnetic radiation equal, for radiation of a particular frequency, to the energy of one mole of photons. The value depends on the frequency (v) of the radiation and is equal to Nhv, where N is the Avogadro constant and h is the Planck constant. [After Albert Einstein (1879–1955), German-born physicist.]

einsteinium. Symbol: Es. A transuranic *actinide element made by neutron bombardment of plutonium. The most stable isotope, ^{254}Es, has a half-life of about 320 days.

Einstein photoelectric equation. See photoelectric effect.

Einstein's equation. The equation $E = mc^2$, giving the energy E equivalent to a mass m, where c is the velocity of light. The interconversion of matter and energy occurs in nuclear fission and fusion and in pair production and annhiliation. See also relativity.

Einstein shift. A slight displacement of lines in a star's spectrum towards longer wavelengths, caused by an intense gravitational field. The Einstein shift is distinct from a *red shift caused by the Doppler effect. It is predicted by the general theory of relativity.

Einthoven galvanometer. A type of galvanometer consisting of a thin conducting filament stretched so that it passes midway between the poles of a magnet. A current passed through the filament causes a deflection, which is measured by a high-power microscope let through one of the pole pieces. [After Willem Einthoven (1860-1972), Dutch physician and physicist.]

elastance. The reciprocal of the electrical *capacitance. Elastance is measured in darafs.

elastic. 1. See elasticity.
2. Denoting a collision between bodies in which the total kinetic energy of the bodies is conserved and no energy is converted into internal energies of the bodies. An elastic collision between a particle and an atom occurs when the particle is scattered without causing excitation or ionization. Collisions that do produce energy changes of this type are said to be *inelastic.*

elasticity. The property of substances that return to their original size and shape when a deforming force is removed. The force causing the change is measured by the *stress and the deformation it produces is measured by the *strain. If the material recovers its dimensions completely, i.e. the strain disappears when the stress is removed, it is said to be *elastic.* Above the *elastic limit* (see Hooke's law), permanent deformation is produced and the material is *inelastic.* A substance that can be deformed easily with no recovery when the strain is removed is said to be *plastic.* The elastic properties of a material are measured by its modulus of elasticity (see modulus).

elastomer. An elastic material, such as a natural or synthetic rubber or some similar synthetic resin.

E-layer. See ionosphere.

electret. A solid dielectric that exhibits permanent dielectric *polarization. Such materials maintain their polarization when the electric field is removed. See ferroelectricity.

electrical image. A hypothetical point charge or system of charges used in the solution of electrostatic problems. If a charge $+q$ is placed a certain distance from a conductor, the force between the charge and the surface is the same as that which would occur between the charge and its "image" in the surface—a charge of $-q$ placed as far behind the surface plane as the original charge is in front of it.

electric arc. See arc.

electric charge. See charge.

electric constant. The constant ϵ_0 appearing in the equation of *Coulomb's law for the force between two charges in a vacuum. It arises from the choice of units: in *SI units its value is 8.854 16 $\times 10^{-12}$ F m^{-1}. The electric constant is sometimes called the *permittivity of free space.*

electric current. See current.

electric displacement. See displacement.

electric field. A *field in which a stationary electric charge (q) experiences a force F proportional to the magnitude of the charge. The *electric field strength* E is given by $E = F/q$. E was formerly called the *electric intensity.* It is measured in volts per metre. The field may be produced by other stationary electric charges, in which case it is an *electrostatic field.* The force on a stationary charge as a result of the electric field from another stationary charge is given by *Coulomb's law. Electric fields can also be created by a changing magnetic field. Thus, if a magnetic field is changed in the neighbourhood of a conductor it induces an electromotive force (see induction).

electric hysteresis. See hysteresis.

electric intensity. See electric field.

electricity. The class of phenomena caused by electric charges. Charges that are stationary form *static electricity.*

Electric charges in motion constitute *current electricity.*

electric motor. A device for obtaining mechanical energy from electrical energy by electromagnetic induction. Alternating-current motors are similar to *generators, with the current supplied to the armature through slip rings. Motors of this type rotate at the speed that would generate current of the same frequency as the current supplied. Such devices (*synchronous motors*) run at a speed proportional to the frequency of the power supply.

Direct-current motors are similar to d.c. generators, with current supplied through a segmented commutator. The coils generating the magnetic field are supplied with the same current. In a *series-wound* motor the coils are connected in series with the armature, giving a motor that loses speed as its load increases. A *shunt-wound* motor has its field coils connected in parallel with the armature. Shunt-wound motors do not lose speed at increased loads.

Most large a.c. motors are *induction motors,* which have a stator supplied with alternating current in such a way as to produce an effective rotating magnetic field. One method of doing this is by mounting three coils at 120° and using a three-phase alternating-current supply. The rotor is mounted in the centre of the stator and typically consists of a steel bar with copper bars parallel to the axis connected at each end by copper rings (called a *squirrel cage*). The magnetic field induces a current in the rotor and interaction of this field and the current exerts a torque on the rotor. Induction motors are nonsynchronous.

electric polarization. *See* polarization.

electric potential. *See* potential.

electric susceptibility. *See* susceptibility.

electrochemical equivalent. The mass of a substance that is deposited or liberated from a solution of its ions by passage of one coulomb of electricity.

electrochemical series. *See* electromotive series.

electrochemistry. The branch of chemistry concerned with the study of electrolysis, electrolytic cells, and similar applications or properties of ions in solution.

electrode. A conducting plate, wire, grid, etc., used to emit, collect, or control the flow of charged particles in a liquid, gas, or semiconductor. The positive electrode of a system is the *anode*: the negative electrode is the *cathode.* The charge carriers may be electrons, holes, or ions. Electrodes are used in transistors, electrolytic cells, thermionic valves, gas-discharge tubes, and similar devices.

electrodeposition. The deposition of a dissolved substance on an electrode by electrolysis, particularly the plating of a cathode by neutralization of metal ions. The cathode process may be written $Ag^+ + e = Ag$.

electrode potential. The electric potential developed on a metal electrode when it is in equilibrium with a solution of its ions. For example, if a metal rod is placed in a solution of metal ions two competing processes occur. The atoms can enter solution as ions, leaving electrons in the electrode. Conversely, ions in solution can gain electrons from the metal and be deposited as atoms. At equilibrium these processes occur at equal rates and the potential on the metal rod depends on the excess or deficiency of electrons. In fact, the potential of such a half cell cannot be measured absolutely and electrode potentials are usually measured against a *hydrogen electrode, which is assigned an electrode potential of zero. Under standard conditions the value is called the *standard electrode potential. See also* electromotive series.

electrodynamometer. *See* dynamometer.

electroforming. A technique for producing thin intricate metal articles by electrodeposition of a layer of metal onto a shaped former. The former is first coated with a conducting layer of carbon and made the cathode in an electrolytic cell.

electroluminescence. *Fluorescence caused by bombardment with electrons. Electroluminescence is the process occurring in neon lights and other types of *fluorescent lamp.

electrolysis. Chemical change produced by passing an electric current through a conducting solution or fused ionic substance. Such liquids (*electrolytes) contain ions of opposite charge. When a voltage is applied across two electrodes in the electrolyte, positive ions drift towards the cathode and negative ions towards the anode, thus carrying an electric current. At the electrode, positive ions can gain electrons and negative ions can lose electrons, forming neutral atoms or radicals. Alternatively, atoms of the electrode material can ionize and pass into solution in the electrolyte.

electrolyte. A liquid that conducts electricity. Electrolytes may either be solutions or molten solids. In both cases the current is carried by ions drifting under the influence of an electric field produced between the electrodes. Electrolytes are of two types. *Strong electrolytes* have a high conductivity. They are made by dissolving electrovalent compounds, such as sodium chloride, in water or by melting ionic crystals. They are simply mixtures of the ions present in the solid. *Weak electrolytes* are produced by dissolving covalent compounds that partially dissociate in water to give ions. Examples of this type are weak acids and bases, such as acetic acid and ammonia.

electrolytic. Pertaining to electrolysis or electrolytes.

electrolytic capacitor. *See* capacitor.

electrolytic gas (detonating gas). An explosive mixture of hydrogen and oxygen in the proportions 2:1 by volume. It can be produced by electrolysis of water.

electrolytic separation. A method of separating isotopes by using electrolysis. The rate at which ions are discharged at an electrode depends on the ion's mass. The technique is used for separating hydrogen from deuterium by electrolysing water. At the cathode, the reaction $H^+ + e = H$ is slightly faster than $D^+ + e = D$. Consequently, over a long period of time the water becomes enriched with heavy water.

electromagnet. A *magnet made by winding a large coil of wire around a core of soft iron, so that a magnetic field is produced when a current is passed through the coil.

electromagnetic induction. The production of an electric current in a conductor by a changing magnetic field. The phenomenon was originally observed by Faraday when a magnet was moved in and out of a coil of wire connected to a galvanometer. It is found that an electromotive force is induced when a conductor cuts a magnetic flux. This can happen in a variety of ways. For example, a circuit placed in a changing magnetic field has an induced e.m.f. as the magnetic flux linked with the circuit changes. This is the origin of the current in the secondary coil of a transformer. A similar effect is produced by moving a magnet towards a coil of wire. Electromagnetic induction also occurs when a conductor moves through a magnetic field, thus cutting the lines of force. This is the mechanism producing current in a generator. The magnitude of the induced e.m.f. is proportional to the rate of cutting of magnetic flux.

In any circuit in which there is a changing electric current there is an associated changing magnetic field and this in turn may induce an e.m.f. in the circuit. The influence of such a field on the circuit in which the current is flowing is called *self-induction*. The flux ϕ linked with the circuit is proportional to the current flowing: i.e. $\phi = LI$, where L is a constant depending on the geometry of the circuit. When the current is changing the induced e.m.f. is given by $E = -L\,dI/dt$, dI/dt being the rate of change of current. The minus sign appears because the e.m.f. opposes the current flow (in accordance with Lenz's law). L is the *self-inductance* of the circuit. Similarly, a current flowing in one circuit can induce an e.m.f. in a neighbouring one. This is *mutual induction*. The induced e.m.f. is given by $E_2 = -M\,dI_1/dt$. Here dI_1/dt is the rate of change of current in the first circuit, E_2 the e.m.f. induced in the second circuit, and M is the *mutual inductance* of the circuits. *See also* induction.

electromagnetic interaction. A type of interaction between *elementary particles arising as a result of their electric charge. Stationary charges interact by an electrostatic field and moving charges produce a magnetic field. The electromagnetic interaction is about 200 times weaker than the strong interaction. It operates at all distances and is inversely proportional to the square of distance. The effect is thought to be caused by the exchange of virtual photons (*see* virtual particle).

electromagnetic moment. *See* magnetic dipole.

electromagnetic pump. A device for pumping liquid metals through pipes by passing an electric current across the diameter of the pipe and applying a strong magnetic field at right angles to the direction of current flow. The liquid metal then experiences an induced force along the pipe, causing it to move. Electromagnetic pumps are used for moving the liquid sodium coolant in nuclear reactors. They are simple and have no moving parts.

electromagnetic radiation. Waves of energy (*electromagnetic waves*) consisting of electric and magnetic fields vibrating at right angles to the direction of propagation of the waves. Electromagnetic waves are produced by accelerating charged particles and their properties are described by *Maxwell's

Electromagnetic Spectrum

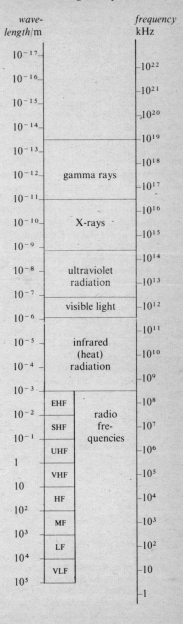

equations. They have a velocity, c, equal to $2.997\ 925 \times 10^8$ metres per second and their properties depend on their frequency (or wavelength). The whole frequency range of electromagnetic radiation is called the *electromagnetic spectrum*, which ranges from radio waves to gamma rays and includes visible radiation (*light). An alternative, and complementary, model of electromagnetic radiation is of streams of *photons.

electromagnetic spectrum. *See* electromagnetic radiation.

electromagnetic units (e.m.u.). A system of electrical units in which the base unit is the unit magnetic pole, defined as a pole that repels a similar pole placed one centimetre away in a vacuum with a force of one dyne. The e.m.u. of current is the current that, flowing in an arc of one centimetre radius and one centimetre length, produces a force of one dyne on a unit magnetic pole placed at the centre of the arc (*see* Ampère's law). Electromagnetic units form part of the system of *CGS units. *See also* ab-.

electromagnetic wave. *See* electromagnetic radiation.

electrometallurgy. The extraction, refining, or shaping of metals by electrolysis.

electrometer. Any of various instruments used to make electrical measurements. The earliest types of electrometer were modified forms of the *electroscope and similar instruments that depend for their action on electrostatic attraction or repulsion. Examples are the *electrostatic voltmeter and the *quadrant electrometer. Most of these are now obsolete except for specialized uses and modern electrometers are electronic devices for amplifying small voltages. The important factor in such instruments is a high input impedance, so that no current is drawn from the circuit investigated. This was formerly achieved by using valves (as in the *valve voltmeter) or by

using a *vibrating-reed electrometer. Now, special field-effect transistors are used. Besides the determination of low voltages, electrometers can be used for small currents (10^{-17} amperes) by measuring the voltage drop they produce across a very high standard resistance. Similarly, they can be used for measuring very high resistances (up to 10^{18} ohms).

electromotive force (e.m.f.). Symbol: E. The energy supplied by a source of current in driving unit charge around an electrical circuit. It is measured in volts. The electromotive force of a current supply is not the same as the potential difference across the terminals when current is flowing because of the internal resistance of the source.

electromotive series. A series of the metals arranged in decreasing order of their tendency to form positive ions by a reaction of the type $M = M^+ + e$. The energy of this process is the *ionization potential for an isolated metal atom. Usually the elements are arranged in order of their *electrode potential, which gives the tendency to form ions in aqueous solution. The series for some of the more common elements is Li, K, Ba, Ca, Na, Mg, Al, Zn, Cr, Fe, Sn, Pb, H, Ti, Cu, Ag, Hg, Au. The more electropositive the element, the higher its position in the series. Metals that have high positions can displace lower metals from their salts. For example, iron displaces metallic copper from copper sulphate: $Fe + Cu^{2+} = Fe^{2+} + Cu$. Similarly, metals such as zinc that are higher than hydrogen in the series will dissolve in dilute acids with evolution of hydrogen: $Zn + 2HCl = ZnCl_2 + H_2$. Metals such as copper will not displace hydrogen. The series is sometimes called the *electropositive* or *electrochemical series*.

electron. An *elementary particle with a negative charge equal to that of the proton and a rest mass of 9.1095×10^{-31} kg (about 1/1836 that of the proton). Electrons are present orbiting the nuclei of all atoms and are also produced by *beta decay. They are produced in electric discharges and in thermionic emission.

electron affinity. Symbol: A. The energy produced by attachment of an electron to an atom or molecule to form a negative ion. The electron affinity of an atom is positive if energy is released. It is usually measured in electronvolts.

electron deficient. Denoting a chemical compound in which there are too few bonding electrons to satisfy the simple rules of valency. The boron hydrides, for example, have molecules in which two boron atoms are linked by an intermediate hydrogen atom with only one pair of electrons shared between the three atoms. This is an example of a multicentre bond.

electron diffraction. Diffraction of electrons by atoms in molecules or crystals. The original example was the *Davisson-Germer experiment in which electrons were diffracted by the surface atoms of a metal. This technique is used for studying solid surfaces. Individual molecules in the gas phase can also produce diffraction patterns from their atoms and these can be used for measuring bond lengths and bond angles of molecules.

electronegative. Tending to form negative ions by gain of electrons.

electronegativity. Symbol: X. The tendency of an atom in a molecule to attract electrons in chemical bonds. The quantity is defined in two ways. In one it is equal to $(I + A)/2$, where I is the ionization potential and A is the electron affinity. The other definition is based on bond energies between two elements Y and Z. If E_{YZ} is the bond energy of a Y-Z bond and E_{YY} and E_{ZZ} are the energies of Y-Y and Z-Z bonds, the difference in electronegativity of Y and Z is given by $X_Y - X_Z = E_{YZ} - E_{ZZ}E_{YY}$. Values are stated with reference to fluorine, which is given the value 4.

electron gun. A device for generating a beam of electrons for use in television tubes, cathode-ray tubes, klystrons, mass spectrometers, etc. The source of the device is usually a wire filament which is electrically heated and which emits electrons by thermionic emission into a surrounding vacuum. The electrons are formed into a narrow beam by arrays of focusing and accelerating electrodes.

electronic. 1. Relating to electronics. **2.** Relating to electrons. Thus the electronic energy levels of a molecule are those defined by the orbital energy of the electrons.

electronics. The study of the properties and applications of semiconductor devices and thermionic valves.

electron lens. An arrangement of coils or electrodes producing a magnetic or electric field used to focus a beam of electrons.

electron microscope. A device in which a magnified image of a sample is produced by using beams of high-energy electrons rather than light. Above a certain magnification, images produced by optical microscopes lack detail because of the limit of resolution. This depends on the wavelength of the light. Electron microscopes have higher resolution because at energies in the keV range the electrons have much shorter wavelengths than light (see de Broglie equation). Magnifications as high as 200 000 can be achieved and objects as small as 0.2-0.5 nm can be resolved. Two main types of electron microscope are used.

In the *transmission electron microscope* a narrow beam of electrons is focused onto the sample in a vacuum by a combination of magnetic and electrical lenses. The sample is thin and electrons passing through it are focused onto a fluorescent screen, thus producing the image.

In the *scanning electron microscope* a thick sample can be used. The primary beam of electrons is deflected by a varying field so that the sample is scanned by the beam and secondary electrons ejected from its surface are focused onto the screen (see secondary emission). Although the scanning microscope has a lower resolution than the transmission instrument it does give a three-dimensional image of the object.

electron multiplier. An electronic device for detecting electrons by using *secondary emission. It consists of an evacuated tube containing a number of electrodes, each held at a higher positive voltage than the one preceding it. An electron hitting the first electrode ejects two or more secondary electrons. These are accelerated to the second electrode where they produce more electrons, and so on. The number of electrons is multiplied by the chain of electrodes (called dynodes). Some types of electron multiplier consist of a narrow spiral glass tube with an insulating coating inside. A very high potential difference is applied between the ends of the coating and electrons hitting one end of the tube are accelerated down it, making succesive collisions with the walls and producing more and more secondary electrons. See also photomultiplier.

electron optics. The study of the focusing and deflection of beams of electrons in cathode-ray tubes, electron microscopes, and similar devices.

electron-probe microanalysis. A technique for the analysis of small quantities of a solid sample by bombarding the specimen with a fine beam of electrons and monitoring the characteristic *X-rays produced. The elements present can be determined and the method can be quantitative, especially for elements of high atomic number. Quantities as small as 10^{-16} kg can be detected with a beam of about 1 micrometre diameter. The technique is particularly useful for investigating the surface structure of solids.

electron spin resonance (ESR). A technique similar to *nuclear magnetic resonance, for investigating the *spins of the electrons in atoms. The radiation absorbed lies in the microwave region.

electron tube. An electronic component in which a current or voltage is controlled by flow of electrons or ions between electrodes in a vacuum or gas. Electron tubes are usually *thermionic valves or types of *discharge tube.

electronvolt. Symbol: eV. A unit of energy equal to the energy acquired by an electron when it is accelerated through a potential difference of one volt. It is used in atomic and nuclear physics, particularly in experimental studies of charged particles.

electrophilic. Having or involving an affinity for negative electric charge. An electrophilic reagent (or electrophile) is one that acts by attacking a negative region of a molecule. Examples are halogen molecules and the nitronium ion (NO_2^+), which occurs in nitrations. Electrophilic reactions are ones involving such a reagent. Thus, in an electrophilic *substitution a hydrogen atom is displaced by an electrophile: this occurs in the substitution of benzene and its compounds. In electrophilic *addition the molecule is initially attacked by an electrophile: this happens in the addition of halogens to alkenes. Compare nucleophilic.

electrophoresis (cataphoresis). The motion of charged colloidal particles through a stationary fluid under the influence of an applied electric field. Electrophoresis is used as a method of separating or analysing mixtures, particularly mixtures of proteins. Typically, two electrodes are placed in contact with a sheet of paper moistened with salt solution. A small amount of sample is applied to the paper and different components of the mixture migrate at different rates.

electrophorus. A simple device for generating an electric charge by friction and electrostatic induction. It consists of a flat insulating plate used in conjunction with a flat metal plate with an insulating handle. The insulating plate is given an electrostatic charge by friction and the metal plate is placed on top of it. Charge on the insulator polarizes the metal so that it has one charge on its lower surface and an equal and opposite charge on its upper surface. The electric charge on the upper surface is then removed by momentarily connecting it to earth: this leaves a net charge on the metal plate.

electroplating. The process of coating a metal object with a thin layer of the same or, more usually, another metal by *electrodeposition.

electropositive. Tending to form positive ions by loss of electrons.

electropositive series. See electromotive series.

electroscope. A simple instrument for detecting an electric charge, consisting of a metal rod with two thin leaves of metal suspended from it passing vertically through an insulating stopper into a glass container. A charge applied to a plate on top of the rod causes the metal leaves to swing apart as a result of the mutual repulsion of their like charges. The leaves are usually made of gold leaf because of its extreme thinness. See also electrometer.

electrostatic. Designating or involving stationary electric *charges or effects produced by such charges. An *electrostatic field*, for example, is an electric field produced by a stationary charge. Electrostatic effects are those caused by forces between stationary charges and are distinguished from effects such as electromagnetism, electrolysis, and the heating effect of a current, all of which are caused by moving charges. Electricity produced by electrostatic charge is known as *static electricity*.

electrostatic field. See electric field.

electrostatic generator. A device, such as a Wimshurst machine or Van de Graaff generator, for generating an electrostatic charge.

electrostatic induction. The separation of charge in a neutral body as a result of the effect of an electric field. If a positively charged body is brought close to a neutral insulating material it attracts it (a plastic comb for example, charged by friction, will pick up small pieces of paper). The electric field produced by the positive charge polarizes the insulator (*see* polarization) and the net effect is that the surface region nearer the positive charge has a negative charge whereas the further part of the insulator has a positive charge. The force of attraction occurs between the original positively-charged body and the induced negative charge on the insulator. Electric charges can also be induced in metal objects: in this case the electrons can flow freely through the metal so one half of the object has a negative charge and the other half a positive charge (the overall charge is zero). See also induction.

electrostatic precipitation. The removal of small solid particles suspended in a gas by electrostatic charging and subsequent precipitation onto a collector in a strong electric field. The process is used in cleaning air in tunnels and effluent gases in industrial chimneys.

electrostatics. The study of electrostatic phenomena. This branch of the study of electricity is also known as *static electricity*.

electrostatic units (e.s.u.). A system of electrical units in which the base unit is the e.s.u. of charge, the statcoulomb, which is defined as the charge that repels an identical charge with a force of one dyne when placed one centimetre away in vacuum (*see* Coulomb's law). Electrostatic units form part of the system of *CGS units. See also stat-.

electrostatic voltmeter. A type of *electrometer used for measuring potential differences in the kilovolt range. A metal vane suspended on a torsion wire is free to move between fixed plates, the angle of deflection being proportional to the square of the potential difference between these and the metal vane. *See also* quadrant electrometer.

electrostriction. A change in the dimensions of a crystal of insulating material when it is subjected to an electric field. *Compare* piezoelectric effect.

electrovalent bond (ionic bond, polar bond). A type of chemical *bond in which atoms or groups of atoms are held together by electrostatic forces between ions. A simple compound between two elements can be formed by transfer of electrons. For example, calcium atoms have the electron configuration 2, 8, 8, 2 and oxygen atoms have the configuration 2, 6. The calcium atom can transfer its two outer electrons to the oxygen and form a calcium ion with a double positive charge, Ca^{2+}. This has the same number of electrons as the inert gas argon. The oxygen atom gains two electrons to form a doubly charged negative ion, O^{2-}. It also has the electron configuration, and thus the stability, of an inert gas (in this case neon). The ions are then held together by electrostatic forces.

Ionic bonding of this type is not confined to simple ions. Ammonium sulphate, $(NH_4)_2SO_4$, contains ammonium ions, $[NH_4]^+$, and sulphate ions, $[SO_4]^{2-}$. Within each ion the bonds are covalent. In many ionic transition-metal compounds the positive ions do not have inert-gas configurations.

No distinct molecules are present in electrovalent compounds. A *crystal of calcium oxide is simply a three-dimensional regular array of ions. Because the forces between ions are strong, ionic solids usually have high melting points. They tend to dissolve in water and other *polar solvents.

electrum. An alloy of gold and silver (15–45%) used in jewellery.

element. A substance consisting of atoms that all have the same atomic number. Elements cannot be broken down into simpler substances. At present, over 104 elements are known although technecium (atomic number 43) and the transuranic elements are made artificially by transmutation. Of the elements in the earth's crust, oxygen (47% by weight) and silicon (28%) are the two most common. Aluminium is the commonest metal and iron, calcium, sodium, potassium, and magnesium are also fairly common. Together these eight elements make up about 99% of the earth's crust. In the universe as a whole, hydrogen is the commonest (about 90% by weight) followed by helium (about 9%). *See also* periodic table.

elementary particle. Any of a number of particles of subatomic size considered as fundamental units of matter. The proton, electron, neutrino, and photon are the only stable particles: the neutron is stable when in the atomic nucleus. Many other unstable particles and *resonances are known from experiments with cosmic rays and accelerators. All decay spontaneously into other particles. Elementary particles are characterized by their rest mass and by other properties such as charge, spin, isospin, hypercharge, parity, and strangeness. In addition to the gravitational and electromagnetic interactions, they partake in *strong and *weak interactions. Those taking part in strong interactions are classified as *hadrons and those taking part in weak interactions as *leptons. Hadrons are further classified into *baryons (including nucleons and *hyperons) and *mesons. Another classification, based on the spin of the particles, divides them into *fermions and *bosons.

Masses of elementary particles are often expressed in units of energy/c^2 (*see* Einstein's equation). 1 MeV/c^2 = 178 × 10^{-30} kg.

Elementary Particles

	Particle	Mass MeV/c^2	Isospin I	Spin J	Parity P	Mean Life/s
photon	γ	0		1	-1	stable
leptons	ν	0		$\frac{1}{2}$		stable
	e	0·511		$\frac{1}{2}$		stable
	μ	105·7		$\frac{1}{2}$		$2·2 \times 10^{-6}$
mesons	$\pi^{\pm}$	139·6	1	0	-1	$2·6 \times 10^{-8}$
	π^0	135·0	1	0	-1	$8·4 \times 10^{-15}$
	$K^{\pm}$	493·8	$\frac{1}{2}$	0	-1	$1·2 \times 10^{-8}$
	K^0	497·7	$\frac{1}{2}$	0	-1	
	K^0_1	497·7	$\frac{1}{2}$	0	-1	$8·6 \times 10^{-10}$
	K^0_2	497·7	$\frac{1}{2}$	0	-1	$5·2 \times 10^{-8}$
	η	548·8	0	0	-1	
baryons	p	938·3	$\frac{1}{2}$	$\frac{1}{2}$	1	stable
	n	939·6	$\frac{1}{2}$	$\frac{1}{2}$	1	932
	Λ	1115·6	0	$\frac{1}{2}$	1	$2·5 \times 10^{-10}$
	Σ^+	1189·4	1	$\frac{1}{2}$	1	$8·0 \times 10^{-10}$
	Σ^0	1192·5	1	$\frac{1}{2}$	1	$1·0 \times 10^{-14}$
	Σ^-	1197·4	1	$\frac{1}{2}$	1	$1·5 \times 10^{-10}$
	Ξ^0	1314·7	$\frac{1}{2}$	$\frac{1}{2}$	1*	$3·0 \times 10^{-10}$
	Ξ^-	1321·3	$\frac{1}{2}$	$\frac{1}{2}$	1*	$1·7 \times 10^{-10}$
	Ω^-	1672·5	0	$\frac{3}{2}$*	1*	$1·3 \times 10^{-10}$

*Predicted by theory

elevation of boiling point. An increase in the boiling point of a solvent when a substance is dissolved in it. The elevation is a *colligative property: for dilute solutions it is proportional to the number of molecules or ions dissolved. Measurement of the elevation is a method of determining molecular weights.

Elinvar. Ⓡ A type of steel containing chromium and nickel. Its elasticity is not affected by temperature changes and it is used for making the hair-springs of watches.

ellipse. A *conic with an eccentricity less than 1. The ellipse is the locus of a point that moves so that the sum of its distances from two fixed points (the foci) is always constant. It has two axes of symmetry, the major and the minor axis. In Cartesian coordinates, the equation is $(x - n)^2/a^2 + (y - m)^2/b^2 = 1$,

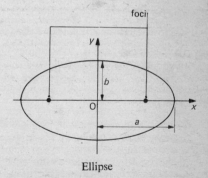

Ellipse

where (n,m) is the centre of the ellipse, $2a$ the length of its major axis, and $2b$ the length of its minor axis. The area is πab.

ellipsoid. A surface or solid all of whose plane sections are ellipses or circles. An ellipsoid generated by rotating an ellipse about one of its axes is a *spheroid*. Rotation about the major axis produces a *prolate* spheroid while rotation about the minor axis gives an *oblate* spheroid.

elliptical. Having the shape of an ellipse or an ellipsoid.

elliptical galaxy. See galaxy.

elliptically polarized light. See polarization.

elongation. The angular distance between the moon or a planet and the sun, measured from the earth.

eluate. See chromatography.

elution. See chromatography.

elutriation. Purification or separation by washing, decanting, and settling. An example of elutriation is the method of panning for gold in river water.

emanation. The radioactive gas given off by radioactive decay of radium, actinium, and thorium. The term is obsolete: the gases are now known to be radon, actinon, and thoron respectively.

e.m.f. See electromotive force.

emission spectrum. A *spectrum produced by emission of electromagnetic radiation by matter. In forming an emission spectrum, energy has first to be supplied to the sample to produce atoms or molecules in excited states. This can be done by heat, as in a flame; by bombardment with particles, as in a discharge tube, electric arc, or X-ray tube; or by irradiation with other electromagnetic radiation. The emitted radiation is dispersed using a prism or grating. Each line in the emission spectrum is produced by decay of excited atoms or molecules to some lower energy level. See also absorption spectrum, spectral series.

emissivity. Symbol: ϵ. A measure of the ability of a body to radiate heat, given by the ratio of the power radiated by the body per unit area to the power radiated per unit area by a *black body at the same temperature. See Kirchoff's law.

emitter. The electrode connected to the central region in a junction *transistor.

empirical. Based on the results of observation rather than on theory.

empirical formula. See formula.

emulsion. A colloidal dispersion of one liquid in another. Oil and water form emulsions: the system may be small globules of oil in a continuous water phase (an oil-in-water emulsion) or small drops of water in oil (a water-in-oil emulsion). Emulsions can be produced by agitating the components together. *Soaps, *detergents, and similar *emulsifiers* are often added to promote the formation of emulsions and, to stabilize them.

emulsoid. See sol.

enantiomer. See optical isomerism.

enantiomorphism. See optical isomerism.

enantiotropy. The existence of a solid in two crystalline forms (see dimorphism) with a definite transition temperature between them. White *tin for example is stable above $13.2°C$: grey tin is stable below $13.2°C$.

endoergic. See endothermic.

endothermic. Denoting a process or chemical reaction that takes place with the absorption of heat. Nuclear processes that absorb heat are said to be *endoergic. Compare* exothermic.

end point. The point in a *titration at which the reaction is complete, so that neither reagent is in excess. It is usually observed using an *indicator.

energy. Symbol: *E*. The capacity of a system to perform *work—i.e. to move a force along its line of action. Energy is divided into *potential energy, due to position, and *kinetic energy, due to motion. Several other forms of energy are distinguished. *Heat* is simply the combined kinetic energies of the atoms and molecules forming the system. *Radiant energy* is energy carried by electromagnetic waves. *Electrical energy* is energy produced by a cell or generator. It is the potential energy of electrons in an electric field. *Chemical energy* is the energy stored in a compound as the kinetic and potential energies of electrons in the chemical bonds. The *law of conservation of energy* states that the total energy of a system is unchanged in any series of processes—energy is converted from one form to another. The principle has been modified by the theory of *relativity in which matter is regarded as a form of energy. *See* mass-energy.

energy band. A band of allowed energies for an electron in a solid. *See* band theory.

energy level. One of a number of discrete energies that may be possessed by a nucleus, atom, molecule, or other system. For example, an electron in a hydrogen atom can only occupy certain orbits around the nucleus, each of a different energy (*see* Bohr atom). The atom can thus have a number of states: one in which the electron is in the inner orbit, another with the electron in the second orbit, etc. A specific energy level exists for each state of the atom. The state with lowest energy is the *ground state*. The higher states are *excited states*. Similarly, according to quantum theory, the vibrations between the atoms in a molecule can only occur so that the molecule has discrete vibrational energy levels.

enol. *See* keto-enol tautomerism.

enrichment. The process of increasing the concentration of one particular *isotope in a mixture of isotopes.

enthalpy. Symbol: *H*. A thermodynamic property of a system equal to the sum of its *internal energy (*U*) and the product of its pressure (*p*) and its volume (*V*): i.e. $H = U + pV$. Changes in enthalpy are useful in chemical thermodynamics because many chemical reactions are carried out at constant (atmospheric) pressure. The measured heat change when a chemical reaction occurs is the change in enthalpy of the system (ΔH): i.e. the sum of its internal-energy change and the work done by a change in volume. By convention, ΔH is negative for an exothermic process.

entropy. Symbol: *S*. A measure of the disorder of a system. Originally, the quantity was introduced in connection with the second law of *thermodynamics. In any system undergoing a reversible change the change of entropy is defined as the energy absorbed divided by the thermodynamic temperature: $dS = dq/T$. The entropy of the system is thus a measure of the availability of its energy for performing useful work. When a practical (irreversible) change occurs, the entropy of a closed system always increases. In fact the entropy of a system, defined in this way, is a measure of the way in which the total energy of the system is distributed amongst its constituent atoms. The statistical idea of entropy has been extended to include changes that do not involve changes in energy. In general the entropy of a system is a measure of its degree of "order"—the more disordered the system, the higher its entropy. For example, two pure liquids in a container will gradually mix together by diffusion. The mixture has a higher entropy than the separated substances because it is more disordered. Similarly, when a liquid freezes part of the entropy change is due to a decrease in entropy caused by a change to a more ordered arrangement of atoms in the crystal.

eosin (bromeosin). A red crystalline solid, $C_{20}H_8Br_4O_8$, showing a greenish fluorescence when in aqueous solution. It is used in colouring textiles and inks.

ephemeris. A table of calculated astronomical data, giving movements and positions of planets and other celestial bodies.

ephemeris time. *See* time.

epicycloid. A plane curve that is the locus of a point on the circumference of a circle as that circle rolls on the outside of another circle. *Compare* hypocycloid.

epidiascope. An apparatus that can function both as a slide projector and an *episcope.

epimerism. A form of *optical isomerism in which a molecule has two or more asymmetric atoms and two isomers (*epimers*) differ in the arrangement about one of these atoms. The term is used in carbohydrate chemistry to describe optical isomerism resulting from different arrangements of the groups attached to the carbon atom next to the aldehyde group.

episcope. An optical apparatus for projecting an image of a flat opaque picture, diagram, etc., onto a screen. The object is illuminated with a high-intensity light and a real image is projected by large lenses. *See also* epidiascope.

epitaxy. The growth of a layer of one solid substance on the surface of another in such a way that the deposited material has the same crystal structure as the underlying substrate. *Epitaxial layers* can be formed by condensation from the vapour: the method is extensively used in making junctions in integrated circuits and other semiconductor devices.

epoxyethane. *See* ethylene oxide.

epoxy group. The group -O- linked to atoms that are themselves linked together, thus forming a ring.

epoxy resin. Any of a class of synthetic resins made by polymerization of compounds containing epoxy groups (see

illustration). Epoxy resins are viscous liquid substances containing ether linkages (-O-) and epoxy groups. They can be mixed with certain curing agents, such as amines or anhydrides, to produce a hard clear resistant plastic. Epoxy resins are also used as adhesives, the resin and the curing agent being mixed together before being applied.

Epsom salt. *See* magnesium sulphate.

equation. 1. A mathematical statement that two expressions have the same value. The equation may hold for particular values of the variables, as in the equation $x^2 + 3x + 2 = 0$ which holds only for $x = -2$ and $x = -1$. Alternatively it may hold for all values of x, as in $x^2 + 3x + 2 = (x + 1)(x + 2)$, in which case it is an *identity. Equations in science are mathematical statements of relationships between physical quantities.
2. A symbolic representation of a chemical reaction using chemical formulae. The right-hand side of the equation represents the products; the left-hand side the reactants. For example, the equation $2NaOH + H_2SO_4 = Na_2SO_4 + 2H_2O$ indicates that two molecules (or moles) of sodium hydroxide combine with one of sulphuric acid to yield one molecule of sodium sulphate and two of water. The numbers of each type of atom are the same on both sides of the equals sign: the equation is then said to be *balanced*. In chemistry a reaction that proceeds completely is often represented using a single arrow (→). One that represents a chemical equilibrium uses the double arrow (⇌).

equation of state. An equation relating the pressure, volume, and temperature of a pure substance, these three quantities serving to define the state of the substance. *See* gas laws, van der Waals' equation.

equilateral. Designating a plane figure with all its sides equal in length.

Epoxy polymer

equilibrium. 1. (in mechanics). The state of a body or system when it has no resultant force or resultant couple acting on it. Its condition is then unchanging: usually the term is applied to a body or structure that is stationary under the influence of balanced forces. Three types of mechanical equilibrium are distinguished depending on the behaviour when the system is subjected to a small momentary displacement from its equilibrium position. In *stable equilibrium*, the system regains its original position. An example of a stable system would be a ball inside a spherical bowl: the ball always rolls back to the lowest point. *Unstable equilibrium* occurs if the initial displacement causes the system to move spontaneously further from the initial position. A ball balanced on top of a sphere, for example, would roll off the sphere if displaced. In *neutral equilibrium* the system takes up a new position as a result of the displacement: an example is that of a ball placed on a plane surface. **2.** (in physics and chemistry). In general, any state of a system in which the properties are not changing with time. A reversible chemical reaction, for example, may reach a condition in which the concentrations of reactants and products are unchanging: the rate of the forward reaction equals that of the back reaction (*see* chemical equilibrium). This is an example of *dynamic equilibrium*—a state in which opposing processes occur at the same rate. Another example is that of a saturated vapour, in which molecules escape from the liquid into the vapour phase at the same rate as they condense. *Thermodynamic equilibrium* is the condition of a system when its free energy

is a minimum. Both the above examples involve states of thermodynamic equilibrium. However, not all unchanging systems are in this state. A supersaturated vapour or solution, for instance, does not have a minimum free energy and can change spontaneously by precipitating liquid or solid. Such systems are said to be in *metastable equilibrium.*

equilibrium constant. In the general reaction $aA + bB = cC + dD$ the rate of the forward reaction, according to the law of *mass action, is $v_f = k_f[A]^a[B]^b$. [A] and [B] are the concentrations of A and B. Similarly the rate of the backward reaction is $v_b = k_b[C]^c[D]^d$. At equilibrium $v_f = v_b$ so that $k_f[A]^a[B]^b = k_b[C]^c[D]^d$. Thus

$$K_c = \frac{[C]^c \, [D]^d}{[A]^a \, [B]^b}.$$

The constant K_c is called the equilibrium constant of the reaction. The subscript c refers to the use of concentration in the derivation. For gasphase reactions it is usual to use *partial pressures instead of concentrations and the constant so derived is written K_p. It can be shown that $K_p = K_c(RT)^{\Delta \nu}$, where $\Delta \nu$ is the number of moles of product less than that of reactants in the stoichiometric equation: i.e. $c + d - a - b$ in the case above.

equinox. One of two points at which the ecliptic crosses the celestial equator. In the Northern Hemisphere, the *vernal equinox* occurs on about March 21 and the *autumnal equinox* occurs on about September 23. At these times the days and nights have equal length. The terms are used both for the points of intersection and for the times at which they occur. *Compare* solstice.

equipartition of energy. The principle that the total energy of molecules is equally distributed amongst their *degrees of freedom. In a gas, the average energy per degree of freedom is $kT/2$, where k is the Boltzmann constant and T the thermodynamic temperature. Thus a monatomic gas, which has three translational degrees of freedom, has an energy per molecule of $3kT/2$.

equipotential. Denoting a line or surface connecting points with the same electric potential.

equivalent circuit. An electrical circuit containing simple resistors, inductors, and capacitors, that is identical in properties to a more complicated electrical device.

equivalent proportions, law of. *See* reciprocal proportions.

equivalent weight. The number of grams of hydrogen that an element could combine with or displace in a chemical reaction. Equivalent weights are sometimes called *chemical equivalents*. They are numbers assigned to substances to indicate the proportions (by weight) in which they combine or react together, using hydrogen as a standard. If an element does not form compounds with hydrogen its equivalent weight can be found by taking the mass that combines with or displaces 8 grams of oxygen or 35.5 grams of chlorine. In general, the equivalent weight of an element is equal to its atomic weight divided by its valency. The equivalent weight of an acid is its molecular weight divided by the number of acidic hydrogen atoms per molecule.

erbium. Symbol: Er. A soft silvery metallic element belonging to the *lanthanide series. A.N. 68; A.W. 167.26; m.pt. 1522 °C; b.pt. 2510 °C; r.d. 9.05; valency 3.

erecting prism or **lens.** A *prism or concave lens used to produce an erect image from an inverted image in a terrestrial telescope or other optical instrument.

E-region. *See* ionosphere.

erg. A CGS unit of energy equal to the work performed when a force of one dyne is moved through a distance of one centimetre in its direction of action. It is equivalent to 10^{-7} joule.

Erlenmeyer flask. A conical laboratory flask with a narrow neck and flat broad bottom. [After Emil Erlenmeyer (1825–1909), German chemist.]

Esaki diode. *See* tunnel diode.

escape velocity. The minimum velocity that would have to be given to an object for it to escape from a specified gravitational field. The escape velocity from the earth is 11 200 metres per second.

ESR. *See* electron spin resonance.

essential oil. A volatile oily liquid mixture extracted from a plant. Essential oils are often fragrant substances containing terpenes and esters.

ester. Any of a class of organic compounds produced by reaction of acids with alcohols. For example, acetic acid and ethyl alcohol give ethyl acetate: $CH_3COOH + C_2H_5OH = CH_3COO\text{-}C_2H_5 + H_2O$
This reaction is called *esterification*: it is reversible and good yields of ester are obtained by distilling off the product as it is formed or by removing the water in some way. The esters of simple acids and alcohols are often volatile fragrant compounds. *Glycerides are naturally occurring esters of glycerol and fatty acids.
Esters can be hydrolysed to give the original acid and alcohol—the reaction is the reverse of esterification and is sometimes called *saponification.

esterification. *See* ester.

etalon. An interferometer in which interference is caused by multiple reflections of light between two parallel semi-reflecting surfaces.

eta meson. Symbol: Gη A type of *meson having zero charge and a mass 280.5 times that of the electron.

ethanal. *See* acetaldehyde.

ethane. A colourless odourless flammable gaseous alkane, C_2H_6, which occurs in natural gas. It is used in organic synthesis and as a fuel. M.pt. -183°C; b.pt. -89°C; r.d. 1.04 (air = 1).

ethanedioic acid. *See* oxalic acid.

ethanediol (ethylene glycol). A colourless hygroscopic viscous liquid, CH_2OHCH_2OH, with a sweet taste. It is made by hydration of ethylene oxide and used as an antifreeze, humectant, and solvent. M.pt. -13°C; b.pt. 197°C; r.d. 1.1.

ethanenitrile. *See* acetonitrile.

ethanoic acid. *See* acetic acid.

ethanoic anhydride. *See* acetic anhydride.

ethanol (alcohol, ethyl alcohol). A colourless volatile flammable liquid, C_2H_5OH. Ethanol is the common alcohol found in intoxicating drinks, in which it is prepared by fermentation. On an industrial scale, ethanol is manufactured by catalytic hydration of ethene or by fermentation of molasses. It has many uses, particularly as a solvent and as a raw material for the manufacture of other chemicals. M.pt. -117°C; b.pt. 78°C; r.d. 0.81.

ethanoyl chloride. *See* acetyl chloride.

ethene (ethylene). A colourless flammable gaseous alkene, C_2H_4, with a sweet taste and odour. It is manufactured by cracking petroleum gases and is used in the preparation of many *polymers. M.pt. -169°C; b.pt. -102°C; r.d. 0.61 (liquid at 0°C).

ethenoid resin. Any synthetic resin produced by polymerizing compounds that contain double bonds. Common examples are *polythene, *polypropylene, the *vinyl resins, and the *acrylic resins.

ethenone. *See* ketene.

ether. 1. *See* diethyl ether.
2. Any of a class of chemical compounds with the general formula ROR′, where R and R′ are alkyl or aryl groups.
3. (aether). An elastic weightless fluid formerly thought to permeate all space and to be the medium in which light and other electromagnetic waves are propagated by vibration. *See also* Michelson-Morley experiment.

ethyl acetate (ethyl ethanoate). A colourless fragrant flammable liquid ester, $CH_3COOC_2H_5$. It is used in flavourings and perfumery and as a solvent for lacquers and plastics. M.pt. -83°C; b.pt. 77°C; r.d. 0.89.

ethyl alcohol. *See* ethanol.

ethylamine (aminoethane). A colourless volatile flammable liquid amine, $C_2H_5NH_2$, made by reacting *chloroethane with alcoholic ammonia. It is used in making dyes and as a stabilizer for rubber latex. M.pt. -81°C; b.pt. 16.6°C; r.d. 0.69 (15°C).

ethyl benzene. A colourless flammable liquid derivative of benzene, $C_6H_5C_2H_5$, made by a *Friedel-Craft's reaction. It is used in the manufacture of styrene. M.pt. -95°C; b.pt. 136°C; r.d. 0.8.

ethyl bromide. *See* bromoethane.

ethyl chloride. *See* chloroethane.

ethylene. *See* ethene.

ethylene diamine. *See* diaminoethane.

ethylene dibromide. *See* dibromoethane.

ethylene dichloride. *See* dichloroethane.

ethylene glycol. *See* ethanediol.

Ethylene oxide

ethylene oxide (epoxyethane). A colourless flammable gaseous ether, C_2H_4O, used in the manufacture of detergents and acrylonitrile. M.pt. -111°C; b.pt. 11°C; r.d. 0.87.

ethyl ethanoate. *See* ethyl acetate.

ethyl group. The organic group C_2H_5-.

ethyl iodide. *See* iodoethane.

ethylxanthic acid. *See* xanthic acid.

ethyne. *See* acetylene.

Euclidean geometry. Geometry based on the assumptions made by Euclid. In his book *Elements* a large number of theorems concerning the properties of figures are deduced from a number of definitions, of concepts such as point and line, and axioms (or postulates). Together the definitions and axioms are considered to be self-evident. Euclid's fifth postulate, the "parallel postulate", says that if a point lies outside a line it is possible to draw only one line through the point that is parallel to the first line. Euclidean geometry is the geometry usually applied to physical measurements involving distance and angle. However, forms of *non-Euclidean geometry exist, in which the parallel postulate is not necessary, and these may be more applicable under special circumstances. [After Euclid (c. 300 B.C.), Greek mathematician.]

eudiometer. An apparatus for measuring volume changes at constant pressure during chemical reactions either in the gas phase or in the solid and liquid phases when evolution of gaseous products occurs. Information about the stoichiometry and in some cases about the mechanism of the reactions can be obtained from these measurements.

Euler's theorem. *See* polyhedron. [After Leonhard Euler (1707–83), Swiss mathematician.]

europium. Symbol: Eu. A soft ductile silvery-white metallic element belonging to the *lanthanide series. A.N. 63; A.W. 151.96; m.pt. 822°C; b.pt. 1597°C; r.d. 5.25; valency 2 or 3.

eutectic. A mixture of two substances in such proportions that the mixture solidifies as a whole when the liquid is cooled, forming a solid solution. In general, if two liquids are mixed and the mixture cooled one or other of the components will solidify until the liquid remaining has the eutectic composition, at which point the whole mixture solidifies. This is the lowest liquid temperature attainable—the *eutectic point*. *Cryohydrates and many *alloys are eutectics.

evaporation. Conversion of a liquid into a vapour, especially when the liquid is below its boiling point. The term is also applied to the process of boiling the solvent from a solution in order to obtain the solid residue.

even-even nucleus. A nucleus that has an even number of protons and of neutrons.

even-odd nucleus. A nucleus that has an even number of protons and an odd number of neutrons.

evolute. A curve that is the locus of the centres of curvature of points on another specified curve (the *involute).

excitation. 1. Change of the energy of an atom, ion, molecule, etc., from one *energy level (usually the ground state) to a higher energy level. The system is then in an *excited state*.
2. Production of a magnetic field by passing a current through the coil of an electromagnet, transformer, or similar device.

exciton. An electron in an excited state in a semiconductor considered as an electron bound electrostatically to a positive hole. The exciton may migrate through the lattice until the hole and electron recombine.

exclusion principle. *See* Pauli exclusion principle.

exoergic. *See* exothermic.

exosphere. The highest region of the earth's *atmosphere, extending from a height of about 400 kilometres.

exothermic. Denoting a process or chemical reaction that takes place with the evolution of heat. Nuclear processes that evolve heat are said to be *exoergic*. *Compare* endothermic.

expanding universe. The universe thought of as expanding in size, as shown by the *red shift of distant celestial objects.

expansion. An increase in the size of a solid, liquid, or gas as a result of an increase in temperature. The expansion results from increased motion of the atoms or molecules. The opposite process—*contraction*—occurs when the solid is cooled. The tendency of a substance to expand is given by its *coefficient of expansion*, α, which is the fractional increase in size per unit rise in temperature: i.e. $\alpha = \Delta s / st$, where s is the original size, Δs the increase, and t the temperature rise. A solid has three coefficients of expansion reflecting increases in length (linear expansion), area (superficial expansion), and volume (volume expansion). Linear expansion obeys the equation: $l_t = l_0(1 + \alpha t)$, where l_0 is the original length, l_t the length at a certain temperature, and t the *rise* in temperature. Similar equations give the area coefficient, β, and the volume coefficient, γ. Coefficients of expansion of solids are measured by the comparator method or by techniques involving the movement of interference fringes.

Liquids have two coefficients of volume expansion: a coefficient of *apparent expansion*, in which the expansion of the container is not corrected for, and a coefficient of *absolute* or *real expansion*, which is based on the true expansion of the liquid. The apparent coefficient is the sum of the absolute coefficient and the volume coefficient of the material of the container. Apparent coefficients can be measured by a *dilatometer and absolute coefficients by Dulong and Petit's method. The expansion of gases is measured at constant pressure and for ideal gases it is 1/273 of the original volume per degree (*see* Charles' law).

exponent. The number indicating the power to which something is raised. Thus in a^n, n is the exponent. Exponents satisfy the laws: $a^n a^m = a^{n+m}$, $(a^n)^m = a^{nm}$, $(ab)^n = a^n b^n$, and $1/a^n = a^{-n}$.

exponential function. A mathematical function e^x, or Ae^{ax}, where A and a are constants. The irrational number e is the base of natural logarithms and has a value of 2.718.... The function e^x is sometimes written exp x. It can be expanded as the infinite series $1 + x + x^2/2! ...$. An *exponential law* is one expressed in the form Ae^{ax}. Thus, exponential decay of a quantity y with time follows an equation of the form $y = Ae^{-at}$.

extender. Any of various substances added to paints, synthetic adhesives, and rubbers to dilute them (for economy) or to modify their physical properties.

extraordinary ray. *See* double refraction.

extremely high frequency (EHF). A *frequency in the range 30 gigahertz to 300 gigahertz.

extrinsic semiconductor. *See* semiconductor.

eye. A section of the human eye is shown in the illustration. Light enters through the transparent *cornea* and is focused by the *crystalline lens* onto the *retina*, the light-sensitive layer of nerve cells on the interior of the eye. The retina is composed of two types of nerve cell—*rods*, which only detect low intensities of light and do not sense different colours, and *cones*, which are the cells detecting normal light intensities. The cones are responsible for colour vision: there are thought to be three types, sensitive to red, green, and blue light and colour sensations are produced by a

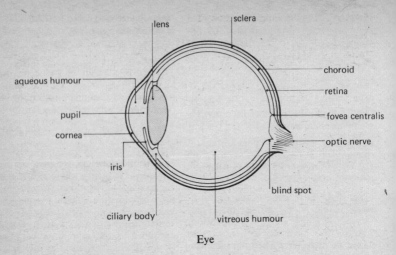

Eye

combination of responses. The most efficient part of the retina is a small region with a high density of cones, the *fovea centralis*. The *blind spot* is a small area with no light sensitivity at the point where the *optic nerve* enters the eye. Besides the retina, the eye has two other layers: the *choroid* and the outer protective *sclera*.

The *iris* is an arrangment of muscle tissue in front of the lens forming a hole, the *pupil*, through which light passes. This increases or decreases in size according to the light intensity. Two fluids are present in the eye: the *aqueous humour* is a water solution between the cornea and the lens and the *vitreous humour* is a gelatinous material behind the lens.

Under normal conditions the lens is held under tension by *ciliary muscles* and focused on distant objects. To observe nearer objects the muscles relax, causing the lens to bulge and shorten its focal length. This process is known as *accommodation*. Defects of vision include *hypermetropia, *myopia, *presbyopia, and *astigmatism.

eye lens. See eyepiece.

eyepiece (eye lens). The lens or lens system that is nearest the eye in a telescope, microscope, or other optical instrument. It is used to view the image formed by the *objective.

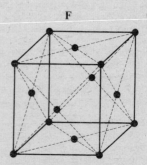

Face-centred cubic unit cell

face-centred. Denoting a crystal structure in which the unit cell has eight atoms at its corners and a further six atoms arranged at the centres of each face. In a *face-centred cubic* lattice (see illustration) the unit cell is a cube. *Compare* body-centred.

factor. Any of two or more integers that give a specified number when multiplied together. Thus, 3 and 5 and 15 and 1 are factors of 15.

factorial. A mathematical function of a number equal to the product of the integers from 1 to the number. Thus, factorial n is $n(n-1)(n-2)...3.2.1$. It is written $n!$ or $\underline{n}$. By definition, $0! = 1$.

faculae. Bright irregular patches on the sun's disc that are much brighter and hotter than the general surroundings. They are apparently associated with *sunspots.

Fahrenheit scale. A *temperature scale in which the melting point of ice is 32° and the boiling point of water is 212° at standard pressure. Temperature is measured in *degrees Fahrenheit* (°F). [After Gabriel Daniel Fahrenheit (1686-1736), German physicist.]

Fajans' rules. Two rules that indicate whether two atoms will be linked by an ionic bond or a covalent bond. They state that:
1. A covalent bond is more likely to be produced if the ions have large charges.
2. A covalent bond is more likely if the cation is small or the anion large.
[After Kasimir Fajans (b. 1887), U.S. chemist born in Poland.]

fall-out. Radioactive material settling on the earth's surface from the atmosphere as a result of a nuclear explosion.

family. A set of curves, surfaces, etc., whose equations are all obtained from one another by replacement of a certain constant by other constants. For example $y = 3x + a$ represents a family of straight lines given by different values of the constant a.

farad. Symbol: F. The SI unit of electric *capacitance, equal to the capacitance of a parallel-plate capacitor with a potential difference of one volt across its plates and holding a charge of one coulomb. The unit is too large for practical use; the microfarad is more commonly employed. [After Michael Faraday (1791-1867), English scientist.]

faraday. *See* Faraday's laws of electrolysis.

Faraday constant. *See* Faraday's laws of electrolysis.

Faraday disc. A simple direct-current generator consisting of a metal disc rotated in a steady magnetic field. The e.m.f. is induced between the centre of the disc and the edge, the current being drawn by brushes at these points.

Faraday effect. Rotation of the plane of *polarization of a beam of plane-polarized light in certain media as a result of the application of a strong magnetic field parallel to the direction of the light. The amount of rotation depends on the nature of the medium, the path length, and the field strength. The effect was originally discovered with light but is also observed for other types of electromagnetic radiation.

Faraday's laws of electrolysis. Two laws used to express the amount of chemical change produced during electrolysis:
1. The amount of chemical change produced is proportional to the quantity of electricity passed.
2. The amount of chemical change produced in substances by a fixed quantity of electricity is proportional to the equivalent weight of the substance.
It follows from Faraday's laws that the charge Q required to deposit or dissolve n moles of an ion with valency v is given by $Q = Fnv$. F is a constant (the *Faraday constant*) having the value $9.648\ 670 \times 10^4$ coulombs per mole. The amount of electricity required to deposit or dissolve one mole of a monovalent ion is $9.648\ 670 \times 10^4$ coulombs. This is sometimes called the *faraday* and used as a unit of charge.

far sight. *See* presbyopia.

fast neutron. A neutron with a high kinetic energy, usually taken to be one with an energy greater than 0.1 MeV. The term is also used for neutrons with sufficient energy to cause *fission in a fast nuclear reactor: i.e. those with energies above 1.5 MeV.

fast reactor. See fission reactor.

fat. Any of various mixtures of *glycerides found in plant and animal tissues, distinguished from natural oils by the fact that they are solid at room temperature.

fathom. A unit of length equal to six feet. It is usually used in expressing the depth of the sea.

fatty acid. See carboxylic acid.

feedback. Any process in which the output of a device or system controls the input in some way. The term is commonly applied to the coupling of a portion of the output of electronic amplifiers to the input. The phase of the returned signal relative to that of the input determines whether the amplification is reduced (*negative feedback*) or reinforced (*positive feedback*). Negative feedback stabilizes the amplifier and cuts down noise. Positive feedback is used in *oscillators.

Fehling's solution. A solution containing a mixture of copper sulphate, sodium hydroxide, and potassium sodium tartrate, used as a test for aldehydes. When heated with Fehling's solution they reduce it to give a red precipitate of copper I oxide, Cu_2O. [After Hermann Fehling (1812–85), German chemist.]

feldspar (felspar). Any of a large group of minerals that are aluminosilicates of sodium, potassium, and calcium. They are found in igneous rocks such as granite. See silicate.

femto-. Symbol: f. Prefix indicating 10^{-15}.

Fermat's principle of least time. The principle that a ray of light (or other electromagnetic radiation) moves between any two points in such a way that the time taken is a minimum: i.e. less than the time that would be taken for other possible paths. [After Pierre Fermat (1601–65), French mathematician.]

fermentation. A reaction in which compounds such as sugar are broken down by the action of microorganisms that form enzymes which catalyse the reaction. A common fermentation reaction is the conversion of sugar to alcohol: $C_6H_{12}O_6 = 2C_2H_5OH + 2CO_2$. The reaction is catalysed by the enzyme *zymase*. Industrial fermentation processes using various yeasts, bacteria, and moulds are used to produce such compounds as butanol, glycerol, acetic acid, and citric acid.

fermi. A unit of length equal to 10^{-15} metre. It is used in nuclear and atomic physics. [After Enrico Fermi (1901–54), Italian-American physicist.]

Fermi-Dirac statistics. A type of *statistical mechanics applied to the distribution of particles amongst energy levels when the particles are indistinguishable from each other and only one particle can occupy each energy state. See also fermion.

Fermi level. The highest occupied energy level in a solid. See band theory.

fermion. Any particle that obeys Fermi-Dirac statistics. All baryons and leptons, such as the proton and the electron, are fermions. Compare boson.

fermium. Symbol: Fm. A transuranic *actinide element made by neutron bombardment of plutonium. The most stable isotope, ^{252}Fm, has a half-life of about 23 hours. A.N. 100.

ferrate. Any salt containing the ion FeO_4^{2-}. Ferrates are salts of the hypothetical acid H_2FeO_4 (*ferric acid*).

ferric. See iron.

ferric acid. See ferrate.

ferricyanide. Any salt containing the complex ion $[Fe(CN)_6]^{3-}$, in which the iron is in the +3 oxidation state. Ferricyanides are salts of the hypothetical acid $H_3Fe(CN)_6$ (*ferricyanic acid*).

ferrimagnetism. A type of magnetic behaviour in which the material has a low magnetic susceptibility that increases with temperature. It is found in certain inorganic solids, such as ferrites, and resembles *antiferromagnetism with the difference that neighbouring antiparallel magnetic moments are unequal. *See also* magnetism.

ferrite. 1. Any of a class of compounds with the formula MFe_2O_4, where M is a divalent transition metal. The ferrites are mixed oxides, $MO.Fe_2O_3$, rather than salts and they are generally ceramic materials exhibiting ferromagnetism or ferrimagnetism. Although they are magnetic they are nonconductors of electricity, a combination that makes them useful in transformer cores and other equipment liable to suffer power loss from eddy currents. Ferrites are also used in computer storage units.
2. Iron with its body-centred cubic crystal structure, existing either as the pure element or as a constituent of steels.

ferroalloy. An alloy of iron and one or more other metals made by smelting mixtures of the metal ore with iron ore. Ferroalloys are used to introduce other metals into steel.

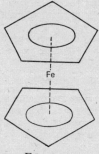

Ferrocene

ferrocene. An orange crystalline solid, $(C_5H_5)_2Fe$, made from iron II chloride and cyclopentadienyl sodium (Na^+-$C_5H_5^-$). Ferrocene was the first *sandwich compound to be prepared. M.pt. 173°C.

ferrochromium. An alloy of iron and chromium (50-70%) used in making steels.

ferrocyanide. Any salt containing the complex ion $[Fe(CN)_6]^{4-}$, in which the iron is in the +2 oxidation state. Ferrocyanides are salts of the hypothetical acid $H_4Fe(CN)_6$ (*ferrocyanic acid*).

ferroelectricity. A property of some dielectrics of exhibiting a permanent electric polarization that is maintained in the absence of a polarizing field. Ferroelectric materials include Rochelle salt and barium titanate. The effect is caused by the presence of electric dipoles in the crystal which line up in domains, just as magnetic dipoles are aligned in the domains of a ferromagnetic material (*see* ferromagnetism). A graph of electric displacement against applied electric field has a hysteresis loop similar to ferromagnetic hysteresis. *See also* electret.

ferromagnetism. A type of magnetic behaviour in which the material has a very high *susceptibility that depends on temperature. Some ferromagnetic substances, such as steel, are also capable of remaining in a magnetized state in the absence of an external magnetizing field (i.e. they form permanent magnets). Ferromagnetism is caused by unpaired electrons, as in paramagnetism. These act as small elementary magnets and in ferromagnetic materials they are aligned parallel to each other within regions of the solid called *domains* by intermolecular forces known as *exchange forces*. Each domain can thus be thought of as a small magnet.
In an unmagnetized sample the domains are orientated at random so the sample has no net magnetic moment. If an external field is applied, the elementary magnets tend to align along its direction and domains whose magnetic moments are directed along the field grow at the expense of neighbouring domains. When the field is large enough, all the elementary magnets point in the direction of the field and the sample is *saturated*. Permanent magnetism occurs when the elementary magnets remain aligned in the absence of the field. Ferromagnetism is found in iron, cobalt,

and nickel, and in certain of their compounds. *See also* Curie-Weiss law, magnetism, hysteresis.

ferromanganese. An alloy of iron and manganese (70-80%) used in making steels.

ferrosilicon. An alloy of iron and silicon (up to 15%) used in making steels.

ferrosoferric oxide. *See* iron oxide.

ferrotungsten. An alloy of iron and tungsten (60-85%) used in making steels.

ferrous. 1. *See* iron.
2. Containing or indicating metallic iron.

ferrovanadium. An alloy of iron and vanadium used in making steels.

fertile. Denoting a material that can be transformed into fissile material for use in a *fission reactor.

FET. *See* transistor.

Fibonacci numbers. The sequence of numbers each of which is the sum of the two preceding. It begins: 1, 1, 2, 3, 5, 8, 13, 21, ... [After Leonardo Fibonacci of Pisa, c. 1170-1250.]

fibre optics. Optical techniques for transmitting images along flexible transparent fibres. Under certain conditions, light falling on the end of a glass fibre is guided along it by a series of total-internal reflections. Bundles of such fibres can be used for viewing or photographing inaccessible objects, inspecting machine parts, medical diagnosis, etc.

field. A region in space in which a force is exerted on an object as a result of its charge (*electric field), magnetic dipole moment (*magnetic field), or mass (*gravitational field). Thus a charged particle can be thought of as producing an electric field in space so that another particle experiences a force at points in the field. Similarly a magnet or a current in a wire creates a magnetic field that can influence a magnetic dipole. A field is often visualized as

lines of force through space. The direction of a line at a point is the direction of the force and the density of lines indicates the strength of the field.

field-effect transistor. *See* transistor.

field emission. Emission of electrons from the surface of a solid as a result of the existence of a high electric field at the surface. The effect is also called *cold-cathode emission*. It is used in the *field-emission microscope*, a device in which a high negative voltage is applied to a fine metal point placed at the centre of a spherical fluorescent screen. Electrons emitted from the metal are accelerated to the screen in straight lines, producing a magnified image of the metal tip. The device has to be used at very low pressures of gas.

field glasses. *See* binocular(s).

field ionization. Ionization of atoms or molecules at the surface of a solid under the influence of a high electric field. Electron transfer occurs between the molecule and the solid to produce a positive ion. The *field-ion microscope* is a similar device to the field-emission microscope with the difference that the metal point has a high positive voltage and a low pressure of helium gas is used. The image is produced by helium ions hitting the screen. The device has a very high resolution: individual atoms can be observed.

field magnet. *See* electric motor.

filler. Any inert substance added to a resin, rubber, or similar substance either to dilute it (for economy) or to modify its physical properties.

film badge. A piece of photographic film in a plastic holder worn by workers exposed to radiation. The amount of blackening of the film indicates the radiation dose received: different types of radiation are detected by the use of metal filters.

filter. 1. A device for separating suspended solid particles from a liquid or gas by passing the fluid through a

porous medium such as paper, sand, sintered glass, etc.
2. A device for passing light or other electromagnetic radiation with a restricted range of wavelengths. Light is filtered by use of pieces of coloured glass or film.
3. An electrical circuit that allows passage of alternating currents within a certain range or ranges of frequency while stopping currents that have frequencies outside the allowed range.

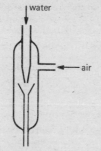

Filter pump

filter pump. A type of vacuum pump in which a jet of water forced through a narrow nozzle traps air molecules and removes them from the system (see illustration). It cannot produce pressures below the vapour pressure of water.

filtrate. The clear liquid obtained by filtration.

filtration. The process of separating solids from a fluid using a *filter.

finder (collimator). A small telescope attached to the side of a large astronomical telescope to enable the large telescope to be directed towards the object to be observed.

fine structure. Structure observed in a spectral line or band when it is viewed at high resolution. Thus a line in an emission spectrum may prove to be a number of closely spaced lines caused by the spin of the electron. *Hyperfine structure* is similar but even higher resolution is required to detect it; the

term is used to designate closely spaced lines produced by emission or absorption in atoms of different isotopes or splitting caused by interaction of the spin of the nucleus with that of the electron.

fire damp. Methane formed in coal mines. It forms an explosive mixture with air. *See also* afterdamp.

Fischer-Tropsch process. A process for producing various hydrocarbons, alcohols, or aldehydes by catalysed reactions between carbon monoxide and hydrogen. [After Franz Fischer (1852-1932), German chemist, and Hans Tropsch (1889-1935), Czech chemist.]

fissile. Capable of undergoing nuclear fission. The term is sometimes restricted to materials that undergo fission by the impact of slow neutrons.

fission. The process in which an atomic nucleus splits into fragment nuclei of comparable size. Nuclear fission is usually accompanied by emission of neutrons and gamma rays and the production of large amounts of energy. It may be spontaneous or induced by the impact of particles or photons. *See also* nuclear weapon, fusion.

fission bomb. *See* nuclear weapon.

fission reactor. A device for producing energy by nuclear fission. In general a fission reactor contains a *core* of fissile material in which a *chain reaction occurs by neutron capture. The reaction is regulated by *control rods*—rods of cadmium, boron, or some other neutron-absorber that can be moved in or out of the core to control the number of neutrons. The heat generated is taken away by a *coolant*, such as liquid sodium, flowing through channels in the core. It is used to heat water to drive a steam turbine for generating electricity.
In the *thermal reactor* the fuel is natural uranium, which is mainly the heavy nonfissile uranium-238 with a small amount of fissile uranium-235. Uranium-238 nuclei capture neutrons to produce stable uranium-239 and under

normal circumstances a chain reaction would not be sustained. A *moderator is used in the core to slow the neutrons to *thermal neutrons, which are less likely to be captured by the uranium-238 isotope. In this way a self-sustaining chain reaction can be maintained by the lighter nuclei. In a *heterogeneous reactor* the core is composed of thin fuel rods embedded in the moderator. A *homogeneous reactor* has a core in which the moderator and fuel are mixed. *Boiling-water reactors* are thermal reactors in which water is used as both moderator and coolant. In *gas-cooled reactors*, the coolant is carbon dioxide.

The other main type of fission reactor—the *fast reactor*—does not have any moderator. Instead the chain reaction occurs by capture of *fast neutrons. For this to be possible the fuel must be enriched, either with uranium-235 or with plutonium-239. The core is surrounded by a *blanket* of natural uranium, in which some neutrons cause the conversion of uranium-238 to plutonium-239, which is used to enrich the fuel. Such devices are called *breeder reactors*.

Fittig reaction. *See* Wurtz reaction. [After Rudolph Fittig (1835–1910), German chemist.]

fixation. *See* nitrogen fixation.

fixed point. A reference point of defined value on a *temperature scale.

fixed star. A true star distinguished from a planet such as Venus.

flame. The hot luminous gas in which chemical reaction is occurring in *combustion. The processes occurring in flames are usually complex, involving free-radical chain reactions. Positive and negative ions are also produced. Light is evolved by fluorescence of excited molecules or ions or, in highly luminous flames, by incandescence of small solid particles of carbon produced in the reactons.

flash photolysis. A technique for investigating the spectra and reactions of free radicals. The sample, usually in the form of a low-pressure gas or vapour, is contained in a glass or quartz tube and subjected to a brief powerful flash of light. *Photolysis of the substance occurs, leading to the formation of atoms and free radicals which can be identified by their absorption spectra using an independent source of radiation. The kinetics of very fast reactions are investigated by observing the change in intensity of spectral lines with time using an oscilloscope.

flash point. The temperature to which a substance must be heated before it can be ignited. This is the temperature at which the sample evolves a flammable vapour.

F-layer. *See* ionosphere.

Fleming's rules. Rules used for remembering the relative directions of motion, magnetic field, and current in dynamos and electric motors. The *left-hand rule* is used for motors. The thumb, forefinger, and second finger are pointed at right angles to each other. If the forefinger points in the direction of the field and the second finger points in the direction of the electric current, then the thumb indicates the direction of motion of the conductor. The *right-hand rule*, which is similar, is used for electrical generators. [After Sir John Ambrose Fleming (1849–1945), British physicist.]

flicker photometer. A type of *photometer in which a screen is illuminated alternately by the test source and a standard source in rapid succession. The distances of the sources are adjusted until the screen does not appear to flicker: the intensities of illumination are then equal. The instrument is particularly useful for comparing sources with different colours.

flint glass. *See* optical flint.

flocculation. The process of coagulating into larger masses.

flotation. *See* froth flotation.

flowers of sulphur. *See* sulphur.

fluid. Any substance that can flow: either a liquid or a gas.

fluidics. The use of fluid flow through narrow pipes in an analgous way to the flow of electric current through circuits. Fluid circuits are much slower than electronic circuits but are unaffected by magnetic fields and ionizing radiation.

fluidity. The reciprocal of the viscosity of a fluid.

fluid ounce. *See Appendix.*

Fluon. Ⓡ *See* polytetrafluoroethylene.

fluorescein. A yellow-red powder, $C_{20}H_{12}O_5$, which exhibits a green fluorescence in alkaline solutions when viewed by reflected light. The solution is green when viewed in transmitted light. It is used as a tracer dye in examining water flow and as an indicator. Decomposes at 290°C.

fluorescence. *Luminescence that ceases as soon as the source of the excitation is removed. It is distinguished from *phosphorescence—in which the emission persists. In scientific usage the term can either denote a luminescence that does not persist long enough to be recognized with the eye or can denote one that lasts less than 10^{-8} second. In general nonscientific usage the term is used for any luminescence under the action of light or electricity, regardless of the persistence.

fluorescent lamp. A lamp in which the light is produced by *luminescence. In devices such as the sodium-vapour and mercury-vapour lamps, an electric current is passed through a low pressure of gas in a glass envelope and light is emitted from excited atoms produced by impact of electrons in the gas phase. This type of fluorescent lighting is also used in neon lights. Domestic fluorescent strip-lights consist of a glass gas-discharge tube that is coated inside with a thin layer of phosphor, which emits light as a result of the impact of electrons from the discharge.

fluoridation. The addition of small quantities (a few parts per million) of a fluorine compound, normally sodium fluoride, NaF, to drinking water for the purpose of decreasing the incidence of tooth decay in the population.

fluoride. Any of certain compounds containing fluorine. *See* halide.

fluorination. A chemical reaction in which one or more fluorine atoms are introduced into a molecule.

fluorine. Symbol: F. A pale yellow corrosive pungent gaseous element occurring chiefly in fluorspar (CaF_2) and cryolite (Na_2AlF_6). It is made by electrolysis of fused hydrogen fluoride and used in producing uranium and in rocket fuels. Fluorine is diatomic, F_2, and is the most reactive of the elements, belonging to the *halogen group. It attacks glass and has to be kept in special steel containers. A.N. 9; A.W. 18.9984; m.pt. –219.64°C; b.pt. –188°C; r.d. 1.7 (air = 1); valency 1.

fluorite. *See* fluorspar.

fluorocarbon. Any of a class of compounds derived from hydrocarbons by replacing some or all of the hydrogen atoms by fluorine atoms.

fluorspar (fluorite). A mineral consisting of calcium fluoride, CaF_2. Fluorspar is the main natural source of fluorine.

flux. 1. A measure of the strength of a field over an area. *See* magnetic flux. **2.** A measure of a flow of energy or particles, usually expressed per unit area. **3.** A substance that can react with an inorganic solid to produce a compound with lower melting point, used in the smelting of metals to form gangue into a molten slag. **4.** A substance used in soldering to keep the surfaces clean and free from oxide.

fluxmeter. An instrument for determining magnetic flux. One type consists of a ballistic galvanometer used in conjunction with a small coil (*search coil*) which is turned in the position investigated. Another type of fluxmeter depends on the voltage produced by the *Hall effect in the field. Fluxmeters are sometimes called *gaussmeters*.

f-number. The ratio of the focal length of a lens to its diameter. The f-number is used in photography, different f-numbers of camera lenses being obtained by use of a variable diaphragm to change the effective lens diameter. Use of higher f-numbers decreases the image brightness and necessitates a longer exposure time. The f-number of a lens is sometimes referred to as its *speed.*

foam. A dispersion of small bubbles of gas in a liquid or solid. The gas is usually air and the dispersion medium is usually a liquid. Foams in liquids are stabilized by the addition of foaming agents, usually surfactants. Solid foams are produced by foaming the substance as a liquid and allowing it to set.

focal length. The distance from the pole of a mirror or centre of a lens to the focal point.

focal point. Either of two points on each side of a *lens, *mirror, or other optical system, such that a ray parallel to the axis converges to or appears to diverge from one of these points.

focus. A point to which rays of light are converged by a *lens, *mirror, or other optical system.

foot. An imperial unit of length equal to one third of a yard. It is equivalent to 0.3048 metres.

force. Symbol: *F.* The agency that can cause a body to alter its state of rest or of uniform motion (*see* Newton's laws). The unit of force is the *newton.

forced convection. *See* convection.

form. *See* crystal form.

formaldehyde (methanal). A colourless poisonous gas, HCHO, with a pungent odour. It is a water-soluble aldehyde forming a solution known as *formalin*, which is used as a preservative. Formaldehyde is manufactured by the catalytic oxidation of methanol or low-boiling petroleum gases and is used in making synthetic resins. M.pt. -92°C; b.pt. -19°C.

formalin. *See* formaldehyde.

formate. Any salt or ester of formic acid.

formic acid (methanoic acid). A colourless liquid, HCOOH, with a pungent odour, made by treating sodium formate with sulphuric acid. It is a strongly reducing carboxylic acid, used in finishing textiles and manufacturing chemicals. M.pt. 8°C; b.pt. 101°C; r.d. 1.2.

formula. A representation of a molecule using symbols for the atoms. The *molecular formula* of a compound indicates the numbers of atoms present, as in $C_2H_4O_2$ for acetic acid. Usually such formulas are written as combinations of functional groups, CH_3COOH. The *empirical formula* indicates the relative numbers of atoms and thus gives the simplest possible structure. Acetic acid has an empirical formula CH_2O. The *structural formula* shows the relative arrangement of the atoms in space.

formyl group. The monovalent group HCO-.

Fortin barometer. A type of mercury *barometer. [After Nicolas Fortin (1750–1831), French instrument maker.]

Fortran. A computer programming language used particularly in scientific work.

Foucault pendulum. A heavy metal ball suspended on a long wire, used to demonstrate the rotation of the earth by the fact that the plane of its motion rotates slowly. [After Jean Bernard Foucault (1819–68), French physicist.]

Fourier series. A series of the form $a_0 + a_1 \cos \omega t + b_1 \sin \omega t + a_2 \cos 2\omega t + b_2 \sin 2\omega t + ...$ Any periodic function $f(t)$ can be expressed in this form, as a sum of harmonic terms.
[After Jean Baptiste Fourier (1763–1830), French mathematician.]

fovea centralis. *See* eye.

FPS units. A system of *units in which the base units are the foot, the pound, and the second.

fraction. A quantity expressed in the form of one quantity divided by another. The dividend in a fraction is the *numerator* and the divisor is the *denominator*: in 7/8, 7 is the numerator and 8 the denominator. If both the numerator and denominator are integers the fraction is a *simple fraction*. Simple fractions are also called *common fractions* or *vulgar fractions*. A fraction in which both denominator and numerator are also fractions, as in (3/4)/(6/7), is a *complex fraction*. If the magnitude of the numerator is less than the denominator, then the expression is a *proper fraction*, otherwise it is an *improper fraction*. Thus, 3/7 is a proper fraction whereas 7/5 is improper.

fractional crystallization. A technique for separating mixtures by utilizing differences in solubility or freezing point of the components. Typically, a solid mixture can be separated by dissolving it in a suitable solvent and lowering the temperature so that one component crystallizes out while the other remains in solution.

fractional distillation. The technique of separating liquid mixtures by *distillation. The term is often used for distillation of a mixture in a container to which is attached a long vertical column filled with loose material, such as glass beads. As the vapour rises in the column, it condenses in the upper parts and the liquid runs back down the column (the *fractionating column*) until it revaporizes. Eventually, the more

volatile components tend to collect in the upper part of the column with less volatile components lower down. This technique enables mixtures of liquids with similar boiling points to be separated efficiently. It is extensively used in the refining of petroleum.

frame of reference. A particular set of coordinates used as a reference system for making physical measurements.

francium. Symbol: Fr. A radioactive metallic element. No stable isotope exists; the most stable, ^{223}Fr, has a half-life of 22 minutes. It is one of the *alkali metals. A.N. 87; valency 1.

Frasch process. A process for obtaining sulphur from underground deposits using a set of three concentric pipes sunk into the ground. Superheated steam is forced down the outer pipe and compressed air is forced down the centre to drive molten sulphur up to the surface.

Fraunhofer lines. Dark absorption lines (*see* absorption spectrum) in the spectrum of radiation from the sun. They are caused by absorption of certain frequencies of light in the cooler outer regions of the sun's atmosphere. [After Joseph Fraunhofer (1787–1826), German instrument maker.]

free energy. A thermodynamic function used to measure the ability of a system to perform work. A change in free energy is equal to the work done. *See* Helmholtz function, Gibbs function.

free radical. An atom or group of atoms that has an independent existence without all its valencies being satisfied. Thus ethane, which contains two methyl groups, can be split into two electrically neutral methyl radicals (*see* homolytic fission). Radicals are usually written with a dot to represent their impaired electron, $CH_3 \cdot$. Atoms produced by breaking diatomic molecules are also regarded as free radicals. Because of their unsatisfied valency species of this type are highly reactive and usually have only a transient existence. They may attack other molecules to abstract an atom and yield another radical. Two radicals can also recombine to give a neutral molecule.

freeze drying. Very rapid freezing and drying of a substance in a high vacuum system, used to preserve foodstuffs, etc.

freezing. The change of a substance from the liquid state to the solid state as a result of lowering the temperature. The temperature at which this occurs is the *freezing point* of the substance. It is the temperature at which the solid and liquid phases are in equilibrium at a specified (usually standard) pressure. The freezing point is equal to the *melting point.

freezing mixture. Two or more substances mixed to give a low temperature. Simple examples of freezing mixtures are ice and sodium chloride (-20°C) and ice and calcium chloride (-40°C).

F-region. *See* ionosphere.

Frenkel defect. *See* interstitial.

Freon. Ⓡ Any of a group of fluorocarbon products that are clear colourless liquids with an ethereal odour. All are chemically inert and of low toxicity. They are used as refrigerants, aerosol propellants, cleaning fluids, and solvents.

frequency. Symbol: f or ν. The number of complete cycles of a periodic process occurring in unit time. The frequency of a wave is equal to c/λ, where c is the velocity and λ the wavelength. Frequency is measured in hertz. *See also* angular frequency.

frequency modulation. *See* modulation.

fresnel. A unit of frequency equal to 10^{12} hertz. [After Augustin Jean Fresnel (1788-1827), French physicist.]

Fresnel lens. A type of lens having a convex surface and a stepped surface (see illustration), used in headlights, searchlights, etc.

Fresnel lens

Fresnel's biprism. An experiment to demonstrate *interference of light by placing a biprism of narrow angle in front of a monochromatic light source. The two images produced by the prism act as coherent sources of equal intensity and produce interference fringes.

friction. A force that tends to oppose the relative motion of two surfaces in contact. The *coefficient of friction* (symbol: μ) is the ratio of the normal force of reaction between the surfaces to the frictional force. In fact the frictional force for a moving body is less than that opposing the movement of a static body, leading to two coefficients: one of *sliding friction* and the other of *static friction*.

Friedel-Crafts reaction. A type of chemical reaction in which an alkyl or acyl group is substituted into a benzene ring by reaction of an alkyl or acyl halide. The two types of reaction can be illustrated by simple examples (see illustration).
All Friedel-Crafts reactions are catalysed by aluminium chloride, boron trifluoride, tin IV chloride, and similar compounds.
[After Charles Friedel (1832-99), French chemist, and James M. Crafts (1839-1917), American chemist.]

benzene   chloromethane   toluene

alkylation

benzene   acetyl chloride   phenyl methyl
                              ketone

acylation

Friedel-Crafts reactions

fringes. Alternate light and dark bands produced by *interference or *diffraction.

froth flotation. Any of several processes in which materials, notably minerals, are separated by agitation of a pulverized mixture of the material with water, oil, and chemicals that cause differential wetting of the suspended particles, the unwetted particles being carried by air bubbles to the surface for collection.

fructose (fruit sugar). A white crystalline simple *sugar, $C_6H_{12}O_6$, that occurs naturally in fruits and honey. M.pt. 103°C; r.d. 1.6.

fruit sugar. See fructose.

fuel cell. A type of electric *cell in which electrical energy is produced directly by electrochemical reactions involving substances that are continuously added to the cell. In the simplest type hydrogen is the fuel and oxygen is the oxidizer. Each gas is fed to a porous metal electrode, usually of nickel or platinum, in an electrolyte of aqueous potassium hydroxide or a molten salt. Electron transfer occurs at the electrodes, the hydrogen electrode being the cathode, and within the cell hydrogen and oxygen combine to produce water.

fulcrum. The pivot about which a *lever turns.

fuller's earth. A type of clay used as a catalyst and absorbent.

full radiator. See black body.

full-wave rectification. Rectification of an alternating current or voltage in such a way that alternate half cycles are reversed in direction, so that the signal repeatedly increases to a maximum and decreases to zero. Compare half-wave rectification.

fulminate. See cyanic acid.

fulminic acid. See cyanic acid.

fumaric acid (trans-butenedioic acid). A colourless crystalline *carboxylic acid with a sharp fruity taste, HOOCCH:- CHCOOH, used in the manufacture of synthetic resins. Fumaric and maleic acids are geometrical isomers. Sublimes at 200°C; r.d. 1.64.

fuming nitric acid. See nitric acid.

fuming sulphuric acid. See sulphuric acid.

function. A mathematical expression involving variables.

functional group. See group.

fundamental. The component with the lowest frequency in a complex wave. See also harmonic.

fur. Calcium carbonate deposited on the inside of kettles, boilers, hot-water pipes, etc. It is precipitated by the decomposition of calcium bicarbonate present in *hard water.

Furan

furan. A colourless liquid, C_4H_4O. M.pt. $-86°C$; b.pt. $31.4°C$; r.d. 0.94.

furanose. *See* sugar.

Furfural

furfural. A colourless liquid cyclic aldehyde, $C_5H_4O_2$, prepared from cereal straw or bran and used as a solvent and in the manufacture of chemicals. M.pt. $-36°C$; b.pt. $162°C$; r.d. 1.2.

fused. Denoting a solid that has been melted and resolidified to a solid mass.

fused ring. A ring of atoms in a molecule attached to another ring in such a way that two atoms are common to both rings. *Naphthalene is a simple example of a fused-ring compound.

fusion. 1. Melting.
2. A nuclear reaction in which two light nuclei join to form a heavier nucleus. An example is the formation of helium from hydrogen: $^{2}_{1}H + ^{3}_{1}H = ^{4}_{2}He + n$. Like all fusion reactions it is accompanied by evolution of large amounts of energy. The nuclei must first have a high kinetic energy to overcome their electrostatic repulsion and this happens at high temperatures. Fusion reactions are often called *thermonuclear reactions*. *See also* nuclear weapon, fission, carbon cycle.

fusion bomb. *See* nuclear weapon.

fusion reactor. A device for obtaining energy by nuclear fusion. In general, a fusion reactor has a configuration of magnetic fields to hold the plasma, which is produced initially by a pulse of high current through the gas. The main problems are in maintaining a stable plasma and in extracting the energy used. So far fusion reactors are only in the experimental stage. *See* ZETA.

G

gadolinium. Symbol: Gd. A silvery-white malleable ductile metallic element belonging to the *lanthanide series. A.N. 64; A.W. 157.25; m.pt. $1311°C$; b.pt. $3233°C$; r.d. 7.9; valency 3.

gain. The increase in signal produced by an amplifier in an electronic circuit. The gain is usually expressed in *decibels as the ratio of the output power to the input power.

galaxy. Any of the innumerable aggregations of stars that, together with gas, dust, and other material, make up the universe. Each galaxy contains, on average, about 100 000 million stars. The vast majority of galaxies are regular systems, either *spiral galaxies*, like our Milky-Way System, or *elliptical galaxies*. The distance between one galaxy and another runs to millions of light-years but galaxies usually form groupings. Our own system has two satellite galaxies, the two *Magellanic Clouds visible in the southern hemisphere.

Galaxy (Milky-Way System). The spiral galaxy in which the sun is located. It is shaped like a Catherine wheel with a convex central nucleus. The Galaxy's diameter is 100 000 light-years and its thickness measured through the nucleus is about 20 000 light-years. It belongs to a local group of 20 or so galaxies including the one in *Andromeda. *See also* galaxy.

galena (lead glance). A grey mineral consisting of lead II sulphide, PbS. It is the principal ore of lead and also contains small amounts of silver.

Galilean telescope. A simple terrestrial telescope with a convex objective lens and a concave eyepiece separated by the difference between the focal lengths of the two lenses. The system gives an erect image and is used in low-powered opera glasses and simple binoculars. [After Galileo Galilei (1564–1642), Italian scientist.]

gallium. Symbol: Ga. A silvery-white metallic element present in small amounts in some ores, including bauxite. It is a low-melting substance and the liquid is more dense than the solid. Gallium arsenide is an important semiconductor. A.N. 31; A.W. 69.72; m.pt. 29.8°C; b.pt. 2403°C; r.d. 5.9°C.

gallon. See Appendix.

galvanic cell. See primary cell.

galvanized iron. Iron or steel coated with a thin layer of zinc to protect it from corrosion.

galvanizing. The process of coating iron or steel with zinc to produce *galvanized iron. Zinc is applied to the metal from a molten bath containing a flux.

galvanometer. Any of several instruments used for detecting or measuring small electric currents. Galvanometers are not usually calibrated in current units (see ammeter). The main types are the *moving-coil and *tangent galvanometers.

gamma iron. The allotropic form of iron stable between 906°C and 1403°C. It is nonmagnetic and has a face-centred cubic crystal structure.

gamma rays. Electromagnetic radiation with wavelengths in the range 10^{-10} – 10^{-13} metre. Gamma rays have shorter wavelengths than X-rays and higher photon energies. They are produced in some radioactive *decay processes.

gangue. The stony material occurring with a metal ore.

gas. A state of matter characterized by ease of flow, compressibility, and spontaneous expansion to fill any container in which the substance is placed. The intermolecular forces in gases are very small and the atoms and molecules show completely random behaviour. See also ideal gas, kinetic theory.

gas carbon. A hard amorphous form of carbon produced as a deposit on the walls of the retort in the manufacture of coal gas.

gas chromatography. See chromatography.

gas constant. The constant R in the equation of state of an ideal gas (see gas laws). Also called the universal gas constant, R has the value 8.314 34 joules per kelvin per mole.

gas-cooled reactor. See fission reactor.

gas laws. Laws describing the behaviour of gases: the two gas laws are Boyle's law ($pV = C$, T constant) and Charles' law ($V = AT$, p constant). The laws strictly apply only to *ideal gases although real gases obey them at low pressure. The laws are combined in the expression $pV = nRT$, where n is the number of moles of gas and R is the *gas constant. This is the equation of state of an ideal gas. Other equations are used for real gases, the best known being *van der Waals' equation.

gas-liquid chromatography. See chromatography.

gas thermometer. A thermometer consisting of a bulb containing gas attached to a pressure measuring device. Two types of gas thermometer are in use. In the constant-volume gas thermometer the volume is maintained at a constant value and the temperature is measured by the pressure of gas in the bulb. In the constant-pressure instrument the temperature is determined by the volume of gas, measured at constant pressure. The temperature is strictly proportional to the pressure or volume only for an ideal gas and corrections are applied in accurate work.

gate. 1. An electronic circuit that gives an output signal only for certain combinations of two or more input signals. The types used in digital computers have two inputs and the signals can be either high or low. Gates can be designed to give a high output if both inputs are high, a high output if either input is low, etc.

2. The electrode that affects the current flow through the channel in a field-effect *transistor.

Gatterman reaction. See Sandmeyer reaction.

gauss. Symbol: G. A CGS-electromagnetic unit of magnetic flux density equal to a flux density of one maxwell per square centimetre. It is equivalent to 10^{-4} tesla. [After Karl Friedrich Gauss (1777–1855), German mathematician and physicist.]

Gaussian units. A system of *CGS units using both electrostatic and electromagnetic electrical units.

Gauss' law. The principle that the total outward *flux of electric field strength E taken over any closed surface is the algebraic sum of the electric charges inside the surface divided by the electric constant.
$$\oint E.ds = \Sigma Q/\epsilon_0$$
The integral is over the whole surface. If $E\epsilon_0$ is replaced by D, the electric displacement, the integral is the electric flux (Ψ), which equals the sum of the internal charges—this is an alternative statement of the law.
Gauss' law can be extended to other fields. In magnetism, the total outward magnetic flux (Φ) is zero.

gaussmeter. An instrument for measuring *magnetic flux density. See fluxmeter.

Gay-Lussac's law. 1. The principle that gases react in volumes that bear a simple ratio to each other and to the volumes of the products if they are gases, all volumes being taken at the same temperature and pressure. Gay-Lussac's law is explained by *Avogadro's hypothesis (equal volumes of different gases contain the same number of atoms).
2. See Charles' law.
[After Joseph Louis Gay-Lussac (1778–1859), French chemist.]

gegenschein (counter-glow). A patch of light seen in the night sky in a position opposite to that of the sun. The phenomenon is rare: like *zodiacal light it is caused by the scattering of sunlight by meteor particles.

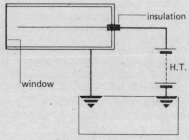

Geiger counter

Geiger counter. A *counter for ionizing radiation, consisting of a cylindrical metallic container with a thin metal window in one end and with a thin wire along the axis (see illustration). The counter contains a low pressure of gas and a high voltage (up to 1000 volts) is applied between the outer metal case (the cathode) and the axial wire (the anode). A particle or photon passing through the tube produces ionization. Electrons are accelerated towards the anode and ions move towards the cathode. On their way they produce further ionization and an *avalanche occurs. The gas in the tube contains halogen compounds or inert gases and the discharge is quickly quenched. The pulse of current is amplified and registered on a counting circuit. [After Hans Geiger (1882–1945), German physicist.]

Geiger-Nuttal law. The principle that the range (R) of a particle emitted in *alpha decay is related to the decay constant (λ) by the equation $\log \lambda = C \log R + B$, where C and B are constants.

Geissler tube. A glass or quartz tube in which a discharge is struck between two electrodes in a low pressure of gas. The electrodes are contained in bulbs connected by a length of capillary tube in which the discharge is intensified. The device is used as a light source for spectroscopy. [After Henry Geissler (1814-79), German instrument maker.]

gel. A colloidal suspension in which the dispersed particles link together, forming a jelly-like mass. Gels can be produced by coagulation of lyophilic *sols. They are often *thixotropic.

gelatin. A colourless or yellow solid obtained in sheets, flakes, or granules. It is a protein, made by boiling skin, tendons, or bones with water, and is used in jellies, films, and adhesives.

gelignite. A high explosive consisting of a mixture of nitroglycerine with nitrocellulose, potassium nitrate, and wood pulp.

gem. See vicinal.

general relativity. The general theory of *relativity.

generator. 1. (dynamo). A machine for producing an electric current by electromagnetic induction. Generators convert mechanical energy into electrical energy and depend on relative motion of a conductor and magnetic field, so that a current is induced. In a simple alternating-current device the coil is rotated in a constant magnetic field produced by a *field magnet*. This coil, the *armature*, is connected to conducting cylindrical *slip rings* on the driving shaft and the current is taken off by carbon *brushes*. The current changes direction, the frequency depending on the rate of rotation. In small a.c. generators the armature is a steel cylinder wound with three coils at 120°. The field is produced by an electromagnet, supplied by a direct current, with curved pole pieces. Such generators produce three a.c. outputs differing in phase by 120°.

Direct-current generators have slip rings made in segments, forming a *commutator*. As the coil rotates the current is taken from different parts of the commutator so that it always flows in the same direction. A single coil gives

a half-wave output. If the armature contains a number of coils, the resulting output can be direct current, with a small ripple. Another type of d.c. generator is the *Faraday disc.

All generators consist of two parts: the *stator*, which remains stationary, and the *rotor*, which moves. Usually, the stator produces the field and the current is induced in the rotor. In very large a.c. generators this is reversed: the magnetic field is produced by rotating field coils and the current is induced in a stationary armature. This arrangement has the advantage that brushes are used to supply the relatively small direct currents to the field coils, and are not required for the large generated currents.

2. Any of various devices for producing electrostatic charge, as in the *Van de Graaff generator.

3. Any of various pieces of chemical apparatus for producing gases for laboratory use, as in *Kipps' apparatus.

4. A line on the surface of a cone or cylinder lying in the same plane as the axis, such that the surface could be generated by rotating the line about the axis.

geocentric. 1. Having the earth at the centre.

2. Measured with reference to the earth.

geochemistry. The study of the chemical composition and reactions of the substances making up the earth.

geodesic. A line on a surface that is the shortest distance between two points on the surface. On a sphere, for example, the geodesics are great circles.

geomagnetism. See terrestrial magnetism.

geometrical isomerism. A type of *stereoisomerism in which two isomers differ in the relative positions of two groups about a double bond. Fumaric and maleic acids are examples of geometrical isomers. In fumaric acid both carboxylate groups are on the same side of the double bond (*cis*) while maleic

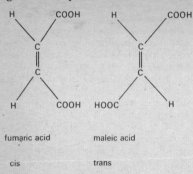

fumaric acid   maleic acid

cis   trans

Geometrical isomers

acid has the groups on opposite sides (*trans*). This type of behaviour is often called *cis-trans isomerism*. It can also be found in certain ring compounds. Square-planar and octahedral inorganic compexes also have geometrical isomers depending on whether two ligands are adjacent or opposite.

geometrical optics. *See* optics.

geometric mean. The *n*th root of the product of *n* numbers. Thus, the geometric mean of *a*, *b*, *c*, and *d* is $(abcd)^{1/4}$.

geometric progression. A *sequence in which each term is the product of the preceding term and a fixed number. A geometric progression has the form a, ar, ar^2, ... ar^{n-1}, where r is the *common ratio* and n is the number of terms. The *geometric series* formed from this sequence has a sum equal to $[a(1 - r^n)]/(1 - r)$. If $-1 < r < 1$ the series converges to $a/(1 - r)$. *Compare* arithmetic progression.

geometry. The mathematical study of shape. *Plane geometry* is the study of figures and curves in two dimensions: *solid geometry* is their study in three dimensions. *See also* Euclidean geometry, non-Euclidean geometry, analytic geometry.

geophysics. The study of the physical properties and behaviour of the earth, including such topics as seismology, oceanography, atmospheric electricity, gravitational properties, and terrestrial magnetism.

germanium. Symbol: Ge. A brittle crystalline grey-white metalloid element found in some zinc ores and some coals. It is extracted by producing the volatile tetrachloride, hydrolysing it to give the oxide, and then reducing this with hydrogen. The principal use is in semiconductors, for which the pure material is obtained by zone refining. The chemistry of germanium is similar to that of silicon, although germanium is rather more reactive. In most of its compounds its valency is 4 although some divalent salts are known, such as $GeCl_2$ and GeS. A.N. 32; A.W. 72.59; m.pt. 937.4°C; b.pt. 2830°C; r.d. 5.32.

getter. A substance used for removing gas from a vacuum tube by adsorption or chemical reaction. Most getters are pure metals. Magnesium, for example, is used in thermionic valves, being heated to combine with residual traces of oxygen and nitrogen after the tube has been sealed. In experimental vacuum equipment at low pressures the getter can be continuously deposited as a clean evaporated film, which adsorbs the gas. A similar process occurs in the getter-ion pump (*see* ion pump).

getter-ion pump. *See* ion pump.

GeV. Gigaelectronvolt; 10^9 electron-volts. In the U.S. *BeV* is used: *B* stands for an American billion.

giant. A giant star; a star of enormous diameter, low mean density, and a high absolute *magnitude.

giant planet. Any of the group of outer planets, Jupiter, Saturn, Uranus, and Neptune, distinguished from the other *terrestrial planets by their large size and mass and by a relatively low density.

gibbous. Denoting the moon when its phase is between a half moon and a full moon.

Gibbs free energy (Gibbs function). Symbol: G. A *free-energy function of a thermodynamic system equal to the *enthalpy minus the product of the *entropy and the thermodynamic temperature: i.e. $G = H - TS$. At constant temperature and pressure the value of G is a minimum when the system is in equilibrium. These are the conditions most usually used in chemical reactions. [After Josiah Willard Gibbs (1839–1903), American physical chemist.]

gilbert. Symbol: Gb. A CGS-electromagnetic unit of *magnetomotive force equal to the magnetomotive force produced by a current of 40π amperes (4π abamperes) passing through a single turn of a coil. It is equivalent to $10/4\pi$ ampere-turns. [After William Gilbert (1544–1603), British physician and scientist.]

glacial acetic acid. Pure acetic acid, free from water.

glass. 1. A hard brittle transparent or translucent substance made by fusing together a mixture of sand (SiO_2), sodium carbonate, and lime. This gives a mixture of calcium and sodium silicates called *soda glass*, which is the common form of glass used in bottles, windows, etc. Other specialized glasses can be made in which part of the sodium or calcium is replaced by metals such as barium and lead. Lead glass, for example, has a high refractivity. Similarly some of the silicon atoms can be replaced by boron atoms forming *borosilicate glass*, which has a lower thermal expansion (see Pyrex). Coloured glass is made by adding coloured metal oxides to the fused mixture.

Glass is an amorphous substance: i.e. the atoms in the solid do not have the regular pattern characteristic of a crystalline solid. Instead they have a fairly random arrangement similar to that in a liquid. Thus glass does not show a sharp melting point but softens over a temperature range. It is best described as a supercooled liquid.

2. Any amorphous solid produced by cooling a liquid.

Glauber's salt. See sodium sulphate.

GLC. See chromatography.

glow discharge. A type of electrical *discharge occurring between two electrodes in a low pressure of gas. The discharge can be maintained in pressures of a few mmHg and consists of a number of dark and bright bands in the region between the electrodes. Most of the potential drop across the discharge occurs between the cathode and a small luminous region close to the cathode called the *cathode glow*. This region is separated from the cathode by a narrow band, the *Crookes dark space*. Another luminous region, the *positive column*, extends from the anode: it is separated from the cathode glow by another dark region, the *Faraday dark space*. At lower pressures the positive column forms into a set of alternate light and dark bands, called *striations*.

The characteristic features of the glow discharge are the nonuniform fall in voltage with distance down the tube, the relatively high current (a few milliamps), and the fact that the voltage drop across the discharge is independent of the current. This last feature makes the glow discharge suitable for use as a voltage stabilizer. *See also* Townsend discharge.

glucose. A colourless crystalline *sugar, $C_6H_{12}O_6$, present in many plants and in sucrose and starch. Carbohydrates and other sugars are broken down to glucose in the body. The natural form is optically active and is also called D-glucose or *dextrose*. M.pt. 146°C; r.d. 1.54.

glyceride. An *ester formed between glycerol and a fatty acid. Glycerol, $CH_2OHCH(OH)CH_2OH$, is a trihydric alcohol and one molecule can react with either one, two, or three molecules of acid. Natural fats and oils are *triglycerides*—esters in which all three hydroxyl groups have been substituted (see illustration). The three fatty acid

O
‖
H₂C — C — R

O
‖
HC — C — R

O
‖
H₂C — C — R

R — hydrocarbon group

Triglyceride

molecules may be identical (*simple glyceride*) or different (*mixed glyceride*). Many different fatty acids are found in natural fats and oils: they are generally classified into saturated and unsaturated acids. Saturated fatty acids, such as stearic and palmitic acids, have a straight-chain alkyl group attached to a carboxyl group. In unsaturated acids, such as oleic and linoleic acids, the hydrocarbon chain contains one or more double bonds. Fats are solid at room temperature and contain a higher proportion of saturated acids: oils are liquid and contain more unsaturated acids.

glycerol (glycerine). A colourless syrupy liquid, $CH_2OHCHOHCH_2OH$, with a sweet taste, obtained as a by-product in the manufacture of soaps. It is used as a solvent, humectant, and plasticizer. M.pt. 18°C; b.pt. 290°C; r.d. 1.3.

glycine (aminoethanoic acid). A white sweet-tasting crystalline solid, NH_2CH_2COOH, made by alkaline hydrolysis of gelatin. The principal amino acid in sugar cane, glycine is used in organic synthesis, medicine, and as a nutrient. M.pt. 233°C (decomposes); r.d. 1.2.

glycol. *See* ethanediol.

glyptal resin. *See* polyester resin.

gold. Symbol: Au. A soft very malleable and ductile heavy yellow *transition element found native and as the telluride. The metal is obtained by sluicing, by amalgamation with mercury, or by dissolving in potassium cyanide followed by reduction of the resulting cyanoaurate. A certain amount is recovered from the anode sludge produced in electrolytic refining of copper. Gold is used in coinage, jewellery, and, in the form of gold leaf, in decorative gilding. Chemically, the metal is less reactive than copper or silver, being unaffected by oxygen, sulphur, and most acids. The element is attacked by the halogens and by *aqua regia. Gold does not have typical ionic salts; most of its compounds are covalently bonded and many complexes exist. It forms both gold I (*aurous*) and gold II (*auric*) compounds. A.N. 79; A.W. 196.97; m.pt. 1063°C; b.pt. 2966°C; r.d. 9.32; valency 1 or 3.

gold chloride (auric chloride, gold trichloride). The only stable chloride of gold, $AuCl_3$. It is a red crystalline solid, prepared by the action of aqua regia on gold. A hydrated form, $HAuCl_4.4H_2O$, is *chloroauric acid*, which is used in photography, gold plating, and ceramics.

golden section. The division of a line or area so that the ratio of the larger to the smaller part is equal to the ratio of the whole to the larger. If a point O divides a line AB such that AO/OB = AB/AO, the ratio AO/OB is the *golden mean*. A rectangle having sides in this ratio is a *golden rectangle*, which can be divided into a square and another golden rectangle. Proportions based on golden sections appear in many paintings: they are considered to be particularly aesthetically pleasing.

gold point. *See* temperature scale.

Goldschmidt process. A process for extracting metals from their oxides by reduction with aluminium. The reaction occurring is the same as that in *thermite.

goniometer. An instrument for measuring the angles between the faces of crystals.

grade. A unit of plane angle equal to one hundredth of a right angle. It is equivalent to 0.9 degree. The symbol is a superscript g used after the value; 1^g = $\pi/200$ radian.

gradient, (of a scalar). The product ∇A, where ∇ is the operator *del and A is a scalar. It is usually written grad A.

graft copolymer. *See* polymer.

Graham's law. The principle that the rate of diffusion of a gas is inversely proportional to the square root of its density. [After Thomas Graham (1805–69), British chemist.]

grain. *See Appendix.*

gram-atom. A unit of mass of an element equal to the element's atomic weight expressed in grams. *See* mole.

gram-equivalent. A unit of mass equal to the *equivalent weight of an element or group expressed in grams.

gram-force. Symbol: gf. *See* gram-weight.

gram (gramme). Symbol: g. A CGS unit of mass equal to one thousandth of a *kilogram.

gramme. *See* gram.

gram-molecule. Symbol: gmol. A unit of mass equal to the molecular weight of a compound expressed in grams. *See* mole.

gram-weight. Symbol: gwt. A unit of force equal to the force that could give a body of mass one gram an acceleration equal to the local *acceleration of free fall. The force in dynes is the force in grams-weight multiplied by the local value of g in centimetres per second per second. The unit varies slightly from place to place. The *gram-force* (symbol: gf) is the force that would give a mass of one gram an acceleration equal to the standard acceleration of free fall (980.665 cm s^{-2}): 1 gf = 980.665 dynes.

graphite (black lead, plumbago). A naturally occurring allotrope of carbon. It is a black solid material with a high

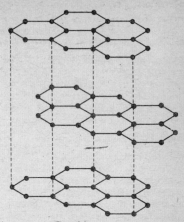

Graphite structure

melting point and a good electrical and thermal conductivity. The crystal structure has layers of covalently bonded hexagonally arranged atoms, with rather weak forces between the layers. This allows the slip of adjacent layers, a property that enables graphite to be used as a lubricant. It is also used in pencils, crucibles, electrodes, and as a moderator in reactors.

graticule. A rectangular network of fine wires or lines in the focal plane of an optical instrument; it is used as a reference system.

grating. *See* diffraction grating.

gravitation. The mutual attraction of bodies as a result of their mass. Gravitation is the effect causing objects to fall to the earth (*see* gravity). *Newton's law of gravitation* states that any two bodies each experience a force of attraction (F) proportional to the product of their masses (m_1 and m_2) and inversely proportional to the square of their separation (x): i.e. $F = Gm_1m_2/x^2$. G is the *gravitational constant. Gravitational interaction is a much weaker force than electromagnetic interaction. A mass has an associated *gravitational field* characterized by the fact that a force is exerted on a second mass placed in this field. According to the general theory of *relativity, gravitation is explained by a curvature of space-time in the neighbourhood of matter.

gravitational constant. Symbol: G. The constant appearing in the formula for Newton's law of *gravitation. It has the value 6.664×10^{-11} newton metres-squared per kilogram squared ($N\,m^2\,kg^{-2}$). *See also* Cavendish's experiment.

gravitational field. *See* gravitation.

gravitational interaction. A type of interaction occurring between bodies or particles as a result of their mass. It is about 10^{40} times weaker than the electromagnetic interaction. *See also* gravitation.

gravitational mass. *See* mass.

gravitational waves. Waves propagated as a result of the acceleration or deformation of a mass. The existence of gravitational radiation travelling at the speed of light is predicted by the general theory of relativity. In principle the radiation can be detected by its effect in exciting oscillations in a solid bar or similar detector. Usually experiments are performed using several detectors, widely separated, and observing coincidences between their signals. It has been suggested that gravitational radiation pulses are reaching earth from some point near the centre of the Galaxy, but the experimental evidence has been disputed.

graviton. A hypothetical elementary particle postulated as responsible for gravitational interaction. Such a particle would have zero charge and rest mass.

gravity. The attraction of the earth or some other celestial body for an object as a result of gravitation. The gravity of a planet is responsible for the *weight of a body near its surface and for the *acceleration of free fall.

grease-spot photometer. A *photometer in which a vertical sheet of white paper with a spot of grease in its centre is illuminated from both sides.

greenockite. A naturally occurring form of cadmium sulphide, CdS.

great circle. A circle on a sphere, whose centre is the centre of the sphere.

green vitriol. *See* iron sulphate.

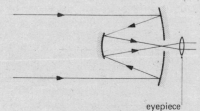

eyepiece
Gregorian telescope

Gregorian telescope. A type of astronomical *telescope similar to the *Cassegrainian telescope but using a second concave mirror rather than a convex mirror (see illustration). [After James Gregory (1638–75), English scientist.]

grid. One of the electrodes controlling the current in a thermionic valve. The *grid bias* is the voltage between the grid and the cathode.

Grignard reagent. Any of a class of organometallic magnesium compounds with the general formula RMgX, where R is an alkyl or aryl group and X is a halogen atom. The Grignard reagents are obtained by reacting alkyl or aryl halides with magnesium metal in dry ether. They are colourless crystalline solids, possibly having the form $R_2Mg.-MgX_2$. Grignard reagents are usually not extracted but are used in ether solution as intermediates in preparing many organic compounds. [After François Grignard (1871–1935), French chemist.]

ground. *See* earth.

ground state. The state of an atom, molecule, nucleus, or other system when it is in its lowest *energy level.

ground wave. A radio wave transmitted directly between points rather than by reflection from the ionosphere. *Compare* sky wave.

group. 1. A group of atoms held together by covalent bonds in a chemical compound, considered as a unit of structure having characteristic properties. A *functional group* is the group in a molecule that is responsible for the compound's chemical behaviour. Thus acetaldehyde, CH_3CHO, contains the methyl group, CH_3-, and the aldehyde group, -CHO. The aldehyde group is the functional group. A group is sometimes called a *radical*, although this term is now usually reserved for a *free radical.
2. A set of elements belonging to the same vertical column in the *periodic table.
3. A set of entities with certain mathematical properties. Members of the set can be combined together by addition, multiplication, or some other operation. They form a group if combination of any two members gives another member and obeys the *associative law. Moreover there must be one member of the set (the *identity*) such that combination with this identity produces no change (as in multiplying a number by 1). Finally, for every member of the group there must be another member that combines with it to give the identity. The members of a group need not be numbers: they can be matrices, vectors, symmetry operations, etc. Group theory is used in spectroscopy, crystallography, and in particle physics (*see* unitary symmetry).

gun cotton. *See* nitrocellulose.

gunpowder. An explosive mixture of potassium nitrate, sulphur, and powdered charcoal.

gypsum. A naturally occurring form of hydrated calcium sulphate, $CaSO_4 \cdot 2H_2O$, used in making plaster of Paris.

gyro-compass. *See* compass.

gyromagnetic ratio. Symbol: γ. The ratio of the *magnetic moment of an atom, electron, nucleus, etc., to its angular momentum.

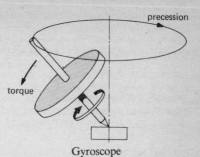

Gyroscope

gyroscope. A heavy wheel mounted so that it can spin freely at high speed and the axis of rotation can change its direction. Usually, a universal mounting is used so that the axle of the wheel can take any direction in space. If a torque is applied tending to change the direction of the axis, the gyroscope precesses at a rate proportional to the size of the torque. In the diagram, the applied torque is produced by the weight of the gyroscope (i.e. by its tendency to fall over) and precession occurs about the vertical axis.
If the wheel has a large moment of inertia and the torque is small, its axis will preserve its direction in space. This effect is used in the *gyro-compass, in stabilizers for ships, guidance systems for spacecraft, and similar devices.

H

Haber process. An industrial process for manufacturing ammonia from nitrogen. A mixture of nitrogen and hydrogen is passed over an iron catalyst at a temperature of 400–500°C and a pressure of 200 atmospheres. The equilibrium is $3H_2 + N_2 = 2NH_3$. The Haber process is the most important method of *nitrogen fixation for use in agricultural fertilizers. *See also* Le Chatalier's principle. [After Fritz Haber (1868–1934), German chemist.]

habit. *See* crystal habit.

hadron. Any *elementary particle that can take part in *strong interactions. The hadrons include the baryons and the mesons: they exclude the photon and the leptons.

haematite (bloodstone). A red or black naturally occurring form of iron III oxide, Fe_2O_3, used as the principal ore of iron and in paint pigments and jeweller's rouge.

hafnium. Symbol: Hf. A highly lustrous silvery metallic element found in zirconium ores. It is usually obtained by the *Kroll process or the *van Arkel-de Boer process. Hafnium is used as a getter and as a neutron absorber in reactors. Its chemistry is similar to that of titanium and zirconium. A.N. 70; A.W. 178.49; m.pt. 2150°C; b.pt. 5400°C; r.d. 13.31; valency 2, 3, or 4.

hair hygrometer. A type of *hygrometer in which the extension or contraction of a hair moves a pointer over a scale. The hair increases in length as the humidity increases and tension is maintained by a spring.

half cell. An electrode dipping into a solution of an electrolyte. The simplest type consists of a metal electrode in contact with a solution of the metal ions of specified concentration. The *electrode potential of such half cells is measured against a *hydrogen electrode.

half-life (half-value period). 1. Symbol: τ. A measure of the stability of a radioactive substance, equal to the time taken for its *activity to fall to one half of its original value. In this time one half of the atoms present disintegrate.
2. A similar measure of the rate of reaction or decay of elementary particles, excited atoms, free radicals, etc.

half-period zones. Zones into which a wavefront is divided so that at a forward point each zone differs from an adjacent one by half of a complete period (i.e. by λ). The construction is used in explaining diffraction.

half-thickness. The thickness of a piece of specified material that will reduce the intensity of a beam of transmitted radiation to one half of its original value.

half-wave plate. A plate cut from a doubly refracting crystal parallel to the optic axis and having a thickness such that ordinary and extraordinary rays of transmitted light differ in phase by π. Half-wave plates rotate the plane of plane-polarized light and are used in polarimeters.

half-wave rectification. Rectification of an alternating signal or current in such a way that current is allowed to pass only in alternate half cycles. *Compare* full-wave rectification.

halide. Any of certain compounds that contain halogen atoms or ions. Such compounds are named *chlorides, bromides, iodides*, or *fluorides*. Simple ionic halides, such as sodium chloride, contain negative halide ions, Cl^-, Br^-, I^-, or F^-. For organic halides the names are usually used when the halogen is bound to a simple alkyl group, as in ethyl bromide, C_2H_5Br, or vinyl chloride, $CH_2{:}CHCl$.

Hall effect. An effect observed when an electric current is passed through a conductor under the influence of a transverse magnetic field. A potential difference is produced between the upper and lower surfaces of the conductor. This potential (the *Hall potential*) is caused by deflection of the moving charged particles. Its size depends on the concentration of charge carriers and its direction is determined by whether the carriers are electrons or positive holes.

Halley's comet. A *comet with an orbital period of 76 years. One of the most spectacular and best-known comets, it last approached the vicinity of the sun in 1910. [After Edmund Halley (1662-1742), English astronomer who predicted its appearance in 1758-9.]

halo. A luminous ring sometimes observed around the sun or moon. It is caused by refraction by small ice particles high in the earth's atmosphere.

haloform. A compound with the formula CHX_3, where X is a halogen atom: i.e. fluoroform (CHF_3), chloroform ($CHCl_3$), bromoform ($CHBr_3$), and iodoform (CHI_3).

halogenation. Any chemical reaction leading to the formation of a bond to a halogen atom, usually used in organic chemistry. Halogenation reactions, which can be either addition or substitution, include *fluorination* (F), *chlorination* (Cl), and *bromination* (Br).

halogens. The nonmetallic elements fluorine, chlorine, bromine, iodine, and astatine, belonging to group VIIA of the periodic table. They are monovalent and strongly electronegative, forming halides.

Hamiltonian. Symbol: *H*.
A function used in mechanics to express the energy of a system. In simple cases it is the sum of the kinetic and potential energies. Thus for a particle of mass m and momentum p: $H = p^2/2m + V$, where V is the potential energy. [After Sir William Rowan Hamilton (1805–65), Irish mathematician and astronomer.]

Harcourt pentane lamp. A lamp burning a mixture of pentane and air under specified conditions: used as a former standard for the international *candle.

hard. 1. Denoting electromagnetic or particle radiation with relatively high energy and, as a consequence, high penetrating or ionizing power. Hard X-rays are X-rays with short wavelength and high photon energy.
2. Denoting a *vacuum in which there is a very low pressure of gas, such that ionization of gas plays little part in phenomena occurring in the vacuum.

hardening. 1. The process of converting liquid vegetable oils, which contain unsaturated glycerides, into saturated compounds with higher melting points by catalytic hydrogenation. Usually nickel powder is the catalyst. The process is widely used in the manufacture of margarines.
2. Any process for making a metal harder. In general, a metal containing small amounts of another substance is less soft than the pure material because the foreign atoms in the lattice make it more difficult for slip of crystal planes to occur. Thus, iron is hardened by introducing carbon to produce steel. *See* case-hardening, strain hardening, precipitation, martensite.

hard solder. See solder.

hardware. The equipment used in a computer, as distinguished from the programs—the *software*.

hard water. Water that does not easily produce a lather with soap due to the presence of soluble calcium salts which react with the soap to form a scum of calcium salts of fatty acids. Two types of hardness are distinguished. *Temporary hardness* is removed by boiling and is caused by the presence of calcium hydrogen carbonate, formed by the action of dissolved carbon dioxide on calcium carbonate in limestone or chalk. Boiling decomposes the soluble hydrogen carbonate and precipitates calcium carbonate, which is insoluble. Softening can also be effected by adding calcium hydroxide to precipitate calcium carbonate (the *Clark process*). Temporary hardness is responsible for the formation of *fur in kettles, boilers, and hot pipes.
Permanent hardness cannot be removed by boiling and is largely caused by calcium sulphate. It can be removed by sodium carbonate (washing soda), which precipitates calcium carbonate, or by use of Permutit or similar *ion-exchange materials.

harmonic. A component of a complex wave with a frequency that is an integral multiple of the base frequency (the *fundamental*). The first harmonic has the frequency of the fundamental, the second harmonic (first overtone) has twice the frequency, etc. The complex wave is made up of the fundamental combined with the overtones. *See also* partial.

harmonic series. A mathematical series in which the reciprocals of each term form an *arithmetic sequence. The simplest example is $1 + \frac{1}{2} + \frac{1}{3} + \frac{1}{4} + ...$

hartley. A unit of information equal to 3.321 93 bits ($\log_2$ 10 bits). It is used in digital computers. [After Sir Walter Noel Hartley (1846–1913), British physicist.]

hartree. A unit of energy equal to e^2/a_0, where e is the atomic unit of charge and a_0 is the atomic unit of length. It is equivalent to 4.8505×10^{-18} joule and is used in spectroscopy and other atomic studies.

H-bomb. *See* nuclear weapon.

HCF. *See* highest common factor.

health physics. The study of the effects of radiation on health and of methods of protecting workers from radioactivity, X-rays, etc.

heat. Symbol: Q. The form of energy that can be transferred from points of high *temperature to points of lower temperature as a result of the temperature difference. The transfer of heat can occur by *conduction, *convection, or *radiation. When the temperature of a system changes, its heat energy changes by an amount equal to the product of its mass, its specific heat capacity, and its change in temperature. The heat of a system is the kinetic energy due to translation, rotation, and vibration of its molecules. Like other forms of energy, heat is measured in joules, although heat units such as calories and British thermal units are sometimes used.

heat capacity (thermal capacity). Symbol: C. The amount of heat required to raise the temperature of a body by one

degree. It is usually measured in joules per kelvin.

heat death. The state of a closed system when its total *entropy has increased to its maximum value. Under these conditions there is no available energy: all the matter is disordered and the temperature is uniform. If the universe is a closed system it will eventually reach this state (the *heat death of the universe*).

heat engine. A device in which heat energy is converted into mechanical energy.

heater. A filament used for heating the cathode in a *thermionic valve.

heat exchanger. Any device for transferring heat from one fluid to another. Heat exchangers are used for cooling, as in a car radiator, or for making use of waste heat, as in the outlet gases of furnaces, chemical plant, etc. They are made of metal and usually have coiled tubes, metal fins, etc., to increase the area and time of contact of the fluids.

heat of atomization. The heat required to decompose one mole of a substance into its atoms.

heat of combustion. The heat evolved when one mole of a substance is burned in oxygen.

heat of formation. The heat liberated or absorbed when one mole of a substance is formed from its elements.

heat of neutralization. The heat evolved when one mole of an acid or base is fully neutralized. All strong acids and bases are fully ionized in solution and the neutralization is the reaction $H^+ + OH^- = H_2O$, which evolves 57 500 joules per mole.

heat of reaction. The heat liberated or absorbed in a chemical reaction, usually expressed for molar amounts of reactants and products. *See* Hess's law.

heat of solution. The heat liberated or absorbed when one mole of substance is dissolved in a large volume of a specified solvent (i.e. enough to give effective infinite dilution).

Heaviside-Kennelley layer. *See* ionosphere. [After Oliver Heaviside (1850-1925) and Arthur Kennelley (1861-1939), English physicists.]

heavy hydrogen. *See* deuterium.

heavy spar. *See* barite.

heavy water. A liquid with the same chemical properties as water but with a higher molecular weight and therefore different physical properties. The term usually applies to *deuterium oxide*, D_2O, in which the hydrogen isotope, deuterium, has an atomic weight of 2. Heavy water may be extracted from natural water by fractional distillation, exchange processes, or by electrolysis. It is used as a moderator in nuclear reactors. M.pt. $3.8°C$; b.pt. $101.4°C$; r.d. 1.11 $(25°C)$.

hectare. A metric unit of area equal to 100 ares: i.e. 10 000 square metres or 2.471 05 acres.

hecto-. Symbol: h. Prefix indicating 100.

Heisenberg uncertainty principle. The principle that it is impossible to measure both the momentum and position of a particle to any desired degree of accuracy: there is always an uncertainty Δp in the momentum and an uncertainty Δx in the position so that their product has the same order of magnitude as the Planck constant h: $\Delta p \Delta x \sim h$. This indeterminacy has nothing to do with the errors in the measuring instruments or with human errors of the observer. It arises from the fact that particles also have a wave-like behaviour (*see* de Broglie wave) and from the observation itself disturbing the system in an unpredictable way. A similar relationship holds for uncertainties in measurements of energy E and time t: $\Delta E \Delta t \sim h$. [After Werner Karl Heisenberg (b. 1901), German physicist.]

heliocentric. 1. Having the sun at the centre. A *heliocentric theory* of the solar system is one in which the planets are considered to move around the sun. **2.** Measured with reference to the sun.

helium. Symbol: He. A colourless odourless tasteless gaseous element found in some natural gas, in radioactive ores such as pitchblende, and in the atmosphere, from which it can be obtained as a by-product in the liquefaction of air. It is used for filling airships and in some fluorescent lamps. Chemically, it is one of the *inert gases and is not known to form any stable compounds. A.N. 2; A.W. 4.0026; m.pt. $-272.2°C$; b.pt. $-268.9°C$; r.d. 0.138 (air $= 1$). *See also* superfluid.

helix. A curve in space that lies on the surface of a cone or cylinder and always makes a constant angle with the generators.

Helmholtz coils. A pair of identical flat coils of wire mounted parallel to each other with a separation equal to the coils' radius. The coils are connected in series and a uniform magnetic field is produced between them when a current is passed. [After Herman Ludwig Ferdinand von Helmholtz (1821-94), German scientist.]

Helmholtz free energy (Helmholtz function). Symbol: A. A *free-energy function of a thermodynamic system equal to the *internal energy minus the product of the temperature and the entropy: i.e. $A = U - TS$. [After Hermann Ludwig Ferdinand van Helmholtz (1821-94), German physicist.]

hemiacetal. *See* acetal.

hemihydrate. A solid compound with one molecule of *water of crystallization for every two molecules of compound, as in calcium sulphate hemihydrate (plaster of Paris), $2CaSO_4.H_2O$.

hemimorphite. A white or yellow mineral consisting of a hydrated zinc silicate, $Zn_4Si_2O_7(OH)_2.H_2O$, used as an ore of zinc.

henry. Symbol: H. An SI unit of inductance equal to the inductance of a closed circuit with a magnetic flux of one weber per ampere of current. [After William Henry (1774–1836), English chemist.]

Henry's law. The principle that the amount of gas dissolved in a certain volume of liquid is directly proportional to the equilibrium pressure of gas above the liquid. The law does not hold if the gas reacts with the solvent.

heptagon. A polygon that has seven sides.

heptahydrate. A solid compound with seven molecules of *water of crystallization per molecule of compound.

heptane. A colourless flammable straight-chain liquid *alkane, C_7H_{16}, obtained from petroleum. Heptane is also called n-*heptane* to distinguish it from its branched chain isomers. It is used as a standard in determining octane rating, as a solvent, and in organic synthesis. M.pt. –906°C; b.pt. 98.4°C; r.d. 0.68.

heptavalent (septivalent). Having a valency of seven.

heptode. A type of thermionic valve with seven electrodes (anode, cathode, and five grids).

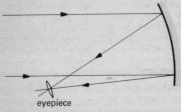

eyepiece

Herschelian telescope

Herschelian telescope. A type of astronomical *telescope in which light is reflected by a concave mirror at a small angle to the direction of the light (see illustration).

hertz. Symbol: Hz. A unit of frequency equal to the frequency of a periodic process with a period of one second. The unit replaces the cycle per second, especially for expressing the frequency of waves or vibrations. [After Heinrich Hertz (1857–94), German physicist.]

Hertzian waves. *See* radio waves.

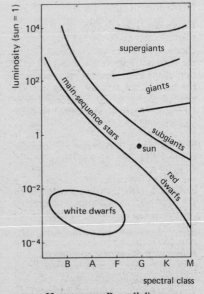

Hertzsprung-Russell diagram

Hertzsprung-Russell diagram. A graph in which the absolute brightness of stars (*see* magnitude) is plotted against their spectral classification according to a system of spectral analysis that reflects decreasing surface temperature. Various classes of stars form concentrations on different parts of the graph. The greatest number of observed stars fall within a narrow band running from top left to bottom right and called the main sequence. The sun is a main-sequence star. At the top of this band lie the brightest and hotter stars, bluish-white in colour. At the bottom lie the cool and relatively dim red stars. *Giants, supergiants and *white dwarfs form isolated concentrations above and below the main sequence. For the much older stars of the centre of the Galaxy and of globular clusters, a second Hertzsprung-Russell diagram may be drawn plotting brightness against colour. The diagram is important as a schematic indication of the evolution of stars. [Named after Ejnar Hertzsprung (1873-1969), Danish astronomer, and the American Henry Norris Russell (1879-1957).]

Hess's law. The principle that if a set of reactants can be converted into a set of chemical products by a series of stages, the algebraic sum of the heats of reaction of the individual stages is equal to the heat of reaction for direct conversion of reactants into products. [After Germain Henri Hess (1802-50), Russian scientist.]

heterocyclic. Denoting a chemical compound that has a *cyclic structure containing two or more types of atom in the ring. *Pyridine, for example, is a common heterocyclic compound, having a hexagonal ring composed of five carbon atoms and one nitrogen atom. The odd atom in the ring, such as the nitrogen atom in pyridine, is termed the *hetero atom. Compare* homocyclic.

heterodyne. Denoting the combination of two radiofrequency signals to give a resultant signal with a frequency equal to the difference in the two radiofrequencies. The phenomenon is the counterpart of *beats produced by sound waves. If the radio signals are close together, the heterodyne frequency is in the audiofrequency range. The heterodyne principle is used in radio reception. The received carrier-wave signal is combined with a fixed signal generated in the receiver. The resulting heterodyne signal, modulated in the same way as the carrier wave, is amplified. *See also* superheterodyne.

heterogeneous. Not uniform in composition; involving or containing more than one phase.

heterogeneous catalysis. *See* catalysis.

heterogeneous reactor. *See* fission reactor.

heterolytic fission. The breaking of a chemical bond with formation of two oppositely charged ions, as in the reaction $AB = A^+ + B^-$. *Compare* homolytic fission.

heteropolar bond. *See* electrovalent bond.

Heusler alloy. Any of a class of alloys of manganese, copper, and aluminium that are ferromagnetic in spite of the fact that they contain no iron, cobalt, or nickel. [After Conrad Heusler, German engineer.]

hexachlorocyclohexane. Any of nine isomeric compounds, $C_6H_6Cl_6$, prepared by the chlorination of benzene in ultraviolet light. The most important is the symmetric isomer *1,2,3,4,5,6-hexachlorobenzene (lindane)*, a white powdery solid used as an insecticide.

hexadecanoic acid. *See* palmitic acid.

hexagon. A polygon that has six sides.

hexagonal. *See* crystal system.

hexahedron. A polyhedron with six faces. A *regular hexahedron* is a cube.

hexahydrate. A solid compound with six molecules of *water of crystallization per molecule of compound, as in iron III chloride hexahydrate, $FeCl_3 \cdot 6H_2O$.

hexane. A colourless flammable straight-chain liquid *alkane, C_6H_{14}, obtained from petroleum. Hexane is also called n-*hexane* to distinguish it from its branched-chain isomers. It is used as a solvent. M.pt. -95°C; b.pt. 69°C; r.d. 0.66.

hexanedioic acid. *See* adipic acid.

hexanoic acid (caproic acid). A slightly soluble oily fatty acid, $CH_3(CH_2)_4COOH$, found, in the form of its glycerides, in cows' and goats' milk and in some vegetable oils. M.pt. -3.4°C; b.pt. 205°C; r.d. 0.93.

hexose. A simple *sugar that has six carbon atoms in its molecules.

hexyl group. The monovalent group $C_6H_{13}-$, existing in several isomeric forms.

HF. See high frequency.

highest common factor (HCF). The largest number that exactly divides into each of a given set of numbers. For example, the HCF of 12, 9, and 15 is 3.

high frequency (HF). A frequency in the range 3 megahertz to 30 megahertz.

high-speed steel. Any steel that retains its hardness at high temperatures, thus being suitable for making high-speed cutting tools, which become heated in use. High-speed steels contain tungsten (10-18%) and smaller amounts of other metals, such as chromium, vanadium, and molybdenum.

high tension. High voltage.

hole. A vacancy in an energy level in one of the energy bands of a *semiconductor.

holmium. Symbol: Ho. A soft silvery metallic element belonging to the *lanthanide series. A.N. 67; A.W. 164.9303; m.pt. 1470°C; b.pt. 2720°C; r.d. 8.78; valency 3.

holography. A technique for recording three-dimensional images on photographic film using laser light. A beam of light from a laser is split by using a semitransparent mirror. One beam of light passes directly through the mirror onto the photographic film. The other is reflected onto the object to be photographed so that it is diffracted onto the film. Interference of the beams produces a pattern of spots on the film (the *hologram*). If the film is then illuminated directly by a beam of the same laser light, it acts as a diffraction grating and gives a real image of the object and a three-dimensional virtual image.

homocyclic. Denoting a chemical compound that has a *cyclic structure in which the ring is composed of identical atoms. Benzene is a common example of a homocyclic compound. *Compare* heterocyclic.

homogeneous. Uniform in composition; involving or containing a single phase.

homogeneous catalysis. See catalysis.

homogeneous reactor. See fission reactor.

homologous series. A series of related chemical compounds for which the formula of each member differs from the preceding one by the same group of atoms, usually a CH_2 group. For example, the alcohols CH_3OH, C_2H_5OH, C_3H_7OH, etc., form a homologous series. Two successive members of such a series are called *homologues*.

homolytic fission. The breaking of a chemical bond with formation of two neutral atoms or radicals, as in the reaction $AB = A\cdot + B\cdot$. *Compare* heterolytic fission.

homopolar bond. See covalent bond.

homopolymer. See polymer.

Hooke's law. The principle that the strain produced in an elastic material is proportional to the stress producing it. The law holds only up to the material's *elastic limit. See also elasticity. [After Robert Hooke (1635-1703), English physicist.]

horizontal component. B_o. The component of the earth's magnetic field in a horizontal direction at a point on the earth. See terrestrial magnetism.

horn silver. A naturally occurring form of silver chloride, AgCl, used as an ore of silver.

horsepower. Symbol: hp. An imperial unit of power equal to 550 foot-pounds-force per second. It is equivalent to 745.700 watts.

hot. 1. Denoting an object or substance that is radioactive.
2. Denoting an atom, ion, or molecule in an excited electronic state, especially one excited by a nuclear decay process.

hot-wire ammeter. *See* ammeter.

Hubble constant. Symbol: *H*. The ratio of the recessional velocity of a star, as measured by its *red shift, to its distance. If it is a constant it is the age of the universe: 7.6×10^9 years. [After Edwin Hubble (1889-1953), American astronomer.]

Hückel rule. *See* aromaticity.

hue. The property of a *colour determined by the wavelength of the light. The hue of a colour is the quality that allows an observer to distinguish it as resembling a spectral colour—a red hue, a violet hue, etc.

humidity. A measure of the concentration of water vapour in the atmosphere. The *absolute humidity* is the mass of water vapour per unit volume of air. The *relative humidity* is the ratio of the mass of vapour per unit volume present in the air to the mass per unit volume that would be present if the air were saturated at the same temperature. *See also* hygrometer.

hybrid computer. A type of *computer in which a continuously variable input is converted into digital form for processing. Hybrid computers have characteristics of both *analog and *digital computers.

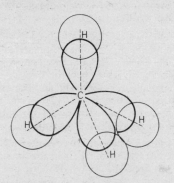

sp^3 hybrid orbitals in methane

hybrid orbital. An orbital formed by combination of two or more *atomic orbitals of an atom. Carbon, for example, has one *s* orbital and three *p* orbitals. These may be thought of as mixing to give four equivalent sp^3 hybrids, each directed towards the corner of a tetrahedron. It is these that overlap with *s* orbitals of hydrogen in methane (see illustration). Other types of *hybridization* can occur: sp^2 hybrids are three coplanar orbitals at 120° to each other; *sp* hybrids are two colinear orbitals directed away from each other. The sp^2 hybrids occur in ethene and *sp* hybrids in acetylene.

hydration. A reaction in which water is added to another substance, as in the conversion of an *anhydride to its acid or the formation of a hydrate. The hydration of ions in solution is an example of *solvation.

hydrazine. A colourless oily toxic explosive liquid, H_2NNH_2. Hydrazine is a strong reducing agent used in rocket fuels, antioxidants, and corrosion inhibitors. M.pt. 2°C; b.pt. 113°C; r.d. 1.0 (25°C).

hydride. A compound formed between an element and hydrogen. Ionic hydrides are formed by electropositive metals and contain the negative *hydride ion* (H^-). An example is lithium hydride (LiH). Interstitial hydrides are the compounds formed by certain transition metals, such as palladium, when they absorb hydrogen. They are generally nonstoichiometric compounds. Nonmetals form covalently bonded compounds with hydrogen.

hydriodic acid. *See* hydrogen iodide.

hydrobromic acid. *See* hydrogen bromide.

hydrocarbon. Any organic compound composed only of carbon and hydrogen. The hydrocarbons include alkanes, alkenes, alkynes, terpenes, and aromatic hydrocarbons, such as benzene and naphthalene.

hydrochloric acid. *See* hydrogen chloride.

hydrochloride. A salt formed by reaction of an organic base with hydrochloric acid: Aniline, for example, forms aniline hydrochloride, $C_6H_5NH_2.HCl$. *See* anilinium.

hydrocyanic acid. *See* hydrogen cyanide.

hydrofluoric acid. *See* hydrogen fluoride.

hydrogel. A *gel in which the dispersion medium is water.

hydrogen. Symbol: H. A colourless odourless flammable gaseous element. It is the lightest element and also the most abundant in the universe, occurring in water and in most organic compounds. The element is manufactured by several methods, including the *Bosch process, the electrolysis of water, and the reaction of natural gas with steam. It is used in the *Haber process for fixing nitrogen, the hydrogenation of organic compounds, the reduction of some metal-oxide ores, and in oxy-hydrogen welding. Chemically, hydrogen reacts directly with many other elements and all elements have *hydrides. It is diatomic, H_2, and has ortho- and para- forms (*see* othohydrogen). There are three isotopes, the heavier ones being *deuterium and *tritium. A.N. 1; A.W. 1.007 97; m.pt. $-259\,°C$; b.pt. $-252\,°C$; r.d. 0.069 (air = 1).

hydrogenation. The reaction of a chemical compound (usually organic) with hydrogen. Common hydrogenations involve the addition of hydrogen to unsaturated molecules using catalysts. An example is the hydrogenation of ethylene: $C_2H_4 + H_2 = C_2H_6$.

hydrogen bomb. *See* nuclear weapon.

hydrogen bond. A weak attraction between an electronegative atom, such as oxygen, nitrogen, or fluorine, and a hydrogen atom that is covalently linked to another electronegative atom. Hydrogen bonds are electrostatic in origin: the electrons in the covalent bond are displaced towards the electronegative atom and electrostatic attraction occurs between the other electronegative atom and the unshielded positive charge of the hydrogen nucleus. Such bonds can be intermolecular: they occur between molecules of water, hydrogen fluoride, etc. They can also be intramolecular: hydrogen bonding is responsible for the structure of proteins and nucleic acids.

hydrogen bromide. A colourless gas, HBr, made by passing hydrogen and bromine vapour over a platinum catalyst. It is a strong acid in aqueous solution (*hydrobromic acid*). M.pt. $-86\,°C$; b.pt. $-66.4\,°C$; r.d. 2.71.

hydrogen chloride. A colourless pungent gas, HCl, prepared by reaction between concentrated sulphuric acid and sodium chloride. It is very soluble in water, forming strongly acidic solutions (*hydrochloric acid*). M.pt. $-114\,°C$; b.pt. $-85\,°C$; r.d. 1.27 (air = 1).

hydrogen cyanide (hydrocyanic acid, prussic acid). An intensely poisonous colourless liquid, HCN, with a faint odour of bitter almonds, made by catalytic reaction between ammonia, air, and methane. It forms weakly acidic solutions in water. Hydrogen cyanide is used in making acrylates. M.pt. $-13\,°C$; b.pt. $26\,°C$; r.d. 0.69.

hydrogen electrode. A half cell in which a platinum electrode is immersed in dilute acid of known concentration and hydrogen gas is bubbled through the solution over the platinum. The electrode reaction is $H^+ + e = H$: the hydrogen electrode is used as a standard for measuring *electrode potentials.

hydrogen fluoride. A colourless highly corrosive liquid, HF, used as a fluorinating agent and catalyst. It is prepared by treating calcium fluoride with sulphuric acid. In solution it forms *hydrofluoric acid*, which, like hydrogen fluoride, attacks glass. M.pt. $-83\,°C$; b.pt. $19.5\,°C$; r.d. 0.99.

hydrogen iodide. A colourless gas, HI, made by passing hydrogen and iodine vapour over a platinum catalyst. It is a reducing agent and a strong acid in aqueous solution (*hydriodic acid*). M.pt $-51°C$; b.pt. $-36°C$; r.d. 4.4.

hydrogen peroxide. A colourless heavy liquid, H_2O_2, which is unstable and usually obtained in aqueous solution. It is a good oxidizing agent and is used as a rocket propellant, antiseptic, and bleach. M.pt. $-2°C$; b.pt. $158°C$; r.d. 1.46.

hydrogen sulphide (sulphuretted hydrogen). A colourless flammable poisonous gas, H_2S, with an offensive odour of rotten eggs. It is a weak acid, made by addition of an acid to iron sulphide and used in metallurgy and analytical chemistry. M.pt. $-84°C$; b.pt. $-60°C$; r.d. 1.2 (air = 1).

hydrolysis. The reaction of a chemical compound with water. Usually the water is the solvent in which the reaction occurs. A common example is the hydrolysis of an *ester to an alcohol and a carboxylic acid. *See also* solvolysis.

hydrometer. An instrument for determining the relative density of a liquid, typically consisting of a weighted bulb with a long graduated stem. The device floats vertically in the liquid and the length of stem immersed is proportional to its density.

hydrophilic. Having an affinity for water.

hydrophobic. Having no affinity for water. *See* sol.

hydroquinone (quinol). A white crystalline solid, $C_6H_4(OH)_2$, used as a photographic developer, antioxidant, and raw material for the manufacture of dyes. M.pt. $170°C$; b.pt. $285°C$; r.d. 1.33. *See also* quinhydrone.

hydrostatics. The branch of mechanics concerned with the equilibrium of fluids.

hydroxide. An ionic base containing the ion OH⁻ (the *hydroxide ion*).

hydroxylamine. An explosive white crystalline solid, NH_2OH, made by reducing ammonium chloride and decomposing the hydroxylamine hydrochloride formed with a base. It is a reducing agent. M.pt. $33°C$; b.pt. $70°C$; r.d. 1.23.

hydroxyl group. The group -OH, as present in alcohols, phenols, and oxy acids.

hygrometer. Any of various instruments for measuring *humidity. Examples are the *wet- and dry-bulb hygrometer, the dew-point hygrometer, and the *hair hygrometer.

hygroscopic. Denoting a substance that absorbs water from the atmosphere. Some solids can absorb enough moisture to form a solution (deliquescence).

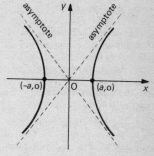

Hyperbola

hyperbola. A *conic with an eccentricity greater than 1. A hyperbola is the locus of a point that moves so that the difference between its distance from two fixed points (the *foci*) is constant. In Cartesian coordinates its equation is $x^2/a^2 - y^2/b^2 = 1$, where a and b are constants and the curves are symmetric about the axes. It approaches the asymptotes $y = bx/a$ and $y = -bx/a$.

hyperbolic. Having the shape of a hyperbola or a hyperboloid.

hyperbolic function. One of a number of functions analogous to the *trigonometric functions. The hyperbolic sine (written sinh x) is defined as $(e^x - e^{-x})/2$ and the hyperbolic cosine (cosh x) is $(e^x + e^{-x})/2$. The hyperbolic tangent (tanh x) is the ratio of sinh x to cosh x and the other three hyperbolic functions—cosech x, sech x, and coth x—are defined as reciprocals in the same way as the trigonometric functions.

hyperbolic logarithm. *See* logarithm.

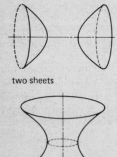

two sheets

one sheet

Hyperboloids

hyperboloid. Either of two types of surface that have hyperbolic cross sections in planes parallel to two of the coordinate planes and elliptical (or circular) cross sections in planes parallel to the third coordinate plane. A *hyperboloid of one sheet* ($x^2/a^2 + y^2/b^2 - z^2/c^2 = 1$) is a single surface with its elliptical cross sections in the *xy* plane. A *hyperboloid of two sheets* is made up of two surfaces and the elliptical cross sections are in the *yz* plane. Simple forms of these hyperboloids, with circular cross sections, can be generated by rotating a standard hyperbola about its *y* or *x* axis respectively.

hypercharge. Symbol: *Y*. A property of elementary particles described by a quantum number equal to the sum of the particle's baryon number and its *strangeness. Hypercharge is conserved in strong and electromagnetic interactions: not in weak interactions.

hyperconjugation. Electronic interaction occurring between a double bond in a compound and a carbon-hydrogen single bond in an attached alkyl group. Under certain circumstances the double bond and the single C-H bond appear to behave rather like conjugated double bonds.

hyperfine structure. *See* fine structure.

hypermetropia. A defect of the eye in which parallel rays are focused to a point behind the retina when the eye is at rest so that distant objects are indistinct and the eye has to accommodate to bring them to focus. In hypermetropia the distance between the front and back of the eyeball is too short and the condition can be corrected by convex spectacle lenses. Hypermetropia is sometimes called *long sight*, which should not be confused with far sight (*presbyopia). *See* eye.

hyperon. Any of a class of *elementary particles that have half-integral spin, take part in *strong interactions, and have masses larger than those of the proton and the neutron. The *hyperons are thus all *baryons that are not nucleons. They include the lambda, xi, sigma, and omega-minus particles. All decay into nucleons.

hypersonic. Relating to a speed in excess of Mach 5. *See* Mach number.

hypo. *See* sodium thiosulphate.

hypochlorite. Any salt or ester of hypochlorous acid.

hypochlorous acid. A greenish-yellow liquid, .HOCl, known only in aqueous solution and in the form of hypochlorite salts. It is an oxidizing agent formed by the action of chlorine on water.

hypocycloid. A curve that is the locus of a point on the circumference of a circle as that circle rolls on the inside of another circle. *Compare* epicycloid.

hypophosphite. *See* phosphite.

hypophosphorous acid. *See* phosphorous acid.

hypothesis. A proposition explaining observed facts. A hypothesis may be a highly probably explanation of the facts. Alternatively it may be a *working hypothesis*, entertained as a guide to further experiment and investigation.

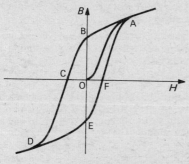

Hysteresis loop

hysteresis. An apparent lag of an effect with respect to the magnitude of the agency producing the effect. *Magnetic hysteresis* is the lag of the magnetization of a ferromagnetic material with respect to the external field producing it. It can be displayed by plotting a graph of the magnetic flux density *B* inside the sample produced by an external magnetic field strength *H*. If the sample is initially unmagnetized, the flux density increases steadily up to a point (A) at which the flux density is increasing linearly with field strength. If the field strength is then decreased to zero, the curve OA is not retraced. Instead, the flux density decreases less markedly and when the external magnetizing field is zero, it has a residual value given by the intercept OB. This value of *B* is the *remanence* of the material. As the value of *H* is increased in the opposite direction, the value of *B* falls further until at C it reaches zero again. The external field necessary to remove the mag-

netization is the material's *coercivity*. If *H* is increased still further in this direction a second saturation (D) is reached. The field is again reduced to zero, reversed, and increased in the opposite direction, giving the complete curve ABDCEF. This is the *hysteresis loop* of the material. A graph of magnetization (*M*) against external field has the same form but differs in that at saturation *M* becomes constant with increasing *H*.

The magnetization and demagnetization of a sample according to such a cycle requires energy and it can be shown that the energy necessary is equal to $\mu_o A$, where *A* is the area of the hysteresis loop and μ_o the magnetic constant. The process causes a dissipation of energy, known as *hysteresis loss*, in systems in which a ferromagnetic material is being magnetized and demagnetized by a varying field.

Magnetic materials used in generators, transformer cores, and electromagnets require low coercivity and low hysteresis loss (i.e. their hysteresis loops should have a small area). Permanent magnets require materials with high remanence and coercivity.

Other types of hysteresis are known. *Electric hysteresis* for example is an effect shown by materials exhibiting *ferroelectricity in which a graph of displacement against electric field strength shows a similar form.

I

Iceland spar. *See* calcite.

ice point. The temperature at which ice melts. It is taken to be the temperature at which ice and water are in equilibrium at standard pressure and is used as a fixed point on *temperature scales.

icosahedron. A polyhedron with twenty faces. A *regular icosahedron* has congruent equilateral triangles as its faces.

ideal gas (perfect gas). An idealized gas composed of atoms that have a negligible volume and undergo perfectly elastic collisions with one another and with the walls of their container. Such a gas would obey the *gas laws under all conditions. Real gases behave like ideal gases at low pressures.

identity. An equation that holds for all values of the variables in the expression. For instance,

$$(x + y)^2 \equiv x^2 + y^2 + 2xy$$

is an identity. $x^2 + y^2 = y^2$ is not. It is only true when $x = 0$.

ignis fatuus. *See* Will-o'-the-wisp.

ilmenite. A black mineral consisting of a mixed iron and titanium oxide, $FeO.TiO_2$, used as an ore of titanium.

image. The visual impression of an object produced by a lens, mirror, or optical system. A *real image* is formed by convergence of rays of light and can be produced on a screen. A *virtual image* is one from which the rays of light appear to diverge. Such an image could not be projected onto a screen.

image converter. An electronic device for obtaining a visible image from an image formed with infrared or other nonvisible radiation. The radiation is focused onto a photocathode and the electrons released are accelerated and focused onto a fluorescent screen.

image intensifier. An apparatus for intensifying the brightness of an optical image, consisting essentially of a form of photomultiplier in which the optical image releases photoelectrons which are multiplied and directed onto a fluorescent screen.

image orthicon. A type of television *camera tube in which the image is produced on a *photocathode. Each part of this electrode emits electrons in proportion to the amount of incident light and these in turn hit a target screen, causing *secondary emission.

The process creates an electrostatic image on the target in which the amount of positive charge at each point corresponds to an amount of incident light. The back of the target is scanned by an electron beam and the reflected beam detected. Its intensity depends on the charge on the target at each point.

imaginary. Denoting a number that is the square root of a negative number. Imaginary numbers are expressed in the form ai, where a is a real number and i is the square root of minus one. *See also* complex number.

imide. An organic compound containing the group -CO.NH.CO- (the *imido group*).

imido group. *See* imide.

imine. An organic compound containing the group -NH-, where the nitrogen is not linked to carbonyl groups or hydrogen atoms. The group is the *imino group*.

imino group. *See* imine.

immersion objective. A microscope objective used in an oil-immersion microscope.

immiscible. Denoting liquids that do not mix completely to form a homogeneous mixture. Oil and water, for example, are immiscible—neither component will dissolve in the other.

impedance. Symbol: Z. A measure of the ability of an alternating-current circuit to resist the flow of current, equal to the potential diffence across the circuit divided by the current through it. The voltage and the current will both be varying quantities and, in general, will not be in phase. The impedance is a complex quantity: $Z = R + iX$, where R is the *resistance and X is the *reactance. It can also be put in the form $Z = (V_0/I_0)e$, where V_0 and I_0 are the amplitudes of the voltage and the current and θ is the phase angle between them. The modulus of the impedance, $|Z|$, is given by $(R^2 + X^2)^{1/2}$. Impedance is measured in ohms.

Imperial units. The system of weights and measures used in Britain, based on the pound and the yard.

implosion. Violent collapse inwards of an evacuated container, caused by the external pressure.

improper fraction. See fraction.

impulse. A force acting for a very short time, as in an impact. Its magnitude is the product of the force and the time during which it acts. An impulse on a body is equal to the change of momentum that it produces.

incandescence. The emission of visible light from a hot object. See also luminescence.

inch. Abbrev.: in. An Imperial unit of length equal to one twelfth of a foot. It is equivalent to 0.0254 metre.

incidence. Impingement of a beam of light or other radiation on a surface. The angle of incidence is the angle between the beam and the normal to the surface at the point of incidence.

inclination. See dip.

inclinometer. See dip circle.

Indene

indene. A colourless flammable liquid heterocyclic compound, C_9H_8, obtained from coal tar and used in the manufacture of synthetic resins. M.pt. $-35\,^{\circ}C$; b.pt. $182\,^{\circ}C$; r.d. 1.01.

independent variable. See variable.

index. A number indicating the power to which a number or expression is raised. In x^7, 7 is the index.

indicator. A substance used to indicate the presence of another compound by a characteristic colour change. Starch, for example, indicates iodine by forming a blue complex. The term is usually used for compounds that are suitable for showing the end point of a *titration by having one colour in the presence of excess of one reactant and another colour in excess of the other and a sharp colour change between the two. Indicators employed in acid-base titrations are often dyes that have one molecular structure in excess H^+ ions and another in excess OH^- ions. Examples are litmus, phenolphthalein, and methyl orange.

indium. Symbol: In. A soft silvery metallic element present in small amounts in some zinc ores. It is used in alloys for jewellery and dental fillings. A.N. 49; A.W. 114.82; m.pt. $156.4\,^{\circ}C$; b.pt. $1450\,^{\circ}C$; r.d. 7.31; valency 3.

Indole

indole. A white to yellowish crystalline solid, C_8H_7N, obtained from coal tar. It is used in medicine and in perfumery. M.pt. $52\,^{\circ}C$; b.pt. $254\,^{\circ}C$.

inductance. A measure of the ability of an electric circuit to resist the flow of a changing current as a result of an induced e.m.f. opposing the current flow. See electromagnetic induction.

induction. Any change in a body produced (induced) by the action of a field. See electrostatic induction, magnetic induction, electromagnetic induction.

induction coil. A device for producing a high pulsed voltage. It consists of a laminated iron core around which are wound two coaxial coils, a primary coil consisting of a few turns of wire and a secondary coil having a large number of turns. A low current is supplied to the primary, continuously interrupted by a mechanical vibrator, and a high pulsed e.m.f. is induced in the secondary coil.

induction heating. A method of heating metals by the eddy currents induced in them by an alternating magnetic field.

induction motor. *See* electric motor.

inductive. Involving electromagnetic induction or inductance.

inductor. A choke or other component used to introduce inductance into a circuit.

inelastic. *See* elastic, elasticity.

inequality. A mathematical statement that one quantity is greater or less than another. Inequalities are written in the forms $a > b$ (a is greater than b) and $a < b$ (a is less than b).

inert. Denoting a chemical substance that is very unreactive.

inert gas (rare gas, noble gas). Any of a class of elements belonging to group O of the periodic table. The inert gases have a filled shell of electrons and were once thought to form no compounds. Xenon and, to a lesser extent, krypton are now known to form compounds, especially with fluorine.

inertia. The tendency of a body to resist changes in its motion. *See* mass.

inferior planet. Either of the two planets, Mercury and Venus, whose orbits around the sun lie within that of the earth. *Compare* superior planet.

infrared radiation. *Electromagnetic radiation with wavelengths in the range 730 nm to about 1 mm. Infrared radiation lies between visible radiation and radio waves (or microwaves) in the electromagnetic spectrum. It is roughly classified into the *near infrared* (730 nm–75 μm) and the *far infrared* (>75 μm). Infrared radiation is emitted by hot bodies: the emission involves changes in vibrational energy of atoms and molecules. When absorbed by bodies it causes an increase in the vibrational energy: i.e. it is converted directly into heat. It is sometimes called *heat radiation* or *radiant heat*.

infrared spectrum. An emission or absorption spectrum in the infrared region. Absorption of infrared radiation occurs by a change of a molecule from one vibrational energy level to another. The spectrum gives information on both the molecule's vibrational frequencies and its shape. Infrared spectroscopy is widely used in organic chemistry to identify the presence of particular groups of atoms by their frequency of vibration.

infrasonic. Below audible frequencies: i.e. below about 20 Hz.

infrasound. Propagated vibrations with a frequency below that of audible sound: i.e. below about 20 Hz.

inorganic. Relating to compounds that do not contain carbon: i.e. compounds that are not *organic. The study of such compounds is the province of *inorganic chemistry*.

input. 1. The signal, current, voltage, etc., fed into an electrical circuit or device.
2. The information fed into a computer. The term also refers to the part of the computer system that converts the information into the form in which it is stored. Input may be by punched cards or paper tape—the card or tape reader converts the information by a light beam and photocell arrangement which produces a series of current pulses. Magnetic tape can also be used as the input medium. In *optical character recognition the information is in the form of conventional printed characters. *See also* output.

insoluble. Denoting a substance that does not dissolve. Unless the solvent is specified it is understood to be water.

instantaneous. Denoting a value of a changing physical quantity at a given instant of time.

insulator. A solid that is a poor conductor of electricity or of heat.

integral calculus. The branch of calculus concerned with *integration.

integrated circuit. An electronic circuit made in a single small unit. Typically, an integrated circuit may be constructed on a small slice of silicon (called a *chip*) by forming a network of junction or field-effect transistors on its surface. Interconnections between the elements are made by thin metal films.

integration. A mathematical operation involving determination of the sum of a number of terms as each term becomes infinitesimally small and the number of terms increases to infinity. A common application is in finding the area under a curve. For instance the area between the curve $y = f(x)$, the x axis, and two ordinates $x = a$ and $x = b$, can be found by dividing it into small rectangles by a number of equally spaced ordinates. If these are separated by a distance Δx, each has an area $y\Delta x$. The limit of the sum of these areas, as Δx tends to zero, is the area under the curve, written $\int y\,dx$. This is the *integral* of y with respect to x. Integration is the inverse operation of *differentiation.

interaction. *See* electromagnetic interaction, strong interaction, weak interaction, gravitational interaction.

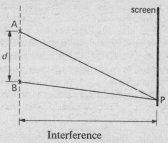

Interference

interference of light. Interaction of two sets of light waves with the same frequency and phase to produce light and dark bands (called *fringes*). If, in the illustration, A and B are monochromatic coherent sources of light with the same frequency and intensity, interference fringes can be observed on the

screen a distance l away. A light wave can be thought of as a train of alternate maxima and minima. At the point P, if two maximum values or minimum values coincide constructive interference occurs and a bright fringe is formed. If a maximum coincides with a minimum, destructive interference results and a dark fringe is produced. For a bright fringe the path difference BP – AP must equal a whole number of wavelengths $n\lambda$, where n is an integer. The separation of two fringes is $\lambda l/d$, where d is the separation of the sources. Interference fringes of this type can be formed by using a single light source and a pair of slits (*see* Young's slits) or a biprism (*see* Fresnel's biprism).
If the light is not monochromatic coloured fringes are formed. Interference effects of this type are seen in soap bubbles and thin films of oil. The constructive and destructive interference occurs between trains of light waves reflected from the upper and lower surfaces of the film. *Newton's rings are formed in a similar way. *See also* diffraction.

interferometer. Any instrument or apparatus designed to produce interference fringes. Interferometers have a variety of uses including the accurate measurement of wavelength (*see* etalon), the testing of flat surfaces, and the measurement of small changes in length.

intermediate vector boson. *See* virtual particle.

intermetallic compound. A compound formed between two or more metallic elements combined in definite proportions. Intermetallic compounds are held together by metallic bonds and are often found as distinct phases in alloys.

internal conversion. A process in which an atomic nucleus in an excited state decays to its ground state and the energy released is used to eject an electron from an inner shell of the atom. The electron ejected is called a *conversion electron*. Internal conversion is a direct process resulting from coupling between electrons in the inner shells and the atomic nucleus. The ion produced, which is in an excited state, can subsequently emit an X-ray photon or an Auger electron.

internal energy. Symbol: U. The total energy possessed by a system on account of the kinetic and potential energies of its component molecules. In practice the absolute internal energy of the system is never known but changes in U occur when heat is supplied to or abstracted from the system and when work is done. *See* thermodynamics.

interstellar dust. Clouds of dust found in deep space between the stars and frequently observed as dark *nebulae absorbing the light of stars beyond them.

interstellar gas. Clouds of gas situated in deep space between the stars. These clouds, of very low density (some ten million to a thousand million atoms per cubic metre), are made up almost entirely of both neutral and ionized hydrogen, though potassium, sodium, calcium, hydrocarbon compounds, and cyanogen have also been identified by their spectra.

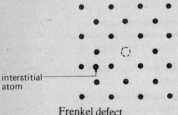

interstitial —
atom

Frenkel defect

interstitial. Pertaining to a hole formed between atoms in their positions in a crystal lattice. The atoms in a lattice can be thought of as spheres packed together in various ways and the interstices between these spheres are *interstitial positions*. Under favourable circumstances smaller atoms (*interstitial atoms*) can occupy these positions with little deformation of the original lattice. This type of structure is found in certain metal *borides, *carbides, and *hydrides.

A *Frenkel defect* is a type of defect in which an atom has moved from its normal position in the crystal to an interstitial position.

intrinsic semiconductor. *See* semiconductor.

Invar. ℝ An alloy of iron and nickel with a low coefficient of expansion, used in making pendulums, etc.

inverse-square law. Any law in which the magnitude of an effect is inversely proportional to the square of the distance from the cause. Examples are Coulomb's law, Ampères law, and Lambert's law.

inversion. The conversion of one *optical isomer of a compound to the other isomer. *See* invert sugar.

inversion temperature. *See* Joule-Kelvin effect.

invert sugar. A mixture of equal parts of glucose and fructose made by hydrolysis of sucrose. The name refers to the fact that the mixture is laevorotatory whereas sucrose is dextrorotatory. *Mutarotation occurs during hydrolysis.

in vitro. Outside the organism: said of biochemical reactions performed in laboratory apparatus (literally "in glass").

in vivo. In the natural living organism: said of biochemical reactions.

involute. A curve that is the locus of the end of a string as it is unwound from another specified curve (the *evolute).

iodate. Any salt of iodic acid.

iodide. Any of certain compounds containing iodine. *See* halide.

iodine. I A bluish-black lustrous nonmetallic element found in several natural bromide salts that occur in sea water, Chile saltpetre, and brines from oil wells. Formerly obtained from seaweed, it is usually made by the action of chlorine on oil-well brine.

Iodine has a violet irritating vapour and gives purple solutions with organic solvents. It is used as a germicide and in the manufacture of organic chemicals. Iodine belongs to the *halogen group of elements. A.N. 53; A.W. 126.9045; m.pt. 113.5°C; b.pt. 184.4°C; r.d. 4.9; valency 1, 3, 5, or 7.

iodoethane (ethyl iodide). A colourless oily liquid alkyl halide, C_2H_5I, made by reacting red phosphorus, iodine, and ethanol. M.pt. -108°C; b.pt. 72°C; r.d. 1.9.

iodoform (tri-iodo methane). A greenish-yellow crystalline powder, CHI_3, with a characteristic penetrating odour. It is made by warming acetone with an alkaline iodine solution and is used as an antiseptic and anaesthetic. M.pt. 115°C; r.d. 4.1.

iodomethane (methyl iodide). A colourless liquid alkyl halide, CH_3I, made by reaction between methanol and iodine in the presence of red phosphorus. M.pt. -66.45°C; b.pt. 42.4°C; r.d. 2.28.

ion. An atom or group of atoms with a net positive or negative charge. Positive ions (*cations*) have a deficiency of electrons and negative ions (*anions*) have an excess. In formulae, the sign and charge of the ion is written as a superscript, as in Ca^{2+} which has a deficiency of two electrons. See also ionization, electrovalent bond.

ion engine. An engine for propelling rockets by ejecting a stream of ions from the exhaust jet using a high electric field.

ion exchange. A process in which positive or negative ions are exchanged between a solution and a solid. Solid materials used in ion exchange are usually polymeric substances having charged sites to which ions are attached. Zeolites and other silicates, for instance, have a lattice of silicon and oxygen atoms with metal ions occupying certain positions in the lattice. If they are placed in contact with a solution of ions, exchange occurs: ions in solution displace ions in the lattice; silicate ion exchanges are often used for removing the calcium ions from hard water and replacing them with sodium ions.

Ion-exchange resins are organic polymers containing ionized side groups. In cation exchange the side groups are carboxyl groups or sulphonic acid groups and positive metal ions are exchanged at these sites. In anion exchange the side groups are positive quaternary ammonium ions and negative ions are exchanged. Ion-exchange resins are extensively used for purifying and separating chemicals and for chromatographic analysis (see chromatography).

ionic bond. See electrovalent bond.

ionic crystal. See crystal.

ion implantation. Bombardment of a semiconductor with high-energy ions, which penetrate the surface, used as a method of controlled doping of transistors.

ionization. The process of forming ions from neutral atoms or molecules. Ionization can occur as a result of several different processes. In solution, molecules may ionize spontaneously: $HCl = H^+ + Cl^-$. The energy necessary to break the bond is compensated by the *solvation of the ions formed. *Heterolytic fission can also occur in the gas phase as a result of the impact of particles or photons. Irradiation can also produce positive ions by ejecting one or more electrons: $A = A^+ + e$. Negative ions can be formed by electron capture: $A + e = A^-$. See also photoionization.

ionization chamber. A chamber containing two electrodes in a gas with a potential difference between them, used to detect ionizing radiation by the current it causes in ionizing the gas between the electrodes.

ionization potential. Symbol: I. The minimum energy that must be supplied to an atom or molecule to remove an electron (to infinity) and produce a positive ion. Ionization potentials are measured in electronvolts. The term *second ionization potential* is used in two ways. It is either the energy required to remove two electrons and form a doubly charged ion or it is the energy necessary to remove the second most strongly bound electron (from an inner shell or sub-shell) to form a singly charged ion in an excited state.

ionizing radiation. Electromagnetic radiation or particles that cause ionization. For this to occur the photon energy of the electromagnetic radiation or the kinetic energy of the particles must exceed the *ionization potential of the substance ionized. The cross section for ionization must also be high. X-rays, ultraviolet radiation, and alpha particles are all forms of ionizing radiation: neutrons and gamma rays are much less effective.

ionosphere. A region of ionized air and free electrons around the earth in the earth's upper atmosphere, extending from a height of about 50 km to 1000 km. It is produced by ionizing radiation, chiefly ultraviolet radiation, coming from the sun. The free electrons in the ionosphere reflect radio waves and the ionosphere can be used for radio transmissions between points that are hidden by the earth's curvature. It is divided into three distinct layers, differing in their concentrations of electrons. The layers are the *D-layer* (50-90 km), the *E-layer* or *Heaviside-Kennelley layer* (90-150 km), and the *F-layer* or *Appleton layer* (150-1000 km). The F-layer contains the highest electron concentration and is the one used for most radio transmission. At night, the concentration of electrons falls in the D-layer due to recombination of atoms and electrons.

ionospheric wave. *See* sky wave.

ion pump. A type of vacuum pump, used at very low pressures, in which gas is removed from the system by ionization and subsequent adsorption of the ions on a surface. The ions can be produced by electrons from a hot metal filament and adsorbed on a metal film. In one type of ion pump ionization occurs in a gas discharge between titanium electrodes. The discharge causes *sputtering of the titanium, thus creating a continuously deposited film of titanium which acts as a *getter. Such devices are called *getter-ion pumps* or *sputter-ion pumps*.

iridium. Symbol: Ir. A silvery-white brittle *transition element occurring associated with platinum. It is used in platinum-iridium and osmium-iridium alloys. The chemical properties of iridium are similar to those of rhodium. Its main compounds are iridium III and iridium IV complexes. A.N. 77; A.W. 192.2; m.pt. 2140°C; b.pt. 4130°C; r.d. 22.42; valency 1–4.

iris. 1. *See* eye.
2. An adjustable aperture for allowing light into an optical instrument.

iron. Symbol: Fe. A silvery-white soft malleable and ductile ferromagnetic *transition element; the fourth most abundant element in the earth's crust, occurring in many minerals, the main ores being haematite (Fe_2O_3), limonite ($FeO(OH).nH_2O$), and magnetite (Fe_3O_4). Iron makes up most of the earth's core and is also found in some meteorites. It is obtained by smelting in a *blast furnace to give pig iron, which is then converted into cast iron, wrought iron, or steel. The pure element is rarely prepared. It exists in several allotropic forms (*see* alpha-, beta-, delta-, gamma iron).

Chemically, the element is reactive, being quickly oxidized in moist air and combining directly with most nonmetals on heating. It dissolves in dilute nonoxidizing acids. Iron forms two series of compounds: iron II (*ferrous*) compounds and iron III (*ferric*) compounds. The element forms numerous ionic salts in both these oxidation states as well as many complexes. Iron also forms compounds in other oxidation states, notably the carbonyls and *ferrates. A.N. 26; A.W. 55.847; m.pt.

1535°C; b.pt. 2750°C; r.d. 7.87; valency 2, 3, 4, or 6.

iron chloride. Either of two chlorides of iron. *Iron II chloride* (ferrous chloride) is a white crystalline solid, $FeCl_2$ or $FeCl_2.4H_2O$, used as a mordant. M.pt. 670°C; r.d. 3.2. *Iron III chloride* (ferric chloride) is an orange crystalline solid, $FeCl_3.6H_2O$, used in the treatment of sewage and industrial waste. It is black when anhydrous. M.pt. 300°C; r.d. 2.9.

iron oxide. Any of three oxides of iron. *Iron II oxide* (ferrous oxide) is a black powder, FeO, obtained by heating iron II oxalate. M.pt. 1420°C; r.d. 5.7. *Tri-iron tetroxide* (ferrosoferric oxide) is a red or black ferromagnetic powder, Fe_3O_4, made by reacting hot iron with air or steam and used in pigments and polishes. M.pt. 1538°C (decomposes); r.d. 5.2. The highest oxide, *iron III oxide* (ferric oxide), occurs naturally as haematite. It is a red powder, Fe_2O_3, used as a pigment and as a metal polish (*jeweller's rouge*). M.pt. 1565°C; r.d. 5.1.

iron sulphate. Either of two salts of iron. *Iron II sulphate* (ferrous sulphate) is a green crystalline solid, $FeSO_4.7H_2O$, prepared by the action of sulphuric acid on iron, and used for making other iron salts, sewage treatment, and as a dietary supplement. It is also known as *green vitriol* and *copperas*. M.pt. 64°C; r.d. 1.9. *Iron III sulphate* (ferric sulphate) is a pale yellow powder, $Fe_2(SO_4)_3$, used in analytic chemistry. R.d. 3.1.

irradiation. Exposure to *radiation, usually with the aim of inducing some change as a result of the impact of particles or photons.

irreversible. *See* reversible.

irreversible reaction. A chemical reaction in which the reactants are completely converted into the products.

isobar. 1. A curve or equation relating quantities measured at the same pressure. An example is a graph of the volume of a gas against its temperature, with the pressure kept constant.
2. Any line joining points that have equal pressures.
3. Any of two or more isotopes that have different atomic numbers but the same mass numbers. For example, the isotopes palladium-105 and silver-105 are isobars. Their atomic numbers are 46 and 47 respectively and their atomic masses are almost identical.

isobutyl group. *See* butyl group.

isochore. A curve or equation relating quantities measured at the same volume. An example is Van't Hoff's isochore.

isocline. A line on the earth connecting points that have the same magnetic *dip. *See* terrestrial magnetism.

isocyanate. *See* cyanic acid.

isocyanic acid. *See* cyanic acid.

isodiapheres. Nuclides that have the same difference between total number of neutrons and total number of protons. Examples are $^{190}_{78}$ Pt and $^{186}_{76}$ Os, the difference being 34 in each case.

isodynamic line. A line on the earth connecting points that have the same horizontal intensity of magnetic field. *See* terrestrial magnetism.

isoelectronic. Denoting atoms or molecules that have the same numbers of electrons. Carbon dioxide and nitrous oxide, for example, have isoelectronic molecules.

isogonal line. A line on the earth connecting points that have the same magnetic *declination. *See* terrestrial magnetism.

isomerism. 1. The existence of two or more chemical compounds with the same molecular formula but different arrangements of atoms in their molecules. Such compounds are said to be *isomers.* Several types of isomerism are distinguished, broadly classified into *structural isomerism and *stereoisomerism.
2. The existence of nuclei with the same

atomic number and mass number but with different energy states. A nucleus in its ground state and one in an excited state constitute a pair of nuclear *isomers*.

isometric. *See* cubic.

isomorphism. The occurrence of two or more different substances with the same crystal structure: the substances are said to be *isomorphous*. The term is often used in a more restricted sense with reference to compounds that have similar compositions and can crystallize together in mixed crystals. *See* Mitscherlich's law.

isopentane (2-methylbutane). A colourless flammable liquid alkane, $CH_3CH(CH_3)C_2H_5$, obtained from petroleum. It is isomeric with *pentane and *neopentaane. M.pt. $-159.9°C$; b.pt. $279°C$; r.d. 0.62.

isopentyl group. *See* pentyl group.

isoprene. A colourless volatile liquid, $CH_2:C(CH_3)CH:CH_2$, made from refinery and coal gases and tars. It polymerizes readily and is used in making synthetic rubber. Isoprene is the structural unit of *terpenes. M.pt. $-146°C$; b.pt. $34.1°C$; r.d. 0.68.

isoscelese triangle. A triangle with two sides equal.

isospin (isotopic spin). A property of elementary particles used to characterize particles in the same *multiplet. The proton and neutron, for example, differ in charge but are identical in their *strong interactions. They can be regarded as two states of the same particle. The property is described by quantum numbers in a similar way to *spin. The multiplet is characterized by the *isospin* (symbol: I) which has values 0, 1/2, 1, etc. The particles within a multiplet have different values of *isospin quantum number* (symbol: I_3) equal to $-I$, $-(I-1)$, ... $(I-1)$, I. Thus in the proton-neutron multiplet, in which $I = 1/2$, the proton has $I_3 = 1/2$ and the neutron has $I_3 = -1/2$.

isotactic. *See* polymer.

isotherm. An equation or a curve on a graph representing behaviour of a system at a constant temperature.

isothermal. Denoting a process in which no change occurs in the temperature of the system. In an isothermal change the system is always in thermal equilibrium with its surroundings. *Compare* adiabatic.

isotones. Nuclides that have the same neutron number but different atomic numbers.

isotonic. Denoting solutions that have the same *osmotic pressure.

isotope. Any of a number of forms of an element, all of which differ only in the number of neutrons in their atomic nuclei. Two isotopes of the same element thus have nuclei with the same atomic number and different mass numbers. Carbon, for instance, is a mixture of two isotopes, one with six protons and six neutrons and the other with six protons and seven neutrons. The respective percentages of carbon-12 and carbon-13 are 98.892 and 1.108, leading to an overall atomic weight of carbon of 12.011 (*see also* carbon-14). All elements also have artificial isotopes produced by bombardment with high-energy particles. Isotopes of an element have identical chemistries because they contain the same number of electrons. They do, however, have different physical properties and chemical reactions involving different isotopes take place at slightly different rates. These two properties are used in separating isotopes. The principle methods of separation are *diffusion, *thermal diffusion, *centrifugal separation, *electrolytic separation, the use of kinetic *isotope effects, and the use of large mass spectrometers. The last method is used only for very small samples.

isotopic number (neutron excess). The difference between the number of neutrons and the number of protons in a nucleus.

isotopic spin. *See* isospin.

isotopic weight. The atomic weight of an isotope. The isotopic weight is almost the same as the *mass number.

isotropic. Having properties that are independent of direction. For example, an isotropic crystal has the same physical properties (refractive index, modulus of elasticity, etc.) along all its axes because of its crystal structure. *Compare* anisotropic.

J

Jaeger's method. An experimental method for measuring surface tension by determining the pressure required to produce a bubble of air from a capillary tube immersed in the sample liquid.

jeweller's rouge. *See* iron oxide.

Joly's steam calorimeter. An apparatus for measuring the specific heat capacities of gases at constant volume. It consists essentially of two equal metal spheres suspended from opposite arms of a balance, one sphere being evacuated and the other filled with a known volume of the sample gas. The spheres are contained in an enclosure through which steam is passed: the specific heat capacity may be calculated by measuring the difference between the amounts of water condensing on the spheres. [After John Joly (1857-1933), Irish physicist and geologist.]

joule. Symbol: J. The *SI unit of energy, equal to the work done when the point of application of a force of one newton is moved one metre, movement being in the direction of the force. It is the work done by a current of one ampere flowing through a resistance of one ohm for one second. [After James Prescott Joule (1818-89), British physicist.]

Joule heating. The heating of a conductor by an electric current flowing through it. The heat produced is I^2Rt,

where I is a steady current flowing for a time t through a resistance R.

Joule-Kelvin effect. A change in temperature occurring in a gas when it undergoes an adiabatic expansion. The effect can be demonstrated by allowing the gas to flow through a porous plug from a region of high pressure into a region of low pressure. Usually a slight fall in temperature is observed, caused by the fact that, in expanding, the gases have to overcome forces between their molecules. Substances have a characteristic temperature—the *inversion temperature*—above which they are heated by expansion and below which they are cooled.

junction transistor. *See* transistor.

Jupiter. The fifth planet in succession outwards from the sun and the first of the *giant planets. Lying beyond the asteroid belt at a distance of 778 million kilometres from the sun, Jupiter is by far the largest of the planets, with an equatorial diameter of 142 800 km and a mass 318 times that of the earth. Despite these dimensions, the planet's relative density is only 1.334—a fifth the size of the earth's. Its gravity is 2.5 times that of the earth.

Jupiter is a typical giant planet, having a dense impenetrable atmosphere of methane, ammonia, and hydrogen. The rotation is not uniform, being slowest at the equator and increasing towards the poles: an average value of 9 hours 55 minutes can be assigned. Jupiter has several interesting features. It has a magnetic field 30 times as strong as the earth's field. It also shows a vast marking in the southern hemisphere—the *Great Red Spot*. While other optical features of Jupiter are transient, this phenomenon, whose origin remains unknown, has been present since serious telescope observation began. Jupiter has 12 satellites.

K

kainite. A mineral consisting of a hydrated mixed salt of magnesium sulphate and potassium chloride, $MgSO_4.$-$KCl.3H_2O$, used as a source of potassium fertilizers.

kaon (K-meson). A type of *meson having either zero charge and a mass 974.2 times that of the electron, or a positive or negative charge and a mass 966.6 times that of the electron.

Kater's pendulum. A pendulum designed for the accurate measurement of the acceleration of free fall. It consists of a metal bar with knife edges mounted near each end and two weights that can be slid up and down the bar between the knife edges. The pendulum is pivoted on each set of knife edges in turn and the positions of the weights are adjusted until the period of oscillation is the same for each pivot. Under these conditions the period is the same as that of a simple pendulum $(2\pi(l/g)^{1/2})$, with a length l equal to the distance between the pivots. [After Henry Kater (1777–1835), British physicist.]

keeper (armature). A small piece of iron or steel placed across the poles of a permanent magnet when it is out of use to complete the magnetic circuit and prevent loss of magnetism.

Kekulé formula. See aromaticity. [After F. A. Kekulé von Stradonitz (1829–96), German chemist.]

kelvin. Symbol: K. The basic *SI unit of *thermodynamic temperature, equal to 1/273.16 of the thermodynamic temperature of the *triple point of water. This unit was formerly the *degree Kelvin* (°K). It is equal to one degree on the centigrade scale of temperature. [After Lord Kelvin (William Thomson) (1824–1907), British physicist.]

Kennelley layer. See ionosphere.

Kepler's laws. Three laws of planetary motion:
1. The planets move in elliptical orbits with the sun at one of the foci of the ellipse.
2. The radius vector (the line joining the centre of the planet to the centre of the sun) sweeps out equal areas of space in the same interval of time.
3. The square of a planet's sidereal period is proportional to the cube of its average distance from the sun.
[Formulated between 1609 and 1619 by Johannes Kepler (1571–1630), German astronomer.]

Kepler telescope. A type of refracting telescope consisting of a convex objective lens and a convex eyepiece. The lenses are separated by a distance equal to the sum of their focal lengths and the magnification is the ratio of the focal lengths. The instrument produces an inverted image. It can be modified for use as a terrestrial telescope by inclusion of an erecting lens or prism.

kerosene. See paraffin.

Kerr effect. 1. The phenomenon in which some liquids or gases show *double refraction in the presence of a strong electric field. The field is applied at right angles to the direction of the light. A *Kerr cell* is a transparent cell containing a suitable liquid and two electrodes. A beam of plane-polarized light passing through the cell can be stopped by applying a voltage across the electrodes: the device acts as a shutter.
2. The phenomenon in which a beam of plane-polarized light is elliptically polarized by reflection from the polished pole face of an electromagnet.
[After John Kerr (1824–1907), British physicist.]

ketal. Any of a class of organic compounds that have the general formula $RR'C(OR'')(OR''')$, where R indicates a hydrocarbon group. Ketals are formed in the same way as *acetals by reaction of a ketone with an alcohol.

ketene (ethenone). A colourless toxic gas, $CH_2:C:O$, with a penetrating odour, made by pyrolysis of acetone. It is an acetylating agent, used in the

$$CH_3-\overset{\overset{O}{\|}}{C}-\overset{\overset{H}{|}}{\underset{\underset{H}{|}}{C}}-\overset{\overset{O}{\|}}{C}-OC_2H_5 \rightleftharpoons CH_3-\overset{\overset{OH}{|}}{C}=\overset{}{\underset{\underset{H}{|}}{C}}-\overset{\overset{O}{\|}}{C}-OC_2H_5$$

keto form (93%)   enol form (7%)

Keto-enol tautomerism in acetoacetic
ester

manufacture of aspirin and cellulose
acetate. M.pt. -150 °C; b.pt. -56 °C.

keto-enol tautomerism. A form of
tautomerism in which transfer of a
hydrogen atom occurs from a carbon
atom next to a carbonyl group to the
oxygen atom of the carbonyl group,
thus producing an unsaturated alcohol.
The two isomeric forms of the compound are called the *keto* and *enol*
forms of the compound.

keto form. *See* keto-enol tautomerism.

ketone. Any of a class of organic compounds with the general formula
RCOR′, where R and R′ are usually
hydrocarbon groups. Ketones can be
prepared by oxidation of secondary
alcohols. They give a negative reaction
with Fehling's solution. They also have
a variety of addition and condensation
reactions.

ketose. A *sugar that is a ketone in its
straight-chain form. *Compare* aldose.

KeV. Kiloelectronvolts: 1000 electronvolts.

kieselguhr. A porous form of silica,
SiO_2, formed from skeletons of microscopic plants. It is used as an absorbent.

killed spirits. A solution of zinc
chloride, $ZnCl_2$, made by adding pieces
of zinc to spirits of salt (hydrochloric
acid). It is used as a soldering flux.

kilo-. Prefix indicating one thousand
(10^3).

kilocycle. A unit of frequency equal to
1000 cycles per second (1000 hertz).

kilogram (kilogramme). Symbol: kg.
The basic *SI unit of mass, equal to the
mass of the *international prototype*
kilogram—a cylindrical piece of platinum-iridium alloy kept in the International Bureau of Weights and Measures
at Sèvres near Paris. It is equivalent to
2.204 62 pounds.

kilohertz. Symbol: kHz. A unit of frequency equal to 1000 hertz.

kilometre. A unit of distance equal to
1000 metres. It is equivalent to 0.6214
mile.

kilowatt. Symbol: kW. A unit of power
equal to 1000 watts.

kilowatt-hour (Board of Trade unit).
Symbol: kWh. A unit of energy equal to
a power of one kilowatt available for
one hour. It is used for the sale of
electrical energy.

kinematics. The branch of mechanics
concerned with motion without
reference to the forces or mass involved
in the motion. *Compare* dynamics.

kinematic viscosity. *See* viscosity.

kinetic energy. Symbol: E_K. The energy
that a system has by virtue of its
motion, determined by the work necessary to bring it to rest. An object of
mass m and velocity v has a kinetic
energy equal to $mv^2/2$. A body rotating
about an axis with moment of inertia I
and angular velocity ω has a kinetic
energy equal to $I\omega^2/2$. *See also* potential
energy.

kinetics. The study of the rates of
chemical reactions, used to study reaction mechanisms.

kinetic theory. Any theory for describing the physical properties of a system
with reference to the motion of its
constituent atoms or molecules. The

most successful is the *kinetic theory of gases*, which is based on the principle that gases are composed of many elastic particles of negligible size in constant random motion. The particles collide with each other and with the walls of the containing vessel, exerting a pressure of $mvC^2/3$, where m is the particle mass, C the *root-mean-square velocity, and v the number of particles per unit volume.

Kipp's apparatus. A laboratory apparatus for generating gas by chemical reaction of a liquid and a solid When the tap is closed the pressure of gas drives the liquid into the top bulb, thus terminating the reaction. [After Petrus Kipp (1808–64).]

Kirchhoff's law. The principle that the emissivity of a body is equal to its absorptance at the same temperature. [After Robert Kirchhoff (1824–87), German physicist.]

Kirchhoff's laws (of electricity). Two laws applied to the flow of electric currents in a network.
1. The algebraic sum of the currents at any point in the network is zero—a current flowing towards the point is taken as having opposite sign to one flowing away from the point.
2. The algebraic sum of the e.m.f.s in any closed loop of the network is equal to the algebraic sum of the products of the currents and the resistances through which they flow: i.e. $E_1 + E_2 + ... = I_1R_1 + I_2R_2 + ...$

Kjeldahl's method. A technique for determining the amount of nitrogen in organic compounds by decomposing the sample with concentrated sulphuric acid to convert the nitrogen into ammonium sulphate. The amount of ammonium sulphate produced is measured by adding an excess of alkali, distilling off ammonia into a standard acid solution, and titrating the excess acid. [After Johan Kjeldahl (1849–1900), Danish chemist.]

klystron. An electron tube for amplifying or generating radiofrequency electromagnetic radiation. A beam of electrons is passed through two consecutive resonant cavities. High-frequency waves generated in the first cavity cause the electrons to "bunch" into a series of pulses. These generate more high-frequency waves, of higher amplitude, in the second cavity, the energy being supplied by the electron beam. The process of bunching the electrons is called *velocity modulation*.

knot. A unit of velocity equal to one nautical mile per hour (1.15 statute miles per hour.) It is used for expressing the speeds of ships and aircraft.

Knudsen flow. *See* molecular flow. [After M. Knudsen (1871–1949), Danish physicist.]

Kroll process. A method of producing metallic elements by reduction of their chloride with magnesium.

krypton. Symbol: Kr. A colourless odourless tasteless nonflammable gaseous element present in the atmosphere (0.5 parts per million) from which it is extracted as a by-product in the liquefaction of air. It is used in fluorescent lamps and lasers. Krypton is one of the *inert gases. A.N. 36; A.W. 83.80; m.pt. $-156.6\,°C$; b.pt. $-152.3\,°C$.

L

label. A radioactive atom in a molecule, used to determine the mechanism of a chemical reaction. *See* tracer.

labile. Denoting a chemical compound that is unstable and likely to break down or change.

lactam. Any of a class of cyclic aliphatic compounds derived from amino acids by elimination of water between the amino group of the molecule and the carboxyl group. Lactams contain the group -NH.CO-.

lactate. Any salt or ester of lactic acid.

lactic acid (2-hydroxypropanoic acid). A colourless hygroscopic syrupy liquid, $CH_3CHOHCOOH$, prepared by fermentation of starch or molasses. Lactic acid is used as an acidulant of foods, preservative, and in making plasticizers and adhesives. M.pt. $18°C$; b.pt. $122°C$ (15 mmHg); r.d. 1.2. *See also* optical isomerism.

lactone. Any of a class of cyclic aliphatic compounds derived from hydroxy carboxylic acids by elimination of water between the hydroxyl group of the molecule and the carboxyl group. Lactones thus contain the group -O.CO- and are internal esters.

lactose (milk sugar). A white crystalline disaccharide sugar, $C_{12}H_{22}O_{11}$, with a sweet taste. It occurs in the milk of mammals and may be prepared from whey. Lactose is used in infant foods, margarine manufacture, and medicine. Decomposes at $203°C$; r.d. 1.5.

Ladenburg benzene. *See* benzene. [After Albert Ladenburg (1842-1911), German chemist.]

laevorotatory (laevorotary). Denoting a chemical compound that is capable of imparting an anticlockwise rotation to the plane of polarization of polarized light, as observed by someone facing the light source. *See* optical activity.

Lagrangian function. Symbol: L. The kinetic energy of a system minus its potential energy. [After Joseph Louis Lagrange (1736-1813), Italian-French mathematician.]

LAH. *See* lithium aluminium hydride.

lake. An insoluble pigment formed by the combination of an organic dyestuff with a metallic compound (salt, oxide, or hydroxide). The organic substance is either chemically combined with the inorganic compound or absorbed by the compound. Lakes are used in paints and printing inks. *See* mordant.

lambda particle. Symbol: Λ. An *elementary particle with zero charge and a mass 2183 times that of the electron. It is classified as a *hyperon.

lambda point. 1. The temperature (2.186 K) at which helium I is in equilibrium with helium II.
2. The temperature at which a second-order phase change occurs, such as the change of an alloy from an ordered to a disordered crystal structure or the change from a ferromagnetic to a paramagnetic state. A graph of specific heat capacity against temperature shows a sharp peak at the lamda point.

lambert. A unit of *luminance equal to the luminance of a surface emitting one lumen per square centimetre. [After Johann H. Lambert (1728-77), German mathematician and physicist.]

Lambert's law. 1. (cosine law). The principle that, for a perfectly diffusing surface, the luminous intensity in any direction is proportional to the cosine of the angle between the direction and the normal.
2. (of illumination). The principle that the illumination of a surface by a point source is universely proportional to the square of the distance from the source to the surface, the illumination occurring normal to the surface. If the normal to the surface is at an angle θ to the light rays, then illumination is proportional to $\cos \theta$.

Lamb shift. A small energy difference observed between two levels in the hydrogen-atom spectrum ($^2S_{1/2}$ and $^2P_{1/2}$). On normal theories of fine structure, the two levels should have the same energy. The shift is explained by interaction between the electron and the radiation field. [After Willis Eugene Lamb (b. 1913), American physicist.]

laminar flow. Steady flow of a fluid in parallel layers, with little or no mixing between adjacent layers. *Compare* turbulent flow.

lamination. The use of thin strips of metal to form the core of a choke, transformer, etc., thus increasing the electrical resistance and reducing *eddy currents. The strips are often oxidized or varnished.

lamp black. The soot (amorphous carbon) produced by burning certain organic compounds, such as oil or turpentine. It is used as a pigment.

lanthanide (lanthanon). Any of a series of fifteen metallic elements with atomic numbers from 57 (lanthanum) to 71 (lutetium) inclusive, in which the elements have an incomplete *f* sub-shell. Barium, which immediately precedes lanthanum in the periodic table, has the outer electron configuration $4s^2$-$4p^64d^{10}5s^25p^66s^2$. The subsequent elements could, in theory, form a fourth transition series (see transition metal) by filling up the $5d$ sub-level. In fact, they form the lanthanide series by filling up the $4f$ level, which can hold a maximum of 14 electrons. Sometimes lanthanum itself is not considered a member of the series.

All are strongly electropositive, mainly trivalent metals which form M^{3+} ions. They all have similar chemical properties and mixtures of their compounds occur naturally, especially in monazite. Lanthanide compounds can be separated by *ion-exchange chromatography. The series of trivalent metal ions show a decrease in their ionic radius as the atomic number increases: the lanthanum ion has a radius of 0.106 nm and lutetium has 0.085 nm. This is known as the *lanthanide contraction.* It is caused by the extra nuclear charges not being shielded by extra electron charges: a consequence of the shapes of *f* orbitals. The lathanides are often called the *rare earths,* or more correctly the *rare-earth elements*—the former term strictly applies to the oxides. Scandium and yttrium, which occupy the same group as lanthanum in the periodic table, are also often classified with the lanthanides.

lanthanon. *See* lanthanide.

lanthanum. Symbol: La. A soft silvery-white ductile metallic element belonging to the *lanthanide series. A.N. 57; A.W. 138.9055; m.pt. 920°C; b.pt. 3454°C; r.d. 6.2; valency 3.

Laplace's equation. The partial differential equation:

$$\nabla^2 U = \frac{\partial^2 U}{\partial x^2} + \frac{\partial^2 U}{\partial^2 y} + \frac{\partial^2 U}{\partial z^2} = 0$$

or a form of it in spherical or cylindrical coordinates. U is a *potential. The operator ∇^2 (read as "del-squared") is the *Laplace operator* or *Laplacian.* [After Pierre-Simon, Marquis de Laplace (1749–1827), French mathematician.]

Larmor precession. The *precession of a charged particle when it moves under the influence of an applied magnetic field and a central force. For instance, the orbit of an electron in an atom will precess about an external field, such that the normal to the plane of the orbit sweeps out the surface of a cone with the field's direction as axis. The frequency of precession (the *Larmor frequency*) is $eH/4\pi mc$, where H is the field strength, m the mass, e the electron's charge, and c the electron's velocity. [After Sir Joseph Larmor (1857–1942), British physicist.]

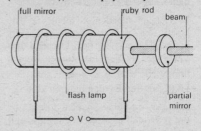

Ruby laser

laser. A device for producing intense light or infrared or ultraviolet radiation by *stimulated emission. The electromagnetic radiation from a laser is coherent and monochromatic and can be obtained in a parallel beam with very little spread. The principle of laser action can be demonstrated by the earliest type of model, the *ruby laser*, which consists of a rod of ruby with a silvered mirror at one end and a partially silvered mirror at the other (see illustration). A flash lamp is used to excite chromium ions in the ruby above their ground state. After a short time each ion spontaneously reverts to the ground state with the emission of a photon. If power is continuously supplied by the external light source, a high proportion of the ions are excited at any time—a situation known as *population inversion*. The process of producing large numbers of excited states in this way is called *optical pumping*. The emitted photons can collide with other excited ions and produce stimulated emission of other photons. The light is reflected backwards and forwards up and down the tube and the beam, which in the ruby laser has a wavelength of 694.3 nm, emerges through the partially reflecting end. It is monochromatic because all the photons correspond to the same energy-level change and it is coherent because in stimulated emission the emitted light has the same phase as the incident light from previous stimulated emissions. Moreover, in stimulated emission the emitted photons move in the same direction as the incident photons—giving the beam a highly directional quality. Other types of solid laser often use lanthanide elements, such as neodymium and yttrium. The name is an acronym for *light amplification by stimulated emission of radiation*. Laser action can also be shown by gases and liquids. In the carbon dioxide laser a mixture of carbon dioxide and nitrogen is used in a gas-discharge tube and population inversion is achieved for vibrationally excited CO_2 molecules. The device gives infrared radiation with a wavelength of 10.6 micrometres. Other types of laser produce population inversion by chemical action.

Lasers can be made with a range of power outputs up to about 100 kilowatts. They are used for cutting metals, surgery, surveying, holography, and scientific research. *See also* maser.

latent heat. Symbol: L. The total heat absorbed or produced during a change of phase (fusion, vaporization, etc.) at a constant temperature. The heat change per unit mass is the *specific latent heat.*

lateral inversion. Inversion from left to right, as seen in an image in a plane mirror.

latex. *See* rubber.

latitude. 1. The distance of a point on the earth from the equator measured as the angle between a normal to the earth at that point and the plane of the equator.
2. (celestial latitude). The angular distance of a body on the *celestial sphere from the ecliptic; considered as positive if the body is north of the ecliptic.

lattice. 1. A regular array of points in two or three dimensions. A crystal lattice is the array of points on which the atoms, molecules, or ions are centred in a *crystal. 2. The regular arrangment of fissile material and moderator in a nuclear reactor.

lattice energy. The energy that would be produced if the ions forming a given crystal were brought together from infinite separation to their positions in the lattice.

laughing gas. *See* nitrogen oxide.

lauric acid (dodecanoic acid). A white crystalline carboxylic acid, CH_3-$(CH_2)_{10}COOH$, obtained from coconut oil. It is used in making alkyd resins, detergents, and insecticides. M.pt. 44°C; b.pt. 225°C; r.d. 0.8.

lawrencium. Symbol: Lr. A transuranic *actinide element obtained in trace amounts by bombarding californium with boron nuclei. The isotope produced, ^{257}Lr, has a half-life of 8 seconds.

A.N. 103. [After E. O. Lawrence (1901-58), U.S. physicist who invented the cyclotron.]

layer lattice. A type of crystal lattice in which the atoms are strongly bound in layers with relatively weak bonding between the layers. Graphite is an example of a substance with a layer lattice.

LCM. *See* least common multiple.

leaching. The process of extracting soluble constituents from a mixture by washing them out with a percolating solvent.

lead. Symbol: Pb. A very soft bluish-white metallic element existing chiefly as galena (PbS), from which it is obtained by roasting. It is used in some alloys, such as solder and Babbitt metal, and in pipes, cable coverings, and accumulators. It is an efficient shielding material for radiation and X-rays. The element is also used in the manufacture of tetraethyl lead, which is used as a petrol additive. Lead is very resistant to corrosion and will not dissolve in dilute sulphuric or concentrated hydrochloric acid. It is attacked by alkalis to form *plumbites. Lead forms covalently bonded lead IV (*plumbic*) compounds and more ionic lead II (*plumbous*) compounds. A.N. 82; A.W. 207.19; m.pt. 327.5°C; b.pt. 1744°C; r.d. 11.35.

lead acetate (sugar of lead). A white crystalline poisonous solid, Pb-$(C_2H_3O_2)_2.3H_2O$, made by the action of acetic acid on lead II oxide. It is a soluble lead salt, used in medicine, textiles, and as an analytical reagent. B.pt. 280°C (anhydrous); r.d. 2.5.

lead carbonate. A white poisonous powder, $PbCO_3$, made by adding sodium hydrogen carbonate to lead nitrate solution. It is used as a pigment in paints. R.d. 6.6.

lead-chamber process. A process for the manufacture of sulphuric acid by oxidizing sulphur dioxide with nitrogen dioxide in large lead chambers. The first step is the production of sulphur trioxide: $SO_2 + NO_2 = SO_3 + NO$. The sulphur trioxide is reacted with water to give sulphuric acid, $SO_3 + H_2O = H_2SO_4$, and the nitrogen dioxide is regenerated by the reaction of nitric oxide with oxygen in the air: $2NO + O_2 = 2NO_2$. The process has been replaced by the contact process.

lead chromate. A bright yellow insoluble salt, $PbCrO_4$, made from a solution of a lead salt and potassium dichromate. It is the basis of various yellow pigments (*chrome yellow*).

lead dioxide. *See* lead oxide.

lead monoxide. *See* lead oxide.

lead nitrate. A white crystalline poisonous water-soluble solid, $PbNO_3$, made by dissolving lead in nitric acid and used in preparing lead salts and as an oxidizer in the tanning industry. Decomposes at 470°C; r.d. 4.5.

lead oxide. Any of three oxides of lead. *Lead II oxide* (plumbous oxide, massicot, litharge, lead monoxide) is a yellow crystalline powder, PbO, made by controlled heating of lead. It is an amphoteric substance, used in storage batteries, ceramics, pigments, and oil refining. M.pt. 888°C; r.d. 9.53. *Trilead textroxide* (red lead, minium) is a red powder, Pb_3O_4, made by heating lead II oxide in air. It is a mixed oxide, used in storage batteries, paints, and ceramics. Decomposes above 500°C; r.d. 8.3-9.2. *Lead IV oxide* (lead dioxide, lead peroxide, plumbic oxide) is a brown powder, PbO_2, prepared by adding bleaching powder to an alkaline solution of lead hydroxide. Its principal use is as an oxidizing agent. Decomposes at 290°C; r.d. 9.4.

lead peroxide. *See* lead oxide.

lead-plate accumulator. A type of *accumulator consisting of two lead plates dipping into dilute sulphuric acid. When the cell is fully discharged both plates are coated with lead sulphate. In charging the cell, a current is passed

.rom the cell's positive electrode to the negative electrode, through the electrolyte. At the positive plate the lead sulphate is converted to a layer of lead oxide, PbO_2, and at the negative plate the lead sulphate is reduced to a layer of spongy lead. During discharge the reverse processes occur: lead oxide changes to lead sulphate with loss of negative charge from the electrode and lead changes to lead sulphate with gain of negative charge. The overall reaction is $PbO_2 + Pb + 2H_2SO_4 = 2PbSO_4 + 2H_2O$.

Lead-plate accumulators are used in car batteries. The relative density of the acid falls as the cell is discharged and should normally be in the range 1.2-1.28. If the battery is allowed to go "flat" an insoluble sulphate may form.

lead sulphate. A white insoluble crystalline salt, $PbSO_4$. M.pt. 1170°C; r.d. 6.2.

lead sulphide. A black insoluble compound, PbS, precipitated from solutions of lead salts by hydrogen sulphide. It occurs naturally as galena. M.pt. 1114°C; r.d. 7.5.

lead tetraethyl. See tetraethyl lead.

leakage. Flow of a small electric current through imperfect insulation.

least common multiple (LCM). The smallest number that every number of a given set of numbers can divide into exactly. For example, the LCM of 2, 7, and 5 is 70.

least squares. A method of obtaining an equation to fit experimental data. A typical case would be one in which a number of measured values of a dependent variable x are known for values of the independent variable y. In general, the values will show a scatter resulting from experimental error and it is necessary to find the best line, say $y = mx + c$, to fit the data. Values of the constants, m and c, are chosen and the deviations of the experimental values of x from calculated values are obtained. The best fit is given for values of m and c for which the sum of the squares of these deviations has a minimum.

Leblanc process. A former process for the manufacture of sodium carbonate by heating sodium chloride with sulphuric acid to give sodium sulphate, which is then heated with coke and limestone to yield sodium carbonate and calcium sulphide. [After Nicolas Leblanc (1742-1806), French chemist.]

Le Chatalier's principle. The principle that if a disturbance is applied to displace a system from chemical equilibrium, the system tends to change in such a way as to compensate for the disturbance. The principle can be illustrated by the reaction: $N_2 + 3H_2 = 2NH_3$. The production of ammonia molecules is exothermic and their decomposition is endothermic. If the temperature is increased, the equilibrium is displaced to the left: ammonia molecules decompose as if to restore the original temperature. Thus, high yields of ammonia are favoured at low temperatures although in practice the rate of attainment of equilibrium must also be considered. If the total pressure is increased, the reaction shifts to the right: more ammonia molecules are produced as if to decrease the total number of molecules and thus restore the total pressure. [After Henri Louis Le Chatalier (1850-1936), French chemist.]

Leclanché cell. A type of primary cell consisting of a carbon anode surrounded by a fabric bag containing carbon powder mixed with manganese dioxide. The electrolyte is ammonium chloride solution and the cathode is a zinc rod. At the cathode zinc atoms change to zinc ions, leaving a negative charge. Ammonium ions from the electrolyte gain electrons at the anode, leaving a net positive charge, and decompose to nitrogen and hydrogen. The manganese dioxide helps to prevent polarization of the anode by oxidizing the hydrogen. Common dry batteries are types of Leclanché cell: the e.m.f. is about 1.5

volts. [After Georges Leclanché (1839–82), French chemist.]

lemma. A result proved as a preliminary to the proof of a theorem.

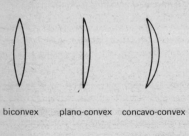

biconvex   plano-convex   concavo-convex

biconcave   plano-concave convexo-concave

Fig. 1 Types of lens

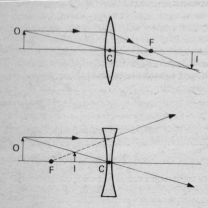

O object

I image

C optical centre

F principal focus

Fig. 2 Ray diagrams

lens. 1. A device for converging or diverging a beam of light by refraction, consisting of a flat piece of transparent material with one or two curved faces. Usually the faces are spherical although other shapes are used for special purposes. Lenses are classified into *converging lenses* (which converge the beam) and *diverging lenses*. Converging lenses are *convex*: i.e. they are thicker at their centre than at their edge. Convex lenses are further classified into *biconvex* lenses, which have both faces curved, *plano-convex* lenses, with one plane face, and *concavo-convex* (or positive *meniscus*) lenses, which have both faces curving in the same direction. Diverging lenses are *concave*: they are thinner at their centre than their edge. Concave lenses are similarly classified into *biconcave*, *plano-concave*, and *convexo-concave* or (negative *meniscus*) lenses. A *thin lens* is one whose thickness is small compared with its *focal length: its properties are described by a simple *lens formula. The formula cannot be applied to a *thick lens*.

The centres of the spheres of which the lens surfaces are part are the *centres of curvature* of the lens. A line through these points is the *optical axis*. A ray of light passing undeviated through the centre of a lens cuts the axis at the *optical centre*. A ray parallel to the axis converges to or appears to diverge from a point on the axis called the *principal focus* or *focal point*. The distance from the optical centre of a thin lens to the principal focus is the *focal length*. See also compound lens, achromatic lens.

2. Any device for diverging or converging a beam. See electron lens.

lens formula. Any equation describing the properties of a lens. The equation $1/v + 1/u = 1/f$, relates the image distance (v), object distance (u), and focal length (f) of a thin lens. Distances are measured from the centre of the lens and the "real is positive" sign convention is used. In the other sign convention, in which distances measured against the light are positive, the plus sign is replaced by a minus. The other equation commonly used for lenses is $1/f = [(n_2 - n_1)/n_1][1/r_1 - 1/r_2]$, where r_1 and r_2 are the radii of the faces and n_2 and n_1 are the refractive indexes of the lens material and of the medium surrounding the lens.

lenticular. Relating to a lens.

Lenz's law. The principle that if the magnetic flux changes with respect to a conductor the current induced in the conductor tends to flow in such a direction as to oppose the change (by producing an opposing field). [After Heinrich Lenz (1804–65), German physicist.]

lepton. Any of a class of *elementary particles that have half-integral spin and take part in *weak interactions. The leptons are *fermions; they include the electron, the muon, and the neutrino, and their antiparticles. It is possible to assign a quantum number (the *lepton number*, l) to each particle, defined to be 1 for the electron, negative muon, and neutrino and –1 for their antiparticles (i.e. for the positron, positive muon, and antineutrino). Particles that are not leptons have lepton number 0. In any reaction the total lepton number is unchanged. For example, in *beta decay: $n \rightarrow p + e + \bar{\nu}$, the neutron has $l = 0$ and the total lepton number of the products, $1 + 0 + (-1)$, is also 0. *Compare* baryon.

Leslie's cube. A large cubic metal container with four of its sides having different finishes or colours, filled with boiling water and used to demonstrate the effect of the surface on the emission of radiant heat. A thermopile or similar detector is used to measure the radiation from each face. [After Sir John Leslie (1766–1832), British physicist.]

level. The ratio of a value of a quantity to a reference value of that quantity. Often a logarithmic scale is used. *See* decibel.

lever. A simple machine consisting of a rigid bar turning about a pivot (the *fulcrum*). There are three types, depending on the relative positions of load, effort, and fulcrum. The mechanical advantage is the ratio of the distance from the load to the fulcrum to the distance from the effort to the fulcrum.

Lewis acid. A compound that can accept a pair of electrons in the formation of a *coordinate bond. According to this extension of the theory of acids and bases the electron acceptor is the acid and the compound donating the lone pair is a *Lewis base*. In the reaction $BCl_3 + :NH_3 = Cl_3B:NH_3$, boron trichloride is the acid and ammonia is the base. [After Gilbert N. Lewis (1875–1946), American chemist.]

Lewis base. *See* Lewis acid.

Lewisite. A colourless volatile liquid, $ClCH:CHAsCl_2$, used as a vesicant war gas. M.pt. –18.2°C; b.pt. 190°C (decomposes); r.d. 1.89. [After Winford Lee Lewis (1878–1943), American chemist.]

Leyden jar. An early form of capacitor, consisting of a glass jar coated with tinfoil on part of its outer and inner surfaces. [After Leyden, the town in Holland where it was invented.]

LF. *See* low frequency.

libration. *See* moon.

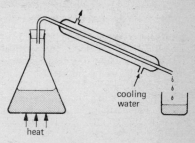

cooling water

heat

Liebig condenser

Liebig condenser. A simple *condenser used in laboratory distillations, consisting of a straight glass tube surrounded by a glass cooling jacket through which water is passed. [After Baron von Liebig (1803–73), German chemist.]

ligand. A molecule, ion, or group that coordinates to a metal atom or ion in a *complex.' See also chelating agent.

light. *Electromagnetic radiation that produces a visual sensation when it strikes the human eye. Light has the wavelength range 400–740 nm. Its velocity, in a vacuum, is $2.997\ 925 \times 10^8$ m s^{-1}. Several theories of light have been used. According to the *corpuscular theory* light consists of small elastic particles emitted by a luminous body. The theory can explain geometrical optics but not interference and polarization. The *wave theory* of light is that light is a wave motion propagated in the ether. Maxwell showed that light is a form of electromagnetic wave motion, consisting of transverse varying electric and magnetic fields. Some phenomena, such as the photoelectric effect, are best explained by considering the light to be streams of *photons.

lightness. The property of a *colour of a pigment, dye, etc., determined by the amount of light it reflects. Colours of the same hue that differ in lightness are *shades*.

lightning. Electrical discharges in the atmosphere, resulting from the formation of regions of positive and negative charge within a cloud. The spark produced can occur to the ground (cloud to ground), between clouds (cloud to cloud), or within a cloud. The potential difference initiating the flash is about 10^8 volts, this being sufficient to cause electrical breakdown of the air. In cloud to ground lightning there is an initial faintly luminous leader stroke passing towards the ground. An upward discharge occurs to meet this leader about 50 metres above the ground, thus inducing the bright return stroke. A typical lightning flash consists of four or five strokes about 40 ms apart, each propagated at about 5×10^7 m s^{-1}. Enormous amounts of energy are dissi-

pated: the return stroke can be 5 km in length dissipating 10^5 joules per metre and currents of 10 000 A and temperatures of 30 000 K are produced. The heating and cooling of the air produces the pressure waves causing thunder. Flashes of this type constitute "forked lightning". "Sheet lightning" is observed when a flash occurs within a cloud, thus lighting up the whole cloud without the channel being visible. *Ball lightning*— the production of a small slowly moving luminous ball of plasma—is the rarest and least understood form of lightning.

lightning conductor. A sharply pointed metal rod attached to the top of a building and connected to earth, used to protect the building from lightning. The electric field induced by an electrically charged cloud has a high gradient in the region of the sharp point. This causes ionization of the air and prevents the build up of the high potential differences necessary for lightning to occur.

light pen. A device connected to the *visual-display unit of a computer. It is capable of sensing the information on the screen and can be used to input information by "drawing" lines on the screen.

light-year. A unit of length equal to the distance travelled by electromagnetic radiation in one year. It is used for expressing astronomical distances and is equal to $9.460\ 7 \times 10^{15}$ metres ($5.878\ 48 \times 10^{12}$ miles).

ligroine. See petroleum ether.

lime. See calcium oxide, calcium hydroxide.

limestone. A rock composed of calcite, $CaCO_3$, used in making carbon dioxide, calcium oxide, cement, and other calcium compounds. It is also used as a flux in smelting metals and as a building stone.

lime water. A clear aqueous solution of calcium hydroxide. It is used as a test for carbon dioxide, which produces a milky precipitate of calcium carbonate.

limit. The value approached by a mathematical function as its independent variable approaches some specified value.

limonite. A dark brown or black mineral consisting of hydrated iron II oxide, $FeO(OH).nH_2O$, used as an ore of iron and a yellow pigment.

linac. See linear accelerator.

lindane. See hexachlorocyclohexane.

Linde process. A process for liquefying air by expansion through a nozzle. The air is compressed by a pump and expanded through a valve into an expansion chamber. The expanding air is cooled as a result of work done against intermolecular forces (see Joule-Kelvin effect). The cooled air is used to cool the incoming compressed air: eventually the temperature falls to the point at which the air is liquefied. [After Carl von Linde (1842-1934), German engineer.]

linear. 1. In a straight line; characterized by one dimension only.
2. Pertaining to a relationship of direct proportionality: i.e. one that would be represented by a straight line on a graph. A *linear equation* is one in which the terms containing the variables are all of the first degree, as in $y = 7x + 5$. A *linear scale* is one in which the intervals are equally spaced. A device, component, or circuit is said to be linear if its output is directly proportional to its input, as in a *linear amplifier*.

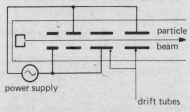

power supply

drift tubes

Linear accelerator

linear accelerator (linac). A type of particle *accelerator in which the particles travel in straight lines along a long

evacuated tube. In low energy machines the particles are accelerated by a series of cylindrical electrodes separated by gaps. Radiofrequency alternating potentials are applied to the electrodes in such a way that there is an accelerating field present when the particles are travelling through the gaps. The lengths and spacing of the electrodes are arranged so that as the particles gain energy in moving along the tube they always reach the inter-electrode gaps at the correct point in time for acceleration to occur. They move at constant velocity inside the cylindrical electrodes (*drift tubes*).

Higher energy linear accelerators do not contain electrodes. Microwave fields are applied and the particles move down the tube accelerated by a travelling wave in a *waveguide. The energy obtained depends on the length, typical energy gains being 7 MeV per metre for electrons and 1.5 MeV per metre for protons. The linear accelerator at Stanford, California, is two miles long and produces particles in the 10-20 GeV range.

line defect. See defect.

line of force. A line in a magnetic or electric *field, whose direction at any point is the direction of the field at that point.

line printer. A device that prints the output of a computer a whole line at a time, rather than printing individual characters. Line printers can work at speeds as high as 3000 lines per minute.

line spectrum. A *spectrum consisting of discrete lines. Line spectra are produced by atoms—the lines correspond to emission or absorption of photons as a result of electron transitions between energy levels. See also spectral series.

linkage. The amount of magnetic flux passing through a coil of wire or other electric circuit.

liquation. The process of separating solid mixtures by heating them to a temperature at which one component melts.

liquefaction. The change of a substance into the liquid state. The term is often applied to the liquefaction of gases. If the substance has a *critical temperature above room temperature, it can be liquefied by pressure alone. Otherwise some cooling process must be used. *See* cascade process, Linde process.

liquid. A state of matter intermediate between a gas and a solid, characterized by ease of flow combined with incompressibility. Thus a liquid, like a gas, will take the shape of its container, offering little resistance to shear stress. Unlike a gas, it will not change its volume to fill the container. The intermolecular forces involved are larger than those in gases and smaller than those in solids, being large enough to prevent spontaneous expansion or significant compression but too small to maintain the order and relative rigidity found in solids. In liquids, molecular order is short range and transient. Theories of liquids are less well developed than theories of solids or gases.

liquid air. A pale blue liquid made by the liquefaction of air and used as a refrigerant. It is mainly liquid oxygen (b.pt. -182.9°C) and liquid nitrogen (b.pt. -195.7°C).

liquid crystal. A state of certain molecules that flow like liquids but have an ordered arrangement of molecules. Substances that form liquid crystals have long molecules and the order results from intermolecular forces. Three types of liquid crystal exist. In *nematic* crystals the molecules are randomly arranged but all their axes are parallel. *Cholesteric* crystals have molecules arranged in layers with their axes all parallel and lying in the plane of the layer. *Smetic crystals* also have molecules arranged in layers; in this case the axes of the molecules are perpendicular to the plane.

liquid-drop model. A model of the atomic *nucleus in which it is visualized as a drop of liquid, with its nucleons behaving like the molecules in a drop of water. The impact of a neutron causes the drop to oscillate and split into two: i.e. nuclear fission occurs. The model can be used to predict fission energies for nuclei.

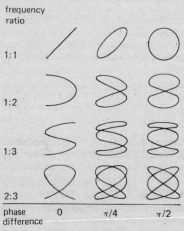

frequency ratio — 1:1, 1:2, 1:3, 2:3

phase difference — 0, π/4, π/2

Lissajous' figures

Lissajous' figures. Figures produced by the path of a point that moves in two dimensions with components that are *simple harmonic motions. The figures can be produced on an oscilloscope by deflecting the spot horizontally with one sinusoidal signal and vertically with another. Two components with the same frequency and phase give a straight line. If they are out of phase, ellipses are formed—a 90° phase difference gives a circle. If the two signals have different phases and frequencies more complicated figures result (see illustration). [After Jules-Antoine Lissajous (1822–80), French mathematician.]

litharge. *See* lead oxide.

lithium. Symbol: Li. A soft silvery metallic element; the lightest solid element. It is made by electrolysis of a fused mixture of lithium chloride and potassium chloride and used in making hydrogenating agents, such as lithium aluminium hydride. Lithium is one of the *alkali metals. A.N. 3; A.W. 6.939; m.pt. 179°C; b.pt. 1317°C; r.d. 0.53; valency 1.

lithium aluminium hydride (LAH). A white or grey powder, $LiAlH_4$, prepared by reacting aluminium chloride with lithium hydride. It is extensively used in preparative organic chemistry as a reducing agent for many functional groups. Decomposes above 130°C; r.d. 0.92.

lithium chloride. A white deliquescent crystalline solid, $LiCl$, used in soldering flux and mineral waters. M.pt. 614°C; b.pt. 1360°C; r.d. 2.07.

lithium hydride. A white crystalline solid, LiH, made by direct combination of the elements at high temperature. It is used as a reducing agent in organic synthesis. M.pt. 680°C; r.d. 0.82.

lithium hydroxide. A soluble basic white solid, $LiOH$, made by the action of calcium hydroxide on lithium salts. M.pt. 450°C; r.d. 1.46.

lithium nitride. A brownish-red powder, Li_3N, used as a nitriding agent in metallurgy and as a reducing agent in organic chemistry. M.pt. 845°C; r.d. 1.3.

litmus. A soluble compound obtained from certain lichens. It has a red colour in acid solutions and a blue colour in alkaline solutions: hence its use as an indicator.

litre. Symbol: l. A metric unit of volume equal to one cubic decimetre. The name is not used in precision measurements. The original definition of the litre, in 1901, was the volume of one kilogram of pure water at the temperature of its maximum density (4°C) and at 760 mmHg pressure, making it equal to 1000.028 cubic centimetres.

liver of sulphur. A brown mixture of potassium polysulphides and potassium thiosulphate made by heating potassium carbonate with sulphur.

lixiviation. The process of separating solid mixtures by washing soluble components out with water.

load. *See* machine.

loaded concrete. Concrete with added amounts of compounds of heavy metals, such as barium or lead. These are efficient absorbers of radiation and the material is used for shielding nuclear reactors.

locus. The curve traced by a point moving so as to satisfy a specified condition. A circle, for example, is the locus of a point that moves so as to be equidistant from a fixed point (the centre).

lodestone. *See* magnetite.

logarithm. The power to which a fixed number (the *base*) must be raised to produce a given number. Thus, if a is the logarithm of c to base b, written $a = \log_b c$, then $c = b^a$. It follows from the laws of *exponents that $\log m + \log n = \log mn$, $\log m - \log n = \log m/n$, and $\log m^n = n \log m$. These rules are used in performing calculations with tabulated values of logarithms of numbers to the base 10 (*common logarithms*). Usually tables only contain the logarithms of numbers between 0 and 10. The logarithm of any number is then found by expressing it as a number between 1 and 10 multiplied by a power of 10. For example, 2570.1 is 2.5701×10^3. $\log_{10} (2.5701 \times 10^3)$ is $\log 2.5701 + 3$, or 3.4099. The integer, in this case 3, is the *characteristic* and the decimal, in this case 0.4099, is the *mantissa*. This system is also used for numbers between 0 and 1. For example, $\log_{10} 0.001670$ is $\log_{10} (1.670 \times 10^{-3}) = 0.2227 - 3$. Usually this is written as $\bar{3}.2227$, where $\bar{3}$ is the mantissa and 0.2227 is the characteristic. *Natural logarithms* use the base e (2.7182...). $\log_e a$ is often written as ln a. They are also called *hyperbolic* or *Naperian* logarithms.

logarithmic. Pertaining to a relationship in which one variable is proportional to the logarithm of another. The scale marked on a slide rule, for example, is logarithmic: the distance of each number from the zero is proportional to the logarithm of the number. Logarithmic scales of this type are found in certain measuring instruments. Some physical measurements are defined by logarithmic relationships. The pH of a solution, for example, is a measure of the reciprocal of the hydrogen-ion concentration, a difference of 1 in pH corresponding to a factor of 10 in concentration.

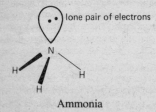

Ammonia

lone pair. A pair of electrons with opposite spins occupying the same orbital in an atom. A lone pair of electrons has a position in space in the same way that a chemical bond is directional. This is used in explaining the shapes of simple molecules. Ammonia (NH_3), for example, is not planar but has a pyramidal structure because of the position of the lone pair (see illustration).

longitude. 1. The angular distance between one of the earth's meridians and the standard meridian through Greenwich.
2. (celestial longitude). The angular distance of a body on the *celestial sphere measured anticlockwise from the first point of Aries (the vernal equinox).
3. See polar coordinates.

longitudinal wave. See wave.

long sight. See hypermetropia.

Lorentz-Fitzgerald contraction. A contraction in length by a factor $(1 - v^2/c^2)$, postulated to occur to a body moving through the ether with a velocity v (c is the velocity of light). The contraction takes place in the direction of the body's motion. The phenomenon was suggested as an explanation of the negative result obtained in the *Michelson-Morley experiment: the dimensions of the interferometer change in the direction of the earth's rotation in such a way that the two light beams take the same time to cover their paths. An explanation of the contraction is given by the theory of relativity. [After Hendrik Antoon Lorentz (1853-1928), Dutch physicist, and George Francis Fitzgerald (1851-1901), Irish physicist.]

Lorentz transformation. A *transformation defined by a set of equations that relate the coordinates of space and time in two frames of reference when one frame moves at a constant velocity relative to the other. They are used in the theory of special *relativity.

Loschmidt's number. Symbol: L. The number of molecules in one cubic centimetre of an ideal gas at standard temperature and pressure. It has the value $2.687\ 19 \times 10^{19}$. [After Joseph Loschmidt (1821-95), Austrian physicist.]

loudness. The sensation that enables a listener to judge the intensity of a sound wave. The loudness is approximately proportional to the logarithm of the sound intensity. See decibel, phon.

loudspeaker. A device for converting changing electric currents into sounds, usually consisting of a small coil that is attached to a cardboard cone and is free to move in the field of a strong permanent magnet. The current produces vibrating motion of the coil, thus vibrating the cone and producing sound. The effect is the opposite of that in the microphone, although loudspeakers work at higher power levels.

low frequency (LF). A frequency in the range 30 kilohertz to 300 kilohertz.

low tension. Low voltage.

lumen. Symbol: lm. An *SI unit of *luminous flux equal to the amount of light emitted per second in a cone of one steradian solid angle by a point source of one candela.

luminance. Symbol: L. The brightness in a particular direction of a surface that is emitting or reflecting light. It is given, at a particular point, by the luminous intensity (I) per unit of area projected onto an area at right angles to the direction. Thus for a uniformly reflecting surface $L = I/A \cos \theta$, where θ is the angle the direction makes with the surface.

luminescence. The emission of light by any mechanism that does not depend on the body having a high temperature: i.e. it is distinguished from *incandescence. Luminescence is the result of transitions of electrons in atoms and ions whereas incandescence is the effect of vibrations of the atoms in a solid. In luminescence, atoms or molecules have to be produced in an excited state. They then change from that state to the ground state and emit light. Many different mechanisms are possible: the excited state may be produced by other electromagnetic radiation (*photoluminescence), by electron bombardment (*electroluminescence), or by chemical or biological reactions (*chemiluminescence and *bioluminescence). Luminescence is also induced by such processes as friction (*triboluminescence), and radioactive decay (*radioluminescence). The term is applied both to the process and to the light emitted and is also used for analogous effects yielding ultraviolet radiation and X-rays. *See also* fluorescence, phosphorescence.

luminosity. 1. The brightness of a source of light.
2. The absolute brightness of a star, depending on the amount of light it emits and independent of the distance from the earth. *See also* magnitude.

luminous flux. The rate of flow of light energy from a source as measured by its visual sensation. The luminous flux is related to the total power emitted corrected according to the sensitivity of the eye to light of different wavelengths.

luminous intensity. Symbol: I. The amount of light emitted per second per unit of solid angle from a point source of light in a given direction. It is measured in candelas. Formerly, the quantity was measured in candles and called *candlepower*.

luminous paint. A type of paint containing a phosphor mixed with a small amount of radioactive material such as radium. Light is emitted by *radioluminescence.

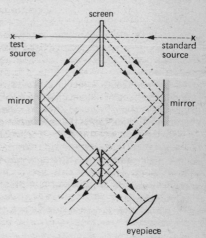

Lummer-Brodhun photometer

Lummer-Brodhun photometer. A type of photometer in which the light source investigated is compared with a standard source using a combination of prisms (see illustration) which allow the illumination from the test source to be seen as a spot in the centre of the illumination from the standard. The positions of the lamps are varied until the field of view is evenly illuminated.

lunar caustic. *See* silver nitrate.

lunar crater. Any of numerous typically saucer-shaped depressions on the surface of the *moon. Their evolution is somewhat mysterious though it is now generally thought that they arose from a combination of meteoroid impacts and lunar volcanic disturbances.

lunar time. Time measured with respect to the *moon. The *lunar month* (which is also called the *lunation* or *synodic month*) is the time between successive new moons. The *lunar year* is defined as 12 lunar months. 1 lunar year = 354.3671 mean solar days. *See* solar time.

lunation. *See* lunar time.

lutetium (cassiopeium). Symbol: Lu. A soft silvery-white metallic element belonging to the *lanthanide series. A.N. 71; A.W. 174.97; m.pt. $1656°C$; b.pt. $3315°C$; r.d. 9.8; valency 3.

lux (metre-candle). Symbol: lx. An *SI unit of intensity of illumination equal to the illumination produced by a flux of one lumen falling uniformly on a surface one square metre in area.

Lyman series. A *spectral series in the ultraviolet emission of hydrogen with wavelengths given by $1/\lambda = R(1/1^2 - 1/m^2)$, where R is the *Rydberg constant and m is 2, 3, 4, etc. [After Theodore Lyman (1874–1954), U.S. physicist.]

lyophilic. Having an affinity for a solvent. *See* sol.

lyophobic. Having no affinity for a solvent. *See* sol.

M

machine. Any of various simple devices in which a small force (the *effort*) is used to overcome a larger force (the *load*). The *velocity ratio* of the machine is the distance moved by the effort divided by the distance moved by the load. The *mechanical advantage* is the load divided by the effort. The *efficiency* is the work done on the load divided by the work done by the effort, often expressed as a percentage. In a perfect machine, with no friction, the efficiency is 1 and the mechanical advantage is equal to the velocity ratio.

Mach number. The ratio of the speed of a fluid or of a body in a fluid to the speed of sound in the fluid. *Mach 1* is the speed of sound in the fluid, *Mach 5* is five times the speed of sound, etc. [After Ernst Mach (1838–1916), German physicist.]

macromolecule. A very large molecule, as found in polymers or in compounds such as haemoglobin.

macroscopic. 1. Large enough to be seen without the use of a microscope.
2. Having properties determined by the statistical behaviour of large numbers of atoms or molecules.

madder. The dyestuff *alizarin, obtained from madder roots.

Magellanic Clouds (Clouds of Magellan, Nubeculae). Two galaxies, generally considered to be satellites of the Milky Way system, visible to the naked eye as conspicuous hazy patches of light in the southern sky. The Larger Cloud (*Nubecula Major*) has a diameter of about 30 000 light-years while the Smaller Cloud (*Nebucula Minor*) is 25 000 light-years in diameter. Both galaxies contain innumerable observable stars (many variable), clusters, and nebulae. The two galaxies are classified as irregular, though the Larger Cloud may be a spiral system like the Milky Way system. Both Clouds lie at a distance of some 120 000 light-years from the sun. [Named after Ferdinand Magellan (1480–1521), Portuguese navigator.]

magic numbers. The numbers 2, 8, 20, 28, 50, 82, and 126. Atomic nuclei that contain these numbers of protons or numbers of neutrons are more stable than other nuclei: elements such as tin and calcium with magic atomic numbers have relatively large numbers of isotopes. The numbers correspond to filled shells in the nucleus. *See* shell model.

magnesia. *See* magnesium oxide.

magnesite. The naturally occurring mineral magnesium carbonate, $MgCO_3$, used in producing magnesium oxide.

magnesium. Symbol: Mg. A silvery malleable metallic element found in several minerals, including carnallite, kainite, magnesite, and dolomite. It is produced by electrolysis of fused magnesium chloride. The metal burns with an intense white flame and is used in pyrotechnics and flash tubes. Other uses include the manufacture of lightweight alloys, the production of some magnesium compounds, such as *Grignard reagents, and the reduction of certain metal oxides to yield the metal. It is an *alkaline-earth metal. A.N. 12; A.W. 24.321; m.pt. 651°C; b.pt. 1107°C; r.d. 1.74; valency 2.

magnesium carbonate. A light white powder, $MgCO_3$, used in heat insulation and as a pigment dentifrice. Decomposes at 350°C; r.d. 3.0.

magnesium chloride. A white deliquescent crystalline solid, $MgCl_2.6H_2O$, extracted from carnallite, sea water, and brines. It is used in the manufacture of metallic magnesium and for fireproofing wood. M.pt. 708°C (anhydrous); b.pt. 1412°C; r.d. 2.3.

magnesium hydroxide. A white powder, $Mg(OH)_2$, manufactured by precipitation from sea water with lime. It is alkaline in water and is used in the refining of sugar and as an antacid. Decomposes at 350°C; r.d. 2.4.

magnesium oxide (magnesia). A white powder, MgO, prepared by heating magnesium carbonate. It is a basic oxide, used in high-temperature refractory materials, electrical insulation,

pharmaceuticals, and cosmetics. M.pt. 2800°C; b.pt. 3600°C; r.d. 3.6.

magnesium sulphate (Epsom salt). A colourless crystalline solid, $MgSO_4.7H_2O$, with a saline bitter taste. It occurs naturally or may be prepared by the action of sulphuric acid on the hydroxide or carbonate. Magnesium sulphate is used in dressing textiles, fireproofing, and as a purgative. Decomposes at 1124°C; r.d. 1.7.

magnet. A device for producing a magnetic field. A permanent magnet is simply a piece of magnetized ferromagnetic material (*see* ferromagnetism). In an *electromagnet the field is produced by a flow of electric current.

magnetic bottle. An arrangement of magnetic fields used to contain the plasma in a controlled thermonuclear reaction.

magnetic circuit. A closed circuit of lines of magnetic flux.

magnetic constant. The constant μ_0 appearing in the equation derived from *Ampere's law for the magnetic flux density produced by an electric current. It arises from the choice of units: in *SI units has the value $4\pi \times 10^{-7}$ henry per metre. It is sometimes called the *permeability of free space*.

magnetic declination. *See* declination.

magnetic dip. *See* dip.

magnetic dipole. A small magnet, either a permanent magnet or a loop carrying an electric current, that can produce a magnetic field and is influenced by the direction of an external magnetic field. In a bar magnet, for example, the field produced comes from two regions near each end—the *poles*. By analogy with an electric *dipole, this can be regarded as a magnetic dipole consisting of two opposite *monopoles separated by a distance. The system will have a *magnetic moment* given by the product of pole strength and distance. In fact, the analogy between magnetic monopoles and isolated electric charges is a weak one because magnetic poles are always found in pairs. It is now more usual to regard the magnetic dipole as the fundamental entity. Its magnetic moment—or *magnetic dipole moment* (symbol: **m**)—is the torque experienced by a magnetic dipole in a field of unit magnetic flux. Its units are weber-metres (Wb m). A small loop of wire carrying a current (*I*) has a magnetic moment given by *IA*, where *A* is the coil's area. This is sometimes called the *electromagnetic moment* and given in units of ampere-metres squared (A m^2), which are equivalent to weber-metres.

magnetic dipole moment. *See* magnetic dipole.

magnetic equator (aclinic line). A line around the earth close to and crossing the true equator, on which all points have zero *dip.

magnetic field. A field characterized by the fact that a small loop of wire carrying a current will experience a force in the field. A magnetic field can be produced by a current flowing in a conductor or by a permanent magnet. A field is characterized by its *magnetic flux density, *B*, or by a *magnetic field strength, *H*.

magnetic field strength. Symbol: *H*. A measure of magnetic field, equivalent to the part of the total field that is not caused by magnetization of the material. It is defined as $H = B/\mu_0 - M$, where *B* is the magnetic flux density, μ_0 the magnetic constant, and *M* the *magnetization. Magnetic field strength was formerly called *magnetic intensity*. It is measured in amperes per metre.

magnetic flux. Symbol: ϕ. The total magnetic field produced by a current in a wire or coil or by a permanent magnet. It is simply the product of the magnetic flux density and the area. The unit is the weber.

magnetic flux density. Symbol: *B*. A measure of the magnetic field at a point by the effect it has on a current-carrying conductor. A small loop of wire carrying a current has a magnetic moment **m** directed along its axis (*see* magnetic dipole). The value of *B* is given by $T = m \times B$, where *T* is the torque produced on the coil when its axis is perpendicular to the field direction. The magnetic flux density is sometimes thought of as the number of lines of force per unit area. It was formerly called *magnetic induction*. The unit is the tesla.

magnetic hysteresis. *See* hysteresis.

magnetic induction. 1. The magnetization of a ferromagnetic material in a magnetic field. An unmagnetized bar of iron, for example, placed close to a bar magnet will itself act like a bar magnet with a north pole close to the south pole of the original magnet (or vice versa) and be attracted to the original magnet. The effect is caused by alignment of magnetic dipoles within the material by the external field. *See also* induction.
2. *See* magnetic flux density.

magnetic intensity. *See* magnetic field strength.

magnetic meridian. One of a set of lines on the earth's surface joining points that have the same *horizontal component of field.

magnetic mirror. A strong magnetic field with a particular configuration suitable for reversing the direction of charged particles. Such fields are used in *magnetic bottles.

magnetic moment. *See* magnetic dipole.

magnetic monopole. A hypothetical particle that would be an isolated north or south magnetic pole. The particle has been postulated by analogy with charged particles. Despite many attempts to detect magnetic monopoles no evidence for their existence has yet been found.

magnetic permeability. *See* relative permeability.

magnetic pole. One of the regions near the ends of a permanent magnet into which the lines of force of the field converge or from which they diverge. If a magnet is freely suspended it will come to rest along the earth's lines of force with its *north pole* (or north-seeking pole) pointing north and its *south pole* (or south-seeking pole) pointing south. The earth's *magnetic North Pole* and *magnetic South Pole* are the points at which the earth's magnetic flux is strongest (*see* terrestrial magnetism). Unlike magnetic poles attract one another and like poles repel in the same way that opposite electric charges attract and similar charges repel. By analogy, it is possible to make use of the idea of *magnetic pole strength* (symbol: *P*). In *electromagnetic units, electrical measurements are based on the concept of a unit magnetic pole. In SI units the pole strength is defined by means of a magnetic dipole (*see* magnetic moment). If a magnetic pole of strength *P* is situated a distance *l* away from an unlike pole of equal magnitude, the magnetic dipole moment is given by $m = P l$. A law of forces between poles can be obtained: $F = \mu_0 P_1 P_2 / 4\pi r^2$, where *r* is the distance between poles of strength P_1 and P_2 and μ_0 is the *magnetic constant.

magnetic quantum number. *See* atom.

magnetic storm. Disturbances of the earth's magnetic field caused by electrical disturbances resulting from sunspot activity.

magnetic susceptibility. *See* susceptibility.

magnetic tape. Plastic tape coated with a layer of iron oxide, which can be magnetized to record information for computers or to record sound in tape recorders.

magnetic variation. *See* declination.

magnetism. The phenomenon produced by magnetic fields, involving interaction at a distance caused by the motion or *spin of charged particles. Several different types of magnetic behaviour are distinguished, depending on the *susceptibility of the material and its

temperature dependence. *See* diamagnetism, paramagnetism, ferromagnetism, antiferromagnetism, ferrimagnetism.

magnetite (lodestone). A black magnetic mineral consisting of ferroferric oxide, Fe_3O_4. It is used as an ore of iron.

magnetization. Symbol: *M*. The effect of an external magnetic field on a sample of material, in which the sample acts as a magnet itself and changes the value of the external field. If the field in free space has a magnetic flux density B_0 and a sample is introduced, then the flux density is changed to *B*, where $B = B_0 + B_M$. B_M is an extra contribution arising in the material and resulting from the alignment of small *magnetic dipoles. The magnetization can be defined as the *magnetic dipole moment per unit volume. It is related to the magnetic field strength (*H*) by the equation $M = B/\mu_0 - H$, where μ_0 is the magnetic constant.

magnetohydrodynamics (MHD). The study of the motion of electrically conducting fluids and their behaviour in magnetic fields.

magnetomotive force (m.m.f.). Symbol: F_m. The integral of magnetic field strength around a closed path $\oint H dx$. Magnetomotive force is analogous to electromotive force.

magneton. A fundamental unit of magnetic dipole moment. The moment of an electron in its orbit in an atom is quantized in values of $eh/4\pi m$, where *e* is the electron charge, *m* is its mass, and *h* is the Planck constant. This is called the *Bohr magneton* (symbol: m_B): its value is $9.274\ 096 \times 10^{-24}$ A m^2. The Bohr magneton is also used for electron spin. The *nuclear magneton*, m_N, is given by $m_e m_B / m_p$, where m_e is the electron mass and m_p is the proton mass. Its value is 5.0503×10^{-27} A m^2.

magnetron. An electron tube for producing microwaves. It contains a hot cathode surrounded by a coaxial cylindrical anode. The device is placed in a magnetic field which causes the electrons emitted by the cathode to spiral around and induce radiofrequency fields in cavities in the anode.

magnification. A measure of the performance of a microscope, telescope, or other optical instrument equal to the ratio of the size of the final image to the size of the object as seen with the unaided eye.

magnitude. The brightness of a celestial object as measured on a scale of numerical values. In ancient times orders of brightness were established for naked-eye objects and Ptolemy fixed the brightest stars as being of the first magnitude and the faintest as the sixth magnitude. In modern times the scale has been fixed so that a difference of five magnitudes gives a brightness ratio of $100:1$. Thus, a star of magnitude m is the fifth root of one hundred (2.512) times brighter than one of magnitude $m + 1$. The largest telescopes have been used to photograph objects as faint as magnitude 22.

The *apparent magnitude* is the observed brightness established by eye (*visual magnitude*) or by photography (*photographic magnitude*). The *absolute magnitude* is the apparent magnitude that a star would have if it were placed at a standard distance of 10 parsecs (32.6 light-years) away from the earth. The absolute magnitude measures the intrinsic brightness or luminosity of a star.

main-sequence star. A star that lies along the narrow diagonal distribution band (the main sequence) on the *Hertzsprung-Russell diagram. The vast majority of observed stars, including the sun, are main-sequence stars.

major axis. *See* ellipse.

majority carrier. The type of carrier responsible for transporting more than one half of the current in a *semiconductor.

malachite. A green basic carbonate of copper, $CuCO_3.Cu(OH)_2$, occurring

naturally and used as an ore and a pigment.

malate. Any salt or ester of malic acid.

maleate. Any salt or ester of maleic acid.

maleic acid (*cis*-butenedioic acid). A colourless crystalline toxic carboxylic acid, $HCOOCH:CHCOOH$, with an astringent taste, used in dyeing and in the manufacture of synthetic resins. Maleic and fumaric acids are *geometrical isomers. M.pt. $130°C$; r.d. 1.59.

Maleic anhydride

maleic anhydride. A colourless crystalline solid, $C_4H_2O_3$, made by the catalytic oxidation of benzene. It is used in the manufacture of polyester resins and dyestuffs. M.pt. $53°C$; b.pt. $202°C$; r.d. 1.5.

malic acid. A colourless crystalline optically active solid, $HOOCCH_2CH(OH)COOH$, occurring naturally in apples and many other fruits. It is used as a food additive and chelating agent. M.pt. $99°C$ (*l*-); b.pt. $140°C$ (decomposes); r.d. 1.6.

malonic acid (propanedioic acid). A colourless crystalline dicarboxylic acid, $CH_2(COOH)_2$, found in sugar beet. M.pt. $135.6°C$; decomposes at $140°C$; r.d. 1.62.

Maltose

maltose (malt sugar). A colourless crystalline sugar, $C_{12}H_{22}O_{11}$, made by the action of enzymes on starch. It is used as a nutrient and sweetener. M.pt. 1-2°C.

manganate. A salt containing the ion MnO_4^{2-}. Manganates are derived from the hypothetical *manganic acid*, H_2MnO_4. The manganate ion is dark green. It is found in the crystalline sodium and potassium salts and exists in strongly basic solutions. In acid or neutral solution it changes to the *permanganate ion.

manganese. Symbol: Mn. A grey-white hard brittle *transition element found in many ores, especially pyrolusite (MnO_2) and rhodochrosite ($MnCO_3$). It is extracted by reduction of the oxide with magnesium or aluminium. Manganese is a ferromagnetic metal and is used in producing ferromagnetic alloys as well as in many important steels. The element is fairly electropositive, reacting with many nonmetals at higher temperatures and dissolving in nonoxidizing acids. It forms manganese II (*manganous*) and the less stable manganese III (*manganic*) compounds, including ionic salts and many complexes. It also exists in higher oxidation states, as in the *manganates and *permanganates. A.N. 25; A.W. 54.938; m.pt. 1247°C; b.pt. 2097°C; r.d. 7.21; valency 1-7.

manganese dioxide. *See* manganese oxide.

manganese oxide. Any of five oxides of manganese. *Manganese II oxide* (manganous oxide) is a green powder, MnO, formed by reducing higher oxides with hydrogen. It is a basic substance. M.pt. 1650°C; r.d. 5.4. *Manganese IV oxide* (manganese dioxide) is a black powder occurring naturally as pyrolusite. It is used in making manganese steel and as an oxidizing agent and a depolarizer in the Leclanché cell. R.d. 5.03. Above 535°C it decomposes to *manganese III oxide* (manganic oxide), Mn_2O_3. The other oxides of manganese are Mn_3O_4 (trimanganese tetroxide) and Mn_2O_7

(manganese heptoxide), which is the anhydride of permanganic acid.

manganic. *See* manganese.

manganic acid. *See* manganate.

manganin. An alloy of copper (70-85%), manganese (15-25%), and nickel. Its high resistivity and low dependence of resistance on temperature make it suitable for use in resistors.

manganous. *See* manganese.

manometer. Any device used for measuring pressure. A simple type consists of a U-shaped glass tube containing mercury. One arm is open to the atmosphere and the other is connected to the system to be measured: the pressure difference is obtained from the difference in heights of the mercury columns in each arm of the tube.

mantissa. *See* logarithm.

marble. A compact form of limestone.

Mars. The fourth planet in the solar system in outward succession from the sun. A characteristically red-coloured planet with polar ice caps that recede and advance with the Martian seasons, it lies at a mean distance of 227 800 000 kilometres from the sun, orbiting it at an average velocity of 24.1 km s⁻¹ once every 687 earth days. Its period of axial rotation is 24 hours 39 minutes. Mars is 6786 km in diameter and has a mean relative density of 3.96. Its gravity is about one third and its mass about one tenth those of the earth. Mars' orange-red surface has dark markings on it and is cratered like the surface of the moon. Its tenuous atmosphere is composed mainly of carbon dioxide and trace amounts of water vapour and carbon monoxide. The pressure at the surface is about 0.01 atmosphere. The polar caps are probably frozen carbon dioxide, though they might have cores of frozen water. Daytime temperatures on Mars reach -40°C at the equator, dropping to -70° at night. At its closest, Mars is less than 56 million km from earth. The eccentricity of Mars' orbit is 0.093 and its inclination is 1.85° to the plane of the ecliptic. It has two satellites, *Phobos* and *Deimos*.

marsh gas. *See* methane.

Marsh's test. An analytical test for arsenic. The sample is mixed with hydrochloric acid and zinc and any arsenic present is reduced to arsine. The gas evolved is passed through a hot tube where the arsine is decomposed to arsenic, which condenses as a brown deposit on the cool part of the tube. Antimony gives a similar result but can be distinguished by the fact that it is insoluble in sodium hypochlorite.

martensite. A solid solution of carbon or iron carbide in alpha iron. It is the main constituent of many *steels, formed from *austenite by rapid quenching from high temperatures. *See also* pearlite.

mascon. One of a number of localized regions of high gravity found on the moon.

maser. A type of amplifier for producing intense monochromatic coherent highly directional beams of microwaves by stimulated emission. Masers work on the same principle as the *laser: the name is an acronym for *microwave amplification by stimulated emission of radiation*.

mass. Symbol: m. The property of a body that determines the acceleration that would be produced by application of a force. According to Newton's law the acceleration is proportional to the force applied: the mass of the body is the constant of proportionality. Thus the mass is a measure of the body's tendency to resist a change in motion: i.e. its *inertia. The quantity is sometimes called the *inertial mass*.
Mass is also the property that determines the mutual attraction of two bodies by gravitational interaction (*see* gravitation). Thus, it is the property determining the weight of the object: the *gravitational mass*, which is

equivalent to the inertial mass. The unit of mass is the kilogram.
The *law of conservation of mass* states that the total mass of a system is constant no matter what changes occur in the system. This principle has been modified by the theory of *relativity in which matter is regarded as a form of energy. *See* mass-energy.

mass action, law of. The principle that the rate of a chemical reaction is proportional to the active masses (concentrations) of the reacting substances. *See* equilibrium constant.

mass defect. The difference between the sum of the masses of the individual nucleons in an atomic nucleus and the mass of the nucleus. This difference in mass is equivalent to the energy holding the nucleons together.

mass-energy. Mass and energy considered as interconvertible, connected by Einstein's equation $E = mc^2$, where c is the velocity of light. Matter can be created or destroyed with corresponding decrease or increase of the energy. Thus the laws of conservation of *mass and of *energy are not strictly true. However, the total energy, calculated by considering matter as a form of energy, *is* conserved. This is known as the *law of conservation of mass-energy*. *See also* relativity, fission, fusion.

massicot. *See* lead oxide.

mass number (nucleon number). Symbol: A. The number of nucleons in the nucleus of an atom.

mass spectrometer. An instrument for producing and identifying ions by deflecting them with magnetic or electric fields. Several types exist. In the simplest a gaseous sample at low pressure is ionized by a beam of electrons and the ions produced are accelerated into the evacuated analyser by an electric field. Here the moving ions are deflected into circular paths by a magnetic field at right angles to their direction of motion. The magnetic or electric field can be continuously varied and successive ions are focused onto a detector: the deflection of an ion depends on its charge-to-mass ratio. A *mass spectrum* is produced consisting of a series of peaks, each corresponding to a different fragment ion. The mass spectrum of a compound can be used to find its formula and chemical structure. The technique is also an accurate method of measuring atomic weights.

matrix. 1. The continuous phase in a heterogeneous solid in which other phases are contained.
2. A mathematical entity consisting of an array of numbers in rows and columns. Matrices obey certain defined mathematical rules. They can be used in solving sets of simultaneous equations.

matte. A mixture of metal sulphides obtained at an intermediate stage in smelting sulphide ores. Copper matte contains a mixture of iron and copper sulphides.

maximum and minimum thermometer. A type of thermometer designed to record the maximum and minimum temperatures reached during a period of time. Two small steel spring-loaded indicators are pushed along the tube by the mercury column and remain in position when the mercury recedes. The instrument is reset by moving the indicators with a magnet.

maxwell. Symbol: Mx. A CGS-electromagnetic unit of magnetic flux equal to a flux of one gauss per square centimetre. One maxwell is equivalent to 10^{-8} weber. [After James Clerk Maxwell (1831–79), Scottish physicist.]

Maxwell's demon. A small imaginary creature who, according to Maxwell, could separate a gas at uniform temperature into a hot region and a colder region by opening and closing a shutter to select the molecules with higher kinetic energy. Such a process would violate the second law of thermodynamics.

Maxwell's equations. Four general differential equations for the varying electric and magnetic fields at a point. The equations are curl $H = \partial D/\partial t + j$; div $B = O$; curl $E = -\partial B/\partial t$; and div $D = \rho$. H is the magnetic field strength, D the displacement, j the current density, E the electric field strength, and ρ the volume charge density. Maxwell showed that electromagnetic waves propagated through free space at a velocity equal to that of light and deduced that light was an electromagnetic wave.

mean. The average value of a set of numbers. *See also* geometric mean.

mean free path. The average distance moved by a molecule in a gas between collisions. It is equal to $1/(2^{1/2}\pi nd^2)$, where n is the number of molecules in unit volume and d is the molecular diameter.

mean life. Symbol: τ. The average lifetime of a nucleus of a radioactive isotope. *See* radioactivity.

mechanical advantage. *See* machine.

mechanical equivalent of heat. Symbol: J. The amount of mechanical energy equivalent to unit amount of thermal energy. It has the value 4.1855 joules per calorie.

mechanism. The complete set of changes occurring in a chemical reaction. In general, a reaction proceeds by a series of simple steps involving the producing of stable molecules, ions, free radicals, or other intermediates, some of which may have only a transient existence. A detailed description of the mechanism also involves specification of the *transition states of the steps.

medium frequency (MF). A frequency in the range 0.3 megahertz to 3 megahertz.

melamine. A white crystalline solid, $C_3N_6H_6$, used together with formaldehyde in producing synthetic resins. M.pt. 354°C (decomposes); r.d. 1.57.

Melamine

melamine resin. A synthetic *amino resin made by reacting melamine with formaldehyde. The process yields a water-soluble syrup which is used in laminates, adhesives, and textile treatment. It can be dried to a powder which is mixed with filler and used to produce the thermosetting plastic. This is hard, strong, and white in colour. It is used in making many moulded household articles and in strong moisture-resistant surface coatings.

melitose. See raffinose.

melting. The process of changing from a solid to a liquid as the temperature is raised. Crystalline solids melt at a fixed temperature, their *melting point*, which is equal to their freezing point.

mendelevium. Symbol: Md. A transuranic *actinide element obtained in trace amounts by bombarding einsteinium with alpha particles. The isotope produced, ^{256}Md, has a half-life of about 1.5 hours. A.N. 101. [After D. I. Mendeleev (1834-1907), Russian chemist who first formulated the periodic table.]

meniscus lens. See lens.

Menthol

menthol. A white crystalline solid, $C_{10}H_{20}O$, with a strong minty taste. It is a terpene alcohol, present in peppermint oil, used in flavourings and perfumes. M.pt. 42°C (*l*-); b.pt. 212°C; r.d. 0.89.

mercaptan. Any of a class of organic compounds of general formula RSH. Mercaptans are analogous to alcohols, with the oxygen atom replaced by sulphur.

mercuric. See mercury.

mercurous. See mercury.

mercury (quicksilver). Symbol: Hg. A heavy silvery liquid metallic element found in cinnabar (HgS), from which it is obtained by heating and condensing the vapour. It is used in mercury-vapour lamps, dental amalgams, thermometers, barometers, diffusion pumps, and other pieces of scientific apparatus. Mercury is a fairly reactive element. Like zinc, it combines with oxygen and other non-metals: however, it is not attacked by nonoxidizing acids. It forms alloys (*amalgams) with most other metals.
The element has two series of ionic compounds. Mercury II (*mercuric*) salts are the more stable, containing the Hg^{2+} ion. They can be reduced to mercury I (*mercurous*) salts, which contain not the Hg^+ ion but the binuclear $^+Hg-Hg^+$ ion. Thus, mercurous fluoride is Hg_2F_2 (not HgF). Mercury also forms covalently bonded compounds, notably organometallic compounds. A.N. 80; A.W. 200.59; m.pt. -38.87°C; b.pt. 356.58°C; r.d. 13.55; valency 2.

Mercury. The closest planet to the sun in the solar system lying at a mean distance of 58 million kilometres from the sun. It is the smallest of the planets, comparable in size with the *moon, and having an equatorial diameter of about 4870 km, a mass only one twentieth that of the earth, and a mean relative density of 5.2. Mercury is probably a rocky planet with little or no atmosphere—some traces of hydrogen and carbon dioxide have apparently been detected—and daytime temperatures that reach 400°C. Its orbital velocity averages 48 km per second and its period of revolution is 87.969 earth days. Its period of axial rotation, originally thought to equal its period of revolution, is about 58 or 59 days. Being the innermost of the planets, Mercury is very difficult to observe, and can only ever be seen as a morning or evening star; its maximum elongation is only $27°45'$ and it never appears high above the horizon.

mercury barometer. See barometer.

mercury chloride. Either of two chlorides of mercury. Both are white powders. *Mercury II chloride* (mercuric chloride), $HgCl_2$, is obtained by direct combination of the elements. It is widely used in making other mercury compounds. M.pt. $276°C$; b.pt. $303°C$; r.d. 5.4 ($25°C$). *Mercury I chloride* (mercurous chloride), Hg_2Cl_2, is prepared by heating mercury with mercury II chloride. It is used in medicine and pyrotechnics and as a fungicide. M.pt. $302°C$; b.pt. $384°C$; r.d. 7.0.

mercury fulminate. A grey crystalline powder, $Hg(CNO)_2$, made by treating mercury with nitric acid and ethanol. It explodes under slight friction or shock and is used in detonators. R.d. 4.4.

mercury oxide. Any of two basic oxides of mercury. *Mercury I oxide* (mercurous oxide) is a black powder, Hg_2O. *Mercury II oxide* (mercuric oxide) is a red or yellow powder, HgO, depending on the particle size. The red form is obtained by heating mercury I nitrate, while the finer yellow form is precipitated by adding an alkali to a solution of a mercury II salt. The compound is used in medicine and paint pigments.

mercury thermometer. A type of *thermometer consisting of a small glass bulb containing mercury, connected to a fine sealed capillary tube. The temperature is measured by the expansion of the mercury.

meridian. Any great circle on the earth's surface passing through both geographical poles.

mesitylene (1,3,5-trimethylbenzene). A colourless liquid, $C_6H_3(CH_3)_3$, extracted from coal tar or petroleum and used as an intermediate in the manufacture of dyes. M.pt. $-52.7°C$; r.d. 0.86.

meso form. See optical isomerism.

meson. Any member of a class of *elementary particles characterized by a mass intermediate between those of the electron and the proton, an integral *spin, and participation in *strong interactions. Mesons are responsible for the forces between nucleons in the *nucleus. They are all unstable particles, undergoing decay in a variety of ways. The particles can have positive and negative charge or can be electrically neutral. Positive and negative mesons are *antiparticles of each other: their charge is the same magnitude as that of the electron. The three types of meson are the *pion, the *kaon, and η-mesons. The muon (or mu-meson) was formerly classified as a meson but is in fact a *lepton.

meta. 1. Indicating that substituents in a benzene ring are attached to carbon atoms separated by one carbon atom: i.e. to the 1,3 positions. See structural isomerism.
2. Denoting the least hydrated acid of a series of acids formed from the same anhydride.

metal. 1. Any of a class of chemical elements that are typically lustrous malleable solids that conduct heat and electricity. There are some exceptions—mercury is a liquid and some metals are brittle. Chemically, metals are electropositive: they have basic oxides and hydroxides and tend to form ionic compounds containing positive ions. In fact, many common metals do not show this behaviour. Tin, for example, also forms covalent compounds, including SnH_4, and has an amphoteric oxide that dissolves in alkalis to form stannates. See also metalloid, nonmetal.
2. An alloy of a metal with one or more other elements.

metaldehyde. A white crystalline polymer of *acetaldehyde, $(CH_3CHO)_n$, where n is either 4 or 6. It is made by acid-catalysed polymerization below 0°C and is used in firelighters and slug killers. Sublimes at 112°C.

metallic bond. The type of chemical bond occurring in metallic crystals. In metallic bonding a regular lattice of positive ions is held together by a cloud of free electrons, which can move freely through the lattice.

metallocene. A *sandwich compound formed between a metal atom and two cyclopentadienyl ions, as in ferrocene.

metalloid. A chemical element with properties characteristic both of *metals and *nonmetals. The term is often used for elements that are difficult to classify in one or other group. Germanium, for example, has a metallic appearance, semiconductivity, and many of the chemical properties of nonmetals. Arsenic has a metallic allotrope and two nonmetallic allotropes. Other metalloids include silicon, tellurium, and antimony.

metallurgy. The study of metals, including industrial processes for smelting, refining, and working metals, and the study of the structure and properties of metals and alloys.

metaphosphoric acid. See phosphoric acid.

metasilicate. See silicate.

metasilicic acid. See silicic acid.

metastable. 1. See equilibrium.
2. Designating an excited atom, molecule, ion, etc., with a relatively long lifetime.

metathesis. See double decomposition.

metavanadate. See vanadate.

metavanadic acid. See vanadate.

meteor (shooting star). The spectacular streak of light caused by the trail of a small meteoroid as it burns up while passing through the earth's atmosphere. When the earth passes through the remains of a comet, there is a considerable increase in the number of meteors. One of the best showers, the *Perseids*, is visible every year at the beginning of August.

meteorite. A large *meteoroid that enters the earth's atmosphere, becoming incandescent as a *bolide (or fireball), and succeeds in reaching the ground. Usually of stone or metal, meteorites range in mass from a few grams to several thousand kilograms. The largest so far recorded was a meteorite found at Grootfontein, South Africa, weighing 65 000 kg.

meteoroid. Any of innumerable extraterrestrial pieces of matter found orbiting the sun and ranging in size from tiny particles to large—sometimes very large—objects. See also meteor, meteorite.

methacrylic acid (2-methylpropenoic acid). A colourless liquid, $CH_2{:}C(CH_3)$-COOH, with an acrid odour, used in the manufacture of acrylic resins. M.pt. 16°C; b.pt. 163°C; r.d. 1.01.

methanal. See formaldehyde.

methane (marsh gas). A colourless odourless tasteless flammable gas, CH_4, that is the chief component of most natural gas. It is the first member of the *alkane series and is used as a fuel and a source of petrochemicals. M.pt. -182.5°C; b.pt. -161°C; r.d. 0.55 (air = 1, 0°C).

methanethiol. A colourless flammable liquid, CH_3SH, with a very powerful unpleasant odour, prepared by reacting methanol with hydrogen sulphide. It is used as a fuel additive and fungicide. M.pt. -121°C; b.pt. 6°C; r.d. 0.87.

methanide. A *carbide that yields methane on hydrolysis.

methanoic acid. See formic acid.

methanol (methyl alcohol, wood alcohol). A colourless poisonous flammable liquid, CH_3OH, usually manufactured by high-pressure catalytic synthesis from hydrogen and carbon monoxide. A typical alcohol, it is used as a solvent, antifreeze, and raw material for the manufacture of other chemicals. Methanol was formerly also called *carbinol*. M.pt. -98°C; b.pt. 64°C; r.d. 0.79.

method of mixtures. A simple method of determining specific heat capacities and latent heats by mixing known amounts of substances (liquids or liquids and solids) at different temperatures and determining the final temperature of the mixture.

methoxybenzene. *See* anisole.

methoxyl group. The organic group CH_3O-.

methyl acetate. A colourless flammable liquid ester, CH_3COOCH_3, with a fragrant odour, used as a solvent for paints and lacquers. M.pt. -98°C; b.pt. 54°C; r.d. 0.92.

methyl alcohol. *See* methanol.

methylamine. A flammable colourless gaseous amine, CH_3NH_2, with a strong ammoniacal odour. It is prepared by reacting methanol and ammonia at high temperature with a catalyst and is an intermediate in preparing dyes, pharmaceuticals, and insecticides. M.pt. -92°C; b.pt. -7°C; r.d. 0.69 (-10.8°C).

methylated spirits. Ethanol denatured by addition of about 9.5% methanol and 0.5% pyridine, with small amounts of blue dye. It is used as a solvent and fuel.

methylation. A chemical reaction in which a methyl group is introduced into a molecule.

methyl bromide. *See* bromomethane.

methyl cellulose. A grey-white powder made by treating cellulose wth an alkali followed by methylation. It swells in water forming a viscous colloidal solution and is used as an adhesive and a thickening and emulsifying agent.

methyl chloride. *See* chloromethane.

methyl cyanide. *See* acetonitrile.

methylene chloride. *See* dichloromethane.

methylene group. The divalent group $H_2C=$.

methyl ethyl ketone. *See* butanone.

methyl group. The organic group CH_3-.

methyl iodide. *See* iodomethane.

methyl methacrylate. A colourless volatile liquid ester, $CH_2:C(CH_3)COOH$. It polymerizes readily forming acrylic resins such as Perspex. M.pt. -48°C; b.pt. 101°C; r.d. 0.94.

methyl orange. An orange powder, $C_{14}H_{14}N_3NaO_3S$, used as an acid-base indicator. It changes colour in the pH range 3.1-4.4, being red below 3.1 and yellow above 4.4.

methyl red. A dark red powder, $C_{15}H_{15}N_3O_2$, used as an acid-base indicator. It changes colour in the pH range 4.4-6.0, being red below 4.4 and yellow above 6.0.

methyl salicylate. A colourless, yellow, or reddish oily ester, $C_6H_4(OH)(COOCH)_3$, with an odour of wintergreen. It is found in some essential oils and is used in perfumes and flavourings. M.pt. -8.3°C; b.pt. 222°C; r.d. 1.18.

metre. Symbol: m. The basic SI unit of length, equal to the length of 1.650 763 73 wavelengths of the radiation from a transition between the $2p_{10}$ and $5d_5$ levels of a krypton-86 atom. The light is produced by a specified type of discharge lamp operating at the triple point of nitrogen (63.15 K).
The metre was originally defined, in 1791, as one ten-millionth of the length of the quadrant of the earth's meridian passing through Paris. This was redefined in 1927 as the distance between two marks on a platinum-iridium bar kept at the International Bureau of Weights and Measures at Sèvres near

Paris. The present definition was introduced in 1960.

metre-candle. *See* lux.

metric system. A system of weights and measures originally based on the metre, the litre, and the kilogram.

metric ton. *See* tonne.

MeV. Million electron volts: 10^6 electronvolts.

MF *See* medium frequency.

MHD. *See* magnetohydrodynamics.

mho. *See* siemens.

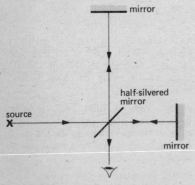

Michelson-Morley interferometer

Michelson-Morley experiment. An experiment designed to measure the velocity of the earth through the ether. The apparatus in the diagram was used. Light from a monochromatic source is split at a half-silvered mirror into two beams at right angles. The beams are reflected back along their path and recombined on reflection at the other side of the mirror, thus producing fringes as a result of *interference.

If the earth is moving through a stationary ether the light would have different velocities when measured in the direction of the earth's rotation and when measured perpendicular to this direction. In this case, rotation of the apparatus through 90° would cause a shift in position of the interference fringes. In fact, no change could be

detected in any position or at any time. The experiment was one of the observations leading to the theory of *relativity. [After Albert A. Michelson (1852–1931) and Edward Morley (1838–1923), American physicists.]

microbalance. An extremely sensitive balance.

micron. Symbol: μm. A unit of length equal to 10^{-6} metre. In SI units it is called the micrometre.

microscope. An optical instrument for producing a magnified image of a small object. A *simple microscope* is a convex lens used as a magnifying glass. A *compound microscope* has a combination of lenses: an objective of short focal length to produce a real image that is viewed by the eyepiece.

microscopic. 1. Visible only with the aid of a microscope.
2. Relating to the behaviour of individual atoms or molecules rather than matter in bulk. *Compare* macroscopic.

microwave background. An isotropic distribution of microwave radiation throughout the universe. The background seems to be black-body radiation corresponding to a temperature of 2.7 kelvin, and it may be a relic of the big bang that was the origin of the universe.

microwaves. *Electromagnetic radiation with wavelengths in the range 1 mm–0.3 m. Microwaves lie in the region between infrared radiation and radio waves.

microwave spectrum. An emission or absorption *spectrum in the microwave region. Microwave radiation is absorbed by transitions between different rotational energy levels of molecules and gives information on the moments of inertia and dimensions of molecules.

mil (thou). 1. One thousandth of an inch.
2. One thousandth of a litre.

3. *See* circular mil.

mile. *See Appendix.*

milk sugar. *See* lactose.

Milky Way. The broad glowing band of faint stars, very conspicuous on a dark moonless night, that arches across the sky in a great circle. It is an effect caused by looking along the equatorial plane of our Galaxy, where the stars appear to be more densely packed. The *Milky-Way system* is another name for the *Galaxy.

minium. *See* lead oxide.

minor axis. *See* ellipse.

minority carrier. The type of carrier responsible for transporting less than one half of the current in a *semiconductor.

minor planet. *See* asteroid.

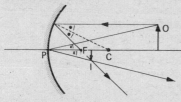

convex

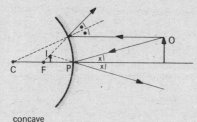

concave

O object

I image   C centre

P pole   F principal focus

Ray diagrams

mirror. A surface that reflects large amounts of light. A *plane mirror* forms a virtual image with the same size as the object. In a plane mirror the image has *lateral inversion and is as far behind the mirror as the object is in front. Curved mirrors are usually spherical, being either *convex* or *concave* depending on whether the reflecting surface lies on the convex or concave side of the sphere. The *centre of curvature* of the mirror is simply the centre point of the sphere of which the reflecting surface forms a part. A line from this to the centre point of the mirror is the *optical axis* and the centre point itself is the *pole*. In a concave mirror, rays of light parallel to the axis are focused to a point on the axis: in a convex mirror they appear to diverge from a point on the axis (see illustration). The point, in both cases, is the mirror's *principal focus* or *focal point*. The distance from the pole to this point is the *focal length*. Spherical mirrors form real or virtual images according to the equation $1/v + 1/u = 1/f = 2/r$, where u is the object distance, v the image distance, f the focal length, and r the radius of the sphere. All distances are measured from the pole, being taken as positive when in front of the mirror and negative when behind it. Spherical mirrors suffer from *spherical aberration and in optical telescopes *parabolic mirrors are used.

mirror image. An image of an object as it would appear in a plane mirror. If one object is the mirror image of the other, it is identical in all respects except that the two could not be superimposed. A left hand is the mirror image of a right hand.

Misch metal. An alloy of cerium (about 25%) and various other lanthanide elements. It is a pyrophoric material, hence its use in lighter flints.

miscible. Denoting two or more liquids that can be mixed together completely to form a homogeneous mixture.

mispickel. *See* arsenopyrite.

mist. A fine suspension of a liquid in a gas.

Mitscherlich's law. The principle that substances that crystallize together in mixed crystals (*see* isomorphism) have similar chemical compositions. Perchlorate salts (such as $KClO_4$) are often isomorphous with the corresponding permanganate ($KMnO_4$). [After Eilhardt Mitscherlich (1794-1863), German chemist.]

mixed crystal. A crystalline compound formed of a mixture of two or more simple compounds that have crystallized together in a homogeneous solid in definite proportions. Mixed crystals are *solid solutions.

mixture. A system containing two or more substances in which one substance is dispersed through the other and there is no chemical bonding between the different components. A *heterogeneous mixture* is one in which small amounts of each component can be distinguished: an example would be a mixture of sugar and sand crystals. In a *homogeneous mixture*, the mixing is on the molecular level and no particles of either component can be distinguished by eye or by a microscope. Examples of such mixtures are systems containing two or more gases or miscible liquids. Solid solutions and solutions of solids and gases in liquids are also homogeneous. Mixtures are distinguished from compounds in that they can be separated by physical means (distillation, magnetic separation, etc.) and that their properties are not widely different from those of their components.

MKS units. A system of *units in which the base units are the metre, the kilogram, and the second. In their original form MKS electrical units used a *magnetic constant of 10^{-7} henries per metre. The modern system of *SI units is a rationalized MKS system in which the magnetic constant is $4\pi \times 10^{-7}$ henry per metre.

m.m.f. *See* magnetomotive force.

mmHg. Millimetre of mercury. The pressure that will support one milli-

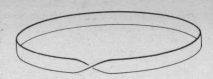

Möbius strip

metre of mercury, equal to 133.322 4 pascals.

Möbius strip. A surface made by twisting a band through 180° and joining the ends together (see illustration). The Möbius strip is topologically interesting in that it is a single surface with a single bounding curve. [After August Ferdinand Möbius (1790-1868), German mathematician.]

moderator. A substance used in *fission reactors to slow down fast neutrons. Moderators are usually light materials, such as graphite and heavy water. The fast neutrons lose energy by colliding with nuclei of the moderator atoms and are then more likely to cause fission of heavy nuclei.

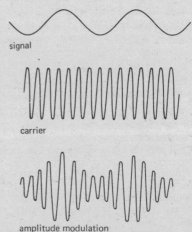

signal

carrier

amplitude modulation

Amplitude modulation

modulation. The process of changing the waveform of one wave by combining it with another wave. Modulation is used in radio transmission to vary the characteristics of a high-frequency carrier wave by impressing the characteristics of the signal upon it. In *amplitude modulation* —the commonest method—the amplitude of the carrier wave is varied. *Frequency modulation* involves changing the frequency of the carrier wave so that it varies in proportion to the amplitude of the signal. *Phase modulation* involves changing the phase of the carrier wave so that its difference from the unmodulated wave is proportional to the signal amplitude. *See also* klystron.

modulus (elastic modulus). The ratio of the *stress applied to a body to the *strain produced for a body obeying Hooke's law. The modulus of elasticity is a property of the material and there are several types depending on the type of applied stress. *Young's modulus* (symbol: E) is the ratio of tensile stress to tensile strain. The *bulk modulus* (symbol: K) is the ratio of the volume stress (or pressure) to the bulk strain. The *rigidity modulus* (symbol: G) is the ratio of the shear stress to the shear strain. *See also* Poisson's ratio.

Mohs' scale. A scale of hardness on which a mineral is compared with ten reference minerals given arbitrary numbers. The members of the scale, in ascending hardness, are 1. talc, 2. gypsum, 3. calcite, 4. fluorite, 5. apatite, 6. orthoclase, 7. quartz, 8. topaz, 9. corundum, 10. diamond. Each member of the scale is capable of scratching a member with a lower number, and any other mineral can be assigned a number by carrying out a similar scratch test. For example, a mineral that is scratched by corundum and not by topaz has Mohs hardness 8-9. [After Friedrich Mohs (1773-1839), German mineralogist.]

molality. The concentration of a solution in moles of solute per kilogram of solution (i.e. 1 kilogram is the mass of solute plus solvent). *Compare* molarity.

molar. 1. Indicating that a specified physical quantity is measured for unit amount of substance. For example, the *molar heat capacity* (C_m) is the *heat capacity of unit amount of substance and is measured in joules per kelvin per mole. Molar quantities are properties of the substance or material rather than properties of objects or systems.
2. Denoting a solution that contains one mole of a specified substance per cubic decimetre of solution. A solution containing x moles per cubic decimetre is said to be x *molar*—written xM.

molar volume. Symbol: V_m. The volume occupied by one mole of a given substance under specified conditions. An *ideal gas at standard temperature and pressure has a molar volume of 22.415 cubic decimetres and real gases approximate to this value.

mole. Symbol: mol. The basic SI unit of amount of substance equal to the amount of substance that contains the same number of entities as there are atoms in 0.012 kilogram of carbon-12. The entities may be atoms, molecules, ions, electrons, or similar elementary units. One mole of any substance contains N entities, where N is Avogadro's number ($6.022\ 52 \times 10^{23}$). One mole of a compound is equivalent to M grams, where M is the molecular weight.

molecular beam. A beam of atoms, ions, molecules, or free radicals in which all the particles in the beam are moving in the same direction and there are few intermolecular collisions. Molecular beams are formed by allowing gas or vapour to escape through a hole in an enclosure and collimating the beam with several apertures, using vacuum pumps to remove molecules that do not pass through the apertures. They are used in research into chemical reactions, spectroscopy, and surface studies.

molecular crystal. *See* crystal.

molecular flow (Knudsen flow). Flow of a gas in which the *mean free path of the gas molecules is large compared with the dimensions of the pipes or containers. Under these conditions, which occur at low pressures, the molecules make very few intermolecular collisions, most of their collisions being made with the walls of the container. The flow characteristics then depend on the molecular weight of the gas rather than its viscosity.

molecular orbital. An *orbital for an electron in a molecule. In the molecular-orbital theory of valence, electrons are considered to move in the combined field of all the nuclei in the molecule. In atoms the electrons are spin-paired and occupy *atomic orbitals. Similarly in molecules they occupy a number of molecular orbitals, which have definite energies and are visualized as regions in space. This approach to valence theory is distinguished from the *valence-bond approach, which uses the idea of definite linkages between atoms. An idea of the possible types and shapes of molecular orbitals is gained by assuming that they result from overlap of the atomic orbitals. The hydrogen molecule, for instance, can be formed by overlap of the 1s orbitals, leading to a molecular orbital containing two electrons. The electrons in the orbital can be found anywhere in the molecule but there is a maximum of electron density mid-way between the atoms, and this holds the atoms together by electrostatic attraction. Two types of molecular orbital are distinguished: *pi orbitals and *sigma orbitals.

molecular sieve. See zeolite.

molecular weight. Symbol: M. The ratio of the average mass per molecule of a compound to one twelfth of the mass of an atom of the isotope carbon-12. The molecular weight is the sum of the atomic weights of the atoms in the molecule.

molecule. The smallest amount of a chemical compound that is capable of independent existence. In covalently bonded liquid or gaseous compounds the molecules are small groups of atoms

held together by covalent bonds. Thus, a molecule of carbon dioxide, CO_2, has a carbon atom bound to two oxygen atoms. Discrete molecules can also be distinguished when such compounds form molecular *crystals. In electrovalently bonded substances discrete molecules do not exist. A crystal of sodium chloride is simply a collection of sodium ions and chloride ions. In such compounds the molecule is considered to be the smallest group of atoms characteristic of the substance: in the case of sodium chloride it is NaCl. Similarly, a covalent crystal, such as boron nitride (BN), does not have distinguishable simple molecules because the covalent bonds extend throughout the whole crystal structure.

mole fraction. The ratio of the number of moles of a specified substance in a mixture to the total number of moles of mixture: i.e. if n_A moles of A are mixed with n_B of B, the mole fraction of A is $n_A/(n_A + n_B)$.

molybdate. Any of a number of oxycompounds of molybdenum formed by dissolving molybdenum trioxide in an alkali. The simplest molybdates have formula M_2MoO_4 (where M is a monovalent metal) and contain MoO_4^{2-} ions. Formally, they are salts of the hypothetical molybdic acid, H_2MoO_4. If alkaline solutions of simple molybdates are acidified, polymolybdates are produced in solution. They contain large anions of the type $(Mo_7O_{24})^{6-}$, and it is possible to obtain crystalline salts and acids (polymolybdenic acids) from such solutions.

molybdenite (molybdenum glance). A grey or black naturally occurring form of molybdenum sulphide, MoS_2, important as the main ore of molybdenum.

molybdenum. Symbol: Mo. A very hard silvery white *transition element found chiefly in molybdenite (MoS_2), from which it can be extracted by roasting to form the oxide and reduction of this with hydrogen. The main uses of molybdenum are in making alloy steels and other molybdenum compounds. The element is chemically unreactive, being inert to most acids and rendered passive by nitric acid. It is oxidized at red heat and it dissolves in molten alkalis. The chemistry of molybdenum is very complicated: it has a variable valency and forms numerous oxides and oxycompounds (see molybdate). A.N. 42; A.W. 95.94; m.pt. 2610°C; b.pt. 5560°C; r.d. 10.22; valency 2–6.

molybdenum glance. See molybdenite.

molybdic acid. See molybdate.

moment. The turning effect of a force or system of forces about an axis. A single force has a moment equal to the product of the force and the perpendicular distance from the axis to the force's line of action. The moment of a system of forces is the algebraic sum of the individual moments. The moment of a force is the same as its *torque.

moment of inertia. Symbol: I. The sum of the products of the masses of particles of a body and the squares of their distances from an axis of rotation. A single point mass m a distance r from an axis has a moment of inertia equal to mr^2. The moment of inertia of a system of such masses is the sum of these products: in the case of a rigid body an integral is used. Moment of inertia is used in place of mass in problems involving rotation. Thus, angular momentum is $I\omega$ and angular kinetic energy is $I\omega^2/2$, where ω is angular velocity. See also radius of gyration.

momentum. 1. Symbol: p. The product of a particle's mass and its velocity.
2. (angular momentum). Symbol: L. The product of the moment of inertia of a rotating body and its angular velocity.
The total momentum (linear or angular) is constant in any system. The principle is the *law of conservation of momentum.*

monatomic. Composed of single atoms. Helium, argon, and other inert gases are monatomic.

monazite. A yellowish-brown mineral consisting of a mixed phosphate of various lanthanide elements together with thorium silicate.

Mond process. An industrial process for purifying nickel by heating the impure metal in carbon monoxide at 50°C. The volatile carbonyl is decomposed at 180°C. [After Ludwig Mond (1839–1909), British industrial chemist born in Germany.]

monochromatic. Denoting light or other electromagnetic radiation that has only one wavelength; not polychromatic.

monochromator. A device for producing monochromatic electromagnetic radiation, usually a spectrometer used to select one frequency of radiation from a polychromatic source.

monoclinic. See crystal system.

monohydrate. A solid compound with one molecule of *water of crystallization per molecule of compound, as in sodium carbonate monohydrate, $Na_2CO_3.H_2O$.

monohydric. Denoting an alcohol or phenol containing one hydroxyl group per molecule.

monolayer. A unimolecular layer of atoms or molecules.

monomer. A simple compound distinguished from a *dimer, *trimer, or *polymer.

monosaccharide. A *sugar that cannot be hydrolysed to simpler sugars.

monosodium glutamate. See sodium hydrogen glutamate.

monoterpene. See terpene.

monotropic. Denoting a substance that exists in only one crystalline form. See polymorphism.

monovalent (univalent). Having a valency of one.

month. The time taken for the moon to move around the earth. It is measured in various ways (see time).

moon. 1. The earth's only natural satellite and, owing to its proximity, the brightest object in the night sky. Its size (diameter: 3476 kilometres; mass: 1/81 that of the earth; mean relative density: 3.34) may be compared to that of the planet *Mercury. Though a solid body like the *terrestrial planets, the moon seems to have no central core. Yet its chemical composition has recently been found to be not altogether dissimilar to that of the earth and chemical and physical data collected first by unmanned and later by manned spacecraft during the 1960s have led scientists to believe that the moon was formed either simultaneously with or only shortly after the earth as part of the same planet-forming process. The moon orbits the earth once every 27.322 days (sidereal period of revolution) at a mean distance of 384 400 km (perigee: 363 000 km; apogee: 406 000 km). Its rotation is captured: it makes one rotation in the same time as it takes to make one orbit and therefore always keeps the same face turned towards us. However, because of an irregularity in the distribution together with the varied velocity of its orbit (faster at perigee and slower at apogee) more than half the moon's surface is visible over a period of time. This phenomenon is called *libration*.

In general, the moon is a cheerless place; rocky, barren, pitted with craters, it has high mountain ranges and huge flat smooth plains. When Galileo first observed them he called them "seas" and the Latin word for sea, *mare* (pl. *maria*) is still applied to these smooth expanses of dark basalt. Lunar craters range in size from tiny pits to huge depressed walled plains hundreds of kilometres across. The moon's smallness means that its gravitation and escape velocity are much lower than those of earth (about 16% and 20% respectively). As a result, any atmosphere that it may have once possessed has almost completely vanished. The fact that the moon's orbit is inclined at some 5° to the ecliptic ensures that lunar and solar *eclipses are rare. Because of the gravitational attraction of the earth, the nodes of the lunar orbit (points of intersection with the ecliptic) move in a westerly direction, completing one revolution every 18.6 years (the *saros*). But the moon's gravity also has a crucial effect on the earth, since it causes the tides. One of the moon's most characteristic features is that because it reflects sunlight, it shares with the inferior planets the capacity to show *phases. The cycle of phases, the synodic month or period, lasts 29.531 days.
2. Any natural satellite of a planet.

mordant. A substance applied to a substrate to allow the attachment of a dye. The mordant is held on the fibres of the substrate and the dye forms a complex with it. Many mordants are inorganic compounds: the most widely used are chromium salts which are employed, together with certain acid dyes (*chrome dyes*), in colouring wool. The colour of the dye is affected by the nature of the mordant: alizarin, for instance, gives a red colour when mordanted with aluminium compounds and a dark purple with iron compounds. *See also* lake.

morphine. A white crystalline poisonous alkaloid, $C_{17}H_{18}NO_3.H_2O$, extracted from opium. Its soluble salts are used in medicine as narcotics.

mosaic. 1. An irregular structure of small crystallites in a crystalline solid.
2. A layer of small particles of light-sensitive material deposited on an insulating support, used in a television *camera for converting an optical image into an electrical image.

mosaic gold. *See* tin sulphide.

Moseley's law. The principle that lines in the X-ray spectra of a set of elements have frequencies that are proportional to the squares of the elements' atomic numbers. The lines must all originate from the same type of transition. For example, the square root of the frequency for characteristic X-rays produced by transition of an electron from an *L* shell to a *K* shell, plotted against atomic number, gives a straight line. [After Henry Gwyn-Jeffreys Moseley (1887–1915), British physicist.]

Mössbauer effect. An effect occurring in the gamma-ray emission from the nuclei of certain isotopes in solids. In the Mössbauer effect the gamma-ray photons have a sharply defined energy (normally a spread of energies would occur because of recoil of the nucleus). The phenomenon is used in a spectroscopic technique for investigating solids (*Mössbauer spectroscopy*). A gamma-ray source is mounted on a moving platform and a similar sample is placed nearby with a detector to measure the scattered gamma rays. The source is moved towards the sample at a varying speed, thus varying the gamma-ray frequency by the *Doppler effect. A minimum in the scattered radiation indicates a resonance absorption of gamma rays by sample nuclei. The method is used for studying nuclear energy levels and for investigating chemical compounds. [After R. L. Mössbauer (b. 1922), German physicist.]

mother liquor. The liquid remaining after a substance has been crystallized from a solution.

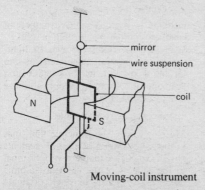

Moving-coil instrument

moving-coil instrument. An electrical measuring instrument consisting of a

flat rectangular coil of wire suspended vertically between the curved pole pieces of a permanent magnet (see illustration). If a current is passed through the coil the vertical sides experience opposite forces, causing the coil to turn. It is restrained by the torsion in a wire suspension or in a flat hair spring: the angle turned depends on the magnitude of the current. It is measured by attaching a small mirror to the suspension wire (in which case the instrument is a moving-coil galvanometer) or by use of a pointer attached to the coil. The pointer instrument is used as an *ammeter or, by including a high resistance to reduce the current drawn from the circuit, as a *voltmeter.

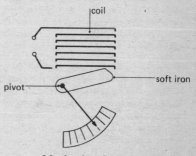

Moving-iron instrument

moving-iron instrument. An electrical measuring instrument consisting of a piece of soft iron pivoted so that it can move into a fixed coil of wire. If a current is passed through the coil, the iron is attracted and the piece of iron rotates. It is restrained by the torsion in a spring: the angle turned through depends on the magnitude of the current and is measured by movement of the pointer (see illustration). The attraction is independent of the direction of the current and the instrument can be used as an *ammeter for both alternating and direct currents. If a high resistance is connected in series with the coil, so that little current is drawn from the circuit, it can be used as a *voltmeter.

multiple proportions, law of. The principle that when two elements combine to form more than one compound, the weights of one element combining with a fixed weight of the other element are in simple proportion. Nitrogen and oxygen, for example form a number of compounds in which the weights of oxygen combining with 14 g of nitrogen are in the ratio 8 : 16 : 24 : 32 : 48 = 1/2 : 1 : 3/2 : 2 : 5/2, corresponding to the oxides N_2O, NO, N_2O_3, NO_2, N_2O_5. The law is one of three laws of *chemical combination.

mu-meson. *See* muon.

Mumetal. ® An alloy containing nickel (about 75%), iron (about 18%), and copper and chromium. It has a high magnetic permeability and is used in applications requiring a low hysteresis loss and in shielding electrical equipment from magnetic fields.

muon. An *elementary particle having a positive or negative charge and a mass equal to 206.77 times the mass of the electron. The muon was formerly called the *mu-meson* but is now considered to be a *lepton.

muriate. A chloride of a metal; a salt of muriatic acid. Thus, muriate of potash is potassium chloride, KCl.

muriatic acid. *See* hydrochloric acid.

mustard gas. A colourless oily liquid, $S(CH_2CH_2Cl)_2$, with a pungent odour, made by reacting ethylene with sulphur chloride. It has been used as a war gas. M.pt. 14°C; b.pt. 217°C; r.d. 1.27.

mutarotation. A change with time in the optical rotation produced by a solution (*see* optical activity). The phenomenon is seen in freshly prepared optically active solutions of some sugars, which change as a result of change of the sugar into another optically active form. Freshly prepared glucose solutions change as a result of conversion of glucose into another isomer of different activity. *See also* invert sugar.

mutual inductance. *See* electromagnetic induction.

myopia. A defect in the eye in which parallel rays of light are focused to a point in front of the retina when the eye is at rest so that distant objects cannot be accommodated. Near objects can be seen. In myopia the distance between the front and back of the eye is too long and it can be corrected by concave spectacle lenses. It is sometimes called *short sight* or *near sight*. *See* eye.

N

nadir. The point opposite the zenith on the *celestial sphere.

Napierian logarithm. *See* logarithm. [After John Napier (1550–1617), Scottish mathematician.]

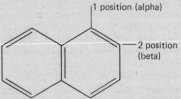

1 position (alpha)

2 position (beta)

Naphthalene

naphthalene. A white flammable crystalline solid, $C_{10}H_8$, with an odour of moth balls. It is the most abundant constituent of coal tar, from which it is extracted by fractional distillation. Naphthalene is an aromatic hydrocarbon and is used in the manufacture of dyes and synthetic resins and as a moth repellent. M.pt. 80.2°C; b.pt. 218°C; r.d. 1.14.

naphthol. Any one of two isomers of hydroxynaphthalene, $C_{10}H_7OH$, both of which are white powders obtained by fusing the corresponding sodium naphthalene sulphonate with caustic soda. The more important is *beta-naphthol* (2-hydroxynaphthalene) which is used in antioxidants for rubber, dyes, and pharmaceuticals. M.pt. 122°C; b.pt. 285°C; r.d. 1.2.

nascent. Denoting a gas that is formed in a reaction mixture and consequently has a higher activity. Nascent hydrogen, for example, can be produced *in situ* by adding zinc and hydrochloric acid to the reaction medium. The hydrogen produced often reduces compounds that are not reduced by "normal" hydrogen. The high activity of the nascent hydrogen can be ascribed to the production of hydrogen atoms or to catalysis involving intermediates formed during the reaction.

natural logarithm. *See* logarithm.

near sight. *See* myopia.

nebula. Any of numerous clouds of interstellar gas and dust. *Galactic nebulae* occur within the Milky Way and fall into a number of classifications. *Bright-emission nebulae* emit light and other forms of radiation, *reflective nebulae* shine by the light of nearby stars, and *dark nebulae* absorb the light of stars lying beyond them. Because of their misty appearance *galaxies were originally referred to as *extragalactic nebulae*, a name still found in modern catalogues. *Planetary nebulae* are also found, consisting of a very hot central star surrounded by an expanding ring or envelope of gas and particles.

nebular hypothesis. Any of a group of theories concerning the origin of the solar system, all of which assume that the sun and planets have been formed by condensation of a nebula.

Néel temperature. *See* antiferromagnetism.

negatron. A negative electron: i.e. an electron as opposed to a *positron.

neodymium. Symbol: Nd. A silvery metallic element belonging to the *lanthanide series. A.N. 60; A.W. 144.24; m.pt. 1010°C; b.pt. 3967°C; r.d. 6.9; valency 3.

neon. Symbol: Ne. A colourless odourless tasteless nonflammable gaseous element present in the atmosphere (0.0012%) from which it is obtained as a by-product in the liquefaction of air. Neon is used in fluorescent lamps. It is one of the *inert gases. A.N. 10; A.W. 20.283; b.pt. -248.6°C; m.pt. -245.92°C; r.d. 0.694 (air = 1).

neopentane (2,2-dimethylpropane). A colourless gas or volatile liquid, $C(CH_3)_4$, found in small amounts in natural gas. It is isomeric with *pentane and *isopentane. M.pt. -16.6°C; b.pt. 9.5°C; r.d. 0.59.

neopentyl group. *See* pentyl group.

neoprene. A synthetic rubber made by polymerizing chloroprene. It is resistant to heat and solvents and is also used in cements and paints.

Neptune. The eighth planet in order of succession from the sun; the outermost of the *giant planets. It orbits the sun once every 164.79 years at a distance of 4.497 million kilometres. Neptune is similar to Uranus, having a diameter of 48 400 km, a mass 17.2 times that of the earth, and a relative density of 1.77. Its rotation period is almost 16 hours and its temperature is about -205°C. As with Uranus, the atmosphere is dense and comprises mainly methane and hydrogen. Neptune has two satellites *Triton* and *Nereid.*

neptunium. Symbol: Np. A silvery metallic *actinide element: the first synthetic transuranic element to be produced. The isotope ^{237}Np is made in nuclear *reactors in the production of plutonium. A.N. 93; m.pt. 640°C; valency 3, 4, 5, or 6.

Nessler's reagent. A colourless alkaline solution of mercury II iodide in potassium iodide used to detect traces of ammonia, with which it gives a brown coloration.

neutrino. An *elementary particle with zero rest mass, a velocity equal to that of light, and a spin of one half. The existence of the neutrino was originally postulated to explain energy conservation in *beta decay. There are two types: one produced in decay processes that yield positrons and the other produced in decays of some mesons. The antiparticles of these two types of neutrino are also called neutrinos.

neutron. An *elementary particle with zero charge and a rest mass of 1.674 92 × 10^{-27} kg, which is similar (but not identical) to that of the proton. Together with protons, neutrons are present in the *nuclei of all atoms except the hydrogen atom. When free, they decay to protons and electrons with a mean-life of 932 seconds. *See also* beta decay.

neutron excess. *See* isotopic number.

neutron star. A type of star, less than 20 kilometres in diameter, that is postulated to result from the gravitational collapse of the matter in a star after the source of nuclear energy has been exhausted. As the matter becomes compressed it reaches a *degenerate state in which its nuclei and electrons are packed together. At densities above about 10^7 kg m^{-3}, protons and electrons may fuse together to give neutrons. A very fast rotational spin is imparted to the new superdense star and energized particles are sent out in a highly directional beam (*see* pulsar). In theory, a neutron star could form a *black hole if it were of sufficient size.

Newlands' law of octaves. The observation made by the British chemist John Newlands (1837–98) that if the elements are arranged in order of their atomic weight, then every eighth element is chemically similar. He compared this to the intervals of the musical scale. *See also* periodic table.

newton. The SI unit of force, equal to the force that will give a mass of one kilogram an acceleration of one metre per second per second. [After Sir Isaac Newton (1647–1727), English physicist.]

Newtonian fluid. A fluid that obeys Newton's law of *viscosity.

Newtonian mechanics. The system of mechanics based on *Newton's laws of motion.

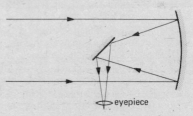

eyepiece

Newtonian telescope

Newtonian telescope. A type of astronomical telescope in which light is reflected from the primary mirror onto a small plane miror on the axis (see illustration).

Newton's law of cooling. The principle that the rate at which a body loses heat is proportional to the difference in temperature between the body and its surroundings. Strictly, it only applies to forced *convection.

Newton's law of gravitation. *See* gravitation.

Newton's law of viscosity. *See* viscosity.

Newton's laws of motion. Three laws formulated by Newton to describe the motion of bodies.
1. Every body will continue in a state of rest or of uniform motion in a straight line unless it is compelled to change that state by an external impressed force.
2. When a force is applied the rate of change of momentum is proportional to the force and takes place in the direction of action of the force: i.e. $F = d(mv)/dt$, where F is the magnitude of the force, m the body's mass, and v its velocity. Usually m is constant and the equation becomes $F = ma$, where a is the acceleration of the body.
3. Every action on a body produces an equal and opposite reaction.
The laws are the basis of *Newtonian*

mechanics, a system of mechanics applicable at velocities that are much less than the velocity of light.

Newton's rings. An interference effect observed when a plano-convex lens of large radius is placed on a flat glass plate and illuminated from above by monochromatic light. When observed from above, a series of concentric light and dark rings is seen centred on the centre of the lens. Coloured rings are produced when white light is used. The phenomenon is caused by interference between light reflected from the glass plate and light reflected at the curved surface of the lens.

nickel. Symbol: Ni. A silvery-white malleable and ductile *transition element found in pentlandite ((Fe,Ni)S), millerite (NiS), and garnierite ((Ni,-Mg)$_6$(OH)$_6$Si$_4$O$_{11}$.H$_2$O). Nickel is also present in many meteorites. The element is obtained by roasting the sulphide to the oxide, which is then reduced or refined by electrolysis or by the *Mond process. It is ferromagnetic and is used in some ferromagnetic alloys, such as Mumetal, as well as alloy steels and protective electroplated coatings. Nickel is also an efficient hydrogenation catalyst, used for hardening oils in making margarine. The element is resistant to corrosion, dissolves in dilute nonoxidizing acids, and is rendered passive by nitric acid. The normal valency is 2: many ionic nickel II (*nickelous*) salts exist as well as numerous complexes. A few complexes (*nickelic*) are formed in higher oxidation states. A.N. 28; A.W. 58.71; m.pt. 2468°C; b.pt. 4927°C; r.d. 8.57; valency 2, 3, or 5.

nickel carbonyl. A poisonous flammable yellow liquid, Ni(CO)$_4$, made by passing carbon monoxide over finely divided nickel. It is used as a catalyst and is also the intermediate in the *Mond process for purifying nickel. M.pt. -25°C; b.pt. 43°C; r.d. 1.32.

nickelic. *See* nickel.

nickelous. *See* nickel.

nickel oxide. Any of two basic oxides of nickel made by heating the corresponding nitrate. *Nickel II oxide* (nickelous oxide) is a green powder, NiO. R.d. 6.6. *Nickel III oxide* (nickelic oxide, nickel peroxide, nickel sesquioxide) is a grey-black powder used in nife cells. R.d. 4.8.

nickel peroxide. *See* nickel oxide.

nickel sesquioxide. *See* nickel oxide.

Nicol prism. A prism made by cutting two pieces of calcite in particular directions and cementing them together with Canada balsam. The arrangement is designed to transmit the ordinary ray and reflect the extraordinary ray at the interface between the two pieces. Nicol prisms are used for producing and analysing plane-polarized light. [After William Nicol (1768–1851), British physicist.]

nicotine. A colourless poisonous hygroscopic oily alkaloid, C$_5$H$_4$NC$_4$H$_7$NCH$_3$, extracted from tobacco. It is used as a horticultural insecticide.

nife cell. A type of *accumulator consisting of a nickel positive plate and an iron negative plate dipping into a solution of sodium hydroxide.

ninhydrin. A white crystalline solid, C$_9$H$_6$O$_4$, used as a test reagent for detecting free amino and carboxyl groups in proteins, with which it gives a blue colour.

niobium. Symbol: Nb. A soft ductile silvery-white metallic element, developing a bluish caste in air, found in columbite-tantalite ((Fe,Mn)(Nb,Ta)$_2$O$_6$) and pyrochlore (NaCaNb$_2$O$_6$F). Niobium is separated from tantalum by fractional crystallization of its complexes. It is extracted by reducing the oxide with calcium or electrolysis of the fused complex fluoride. The metal is superconducting at low temperatures. Its main uses are in alloy steels and vacuum getters.

Chemically, it is not attacked by acids, dissolves in fused alkalis, and reacts with oxygen and other nonmetals at high temperatures. In its compounds it exhibits several valencies, generally behaving like a nonmetal in forming volatile covalent compounds, such as halides and oxyhalides. Niobium was formerly called *columbium* in the U.S. A.N. 41; A.W. 92.906; m.pt. 2468°C; b.pt. 4927°C; r.d. 8.57; valency 2, 3, or 5.

nitration. A chemical reaction in which a nitro group ($-NO_2$) is introduced into a molecule.

nitre. *See* potassium nitrate.

nitric acid (aqua fortis). A colourless or yellowish fuming corrosive liquid, HNO_3, obtained by the catalytic oxidation of ammonia (*see* Ostwald process). It is a strong oxidizing agent, which attacks almost all metals. Nitric acid is used in manufacturing ammonium nitrate for use in fertilizers and explosives. M.pt. -41°C; b.pt. 83°C; r.d. 1.5.

nitric oxide. *See* nitrogen oxide.

nitrile. Any of a class of organic compounds containing the group $-CN$.

nitrobenzene. A poisonous pale yellow oily liquid, $C_6H_5NO_2$, with an almond-like odour. It is prepared by nitrating benzene and used in the manufacture of aniline. M.pt. 6°C; b.pt. 211°C; r.d. 1.2.

nitrocellulose (gun cotton). A cotton or pulp-like polymer of variable composition made by treating cellulose with a mixture of nitric and sulphuric acids. It is highly flammable and is used in explosives and solid rocket propellants. Strictly speaking it is a nitric acid ester of cellulose, not a nitro compound, and the more correct name is *cellulose nitrate.*

nitro compound. A compound with the formula RNO_2, where R is usually an aryl group.

nitrogen. Symbol: N. A colourless odourless tasteless nonflammable gas forming 78% of the air by volume. Pure nitrogen is obtained by liquefying air. The element is used in the manufacture of ammonia and other nitrogen compounds and as an inert atmosphere in welding, forging, and similar operations. The element is diatomic, N_2, and relatively inert: it reacts with hydrogen under the influence of a catalyst (*see* Haber process), forms nitrogen oxides in electric sparks, and combines with magnesium, lithium, and calcium to produce nitrides. Nitrogen compounds are essential to plant growth. A.N. 7; A.W. 14.0067; m.pt. -209.86°C; b.pt. -195.8°C; r.d. 0.967 (air = 1).

nitrogen fixation. The conversion of nitrogen gas from the atmosphere into nitrogen compounds. Nitrogen is fixed by bacteria in the roots of some plants. Industrial nitrogen fixation can be effected by several processes including the combination of calcium carbide and nitrogen to give calcium cyanamide, $CaCN_2$, and the *Birkeland-Eyde process for producing nitric oxide, NO. The most widely used method is the *Haber process for synthesizing ammonia.

nitrogen oxide. Any of seven oxides of nitrogen, all gaseous at room temperature. *Dinitrogen monoxide* (nitrous oxide, laughing gas) is the lowest oxide, with the formula N_2O. It is neutral, colourless, sweet-tasting, and nonflammable. The substance is prepared by heating ammonium nitrate and is used as an anaesthetic. M.pt. -91°C; b.pt. -88°C; r.d. 1.52 (air = 1). *Nitrogen monoxide* (nitric oxide), NO, is colourless, nonflammable, and poisonous. It is prepared by oxidation of ammonia above 500°C and reacts with air to give nitrogen dioxide. M.pt. -164°C; b.pt. -152°C; r.d. 1.3. *Nitrogen dioxide* (nitrogen peroxide) is a brown acidic gas, NO_2, which on cooling is converted to a colourless gas, N_2O_4, *dinitrogen tetroxide.* It is prepared by oxidation of nitrogen monoxide and used as a nitrating and oxidizing agent. M.pt. -11.2°C; b.pt. 21°C; r.d. 1.4 (20°C). *Dinitrogen trioxide*, N_2O_3, *dinitrogen* *pentoxide*, N_2O_5, and *nitrogen trioxide*, NO_3, are the other less stable oxides of nitrogen.

nitrogen peroxide. *See* nitrogen oxide.

nitroglycerine. A pale yellow thick explosive liquid, $C_3H_5(NO_3)_3$, used in dynamite and other explosives.

nitrous oxide. *See* nitrogen oxide.

nitro group. The monovalent group $-NO_2$, characteristic of nitro compounds.

NMR. *See* nuclear magnetic resonance.

nobelium. Symbol: No. A transuranic *actinide element made by bombarding curium with carbon nuclei. The isotope produced, ^{254}No, has a half-life of about 3 seconds. The name *nobelium* was proposed by earlier workers claiming the production of a longer-lived isotope with a half-life of about 10 minutes. A.N. 102.

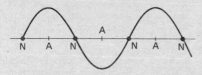

N node

A antinode

Nodes and antinodes

node. A point of zero displacement of a *standing wave.

noise. Unwanted random disturbances in an electrical circuit.

non-Euclidean geometry. Any logical self-consistent mathematical system concerning shape, based on postulates that do not include the parallel postulate of Euclid (*see* Euclidean geometry). Non-Euclidean geometry is used in the theory of *relativity.

nonmetal. A chemical element that is not a metal. Nonmetals are poor electrical and thermal conductors (although graphite does conduct electricity) and are not generally lustrous or malleable when solid. They include hydrogen, the inert gases, the halogens, oxygen, sulphur, nitrogen, and carbon. Chemically, such elements are electronegative, forming either covalent compounds or negative ions in ionic compounds. Their oxides and hydroxides are usually acidic. *See also* metalloid.

non-Newtonian fluid. A fluid that does not obey Newton's law of *viscosity: i.e. the viscosity is not constant but depends on the rate of shear. Fluids that consist of a mixture of different phases show non-Newtonian behaviour: usually the viscosity decreases as the rate of shear increases (*thixotropy), although in some suspensions the fluid becomes more viscous.

nonoxidizing acid. An acid that does not act as an oxidizing agent. The term is used for compounds such as hydrochloric acid and dilute sulphuric acid, which dissolve metals that are higher than hydrogen in the *electrochemical series but do not attack lower metals such as copper. In contrast, *oxidizing acids do dissolve such metals.

nonpolar. Denoting a substance whose molecules do not have a permanent dipole moment. Carbon tetrachloride is an example of a nonpolar compound. Nonpolar solvents are not capable of dissolving ionic solids or other *polar compounds. They do dissolve other nonpolar substances: carbon tetrachloride, for example, is a good solvent for many organic compounds.

nonreducing sugar. *See* sugar.

normal. 1. Denoting a solution containing one gram-equivalent per litre. A solution containing x gram-equivalents per litre is said to be x-*normal*, written xN.
2. Denoting an isomer with a straight chain of atoms. Normal hexane, for example, has the formula CH_3-$CH_2CH_2CH_2CH_2CH_3$, and is distinguished from isomers with branched

chains. The abbreviation *n-* is used, as in *n*-hexane.

3. A line or plane perpendicular to a given line or surface. At any point, the normal to a curved line or surface is perpendicular to the tangent at that point.

normality. The concentration of a solution in gram-equivalents per litre of solution. A two-normal (2N) solution for example has a normality of 2. *See* normal.

normal temperature and pressure (N.T.P.). *See* standard temperature and pressure.

Norway saltpetre. *See* ammonium nitrate.

nova. A star that undergoes an explosion, increasing in luminosity by up to 100 000 times its original value. *See also* supernova.

n-p-n transistor. *See* transistor.

N.T.P. *See* standard temperature and pressure.

n-type. *See* semiconductor.

nuclear fission. *See* fission.

nuclear fusion. *See* fusion.

nuclear magnetic resonance (NMR). A method of investigating the *spins of atomic nuclei. In the presence of a strong magnetic field the magnetic moment of a nucleus precesses around the field direction (*see* precession). Only certain orientations are possible, leading to the existence of a number of discrete energy states for the nucleus. Radiofrequency radiation is supplied to the sample by passing a high-frequency current through a coil and the radiation is detected by a second coil. If the magnetic field is changed slowly, thus changing the difference in energy between levels, radiation is absorbed at certain values of the field when the frequency of the radiation corresponds to the difference between two energy levels. A graph of detector response against magnetic field is an *NMR*

spectrum of the sample. The technique is extensively used in chemistry for studying molecules—the energy levels of the nucleus are affected by the electrons surrounding the nucleus and the frequency at which the nucleus absorbs radiation depends on its position in the molecule. A modification of the method described can be used for the accurate measurement of magnetic fields. *See also* electron spin resonance.

nuclear physics. The branch of physics concerned with the study of the nuclei of atoms.

nuclear weapon. A bomb in which the explosion is the result of uncontrolled nuclear *fission or *fusion. In a *fission bomb* (*atomic* or *A-bomb*) two masses of uranium or plutonium, each below the critical mass, are brought together to form a mass greater than the critical mass. The resulting chain reaction occurs with enormous production of energy. In a *fusion bomb* (*hydrogen* or *H-bomb*) a fission bomb is used to increase the temperature of a hydrogen-containing material to the point at which nuclear fusion occurs.

nucleon. A particle that is a constituent of an atomic nucleus; either a proton or a neutron.

nucleonics. The technology of nuclear physics and its applications.

nucleon number. Symbol: N. The number of nucleons in the nucleus of an atom.

nucleophilic. Having or involving an affinity for positive electric charge. A nucleophilic reagent (or *nucleophile*) is one that acts by attacking a positive region of a molecule. Examples are negative ions such as Cl^- and CN^-. Nucleophilic reactions are ones involving such a reagent. Thus, in a *nucleophilic *substitution* one atom in a molecule is displaced by a nucleophile: this occurs in *displacement reactions involving alkyl halides. In *nucleophilic *addition* the initial attack is by a nucleophile: this occurs in addition to the carbonyl group of aldehydes and ketones. *Compare* electrophilic.

nucleus. The positively charged part of the *atom about which the electrons orbit. The nucleus is composed of neutrons and protons (nucleons) held together by *strong interactions. The number of protons determines the element and the number of neutrons determines the particular *isotope of the element. The principal theoretical models of the nucleus are the *liquid-drop model and the *shell model. *See also* fission, fusion, radioactivity.

null method. A method of measurement in which a zero reading is obtained by balancing one quantity against another, as in the Wheatstone bridge or the Lummer-Brodhun photometer.

numerator. The number placed above the line in a vulgar fraction. In ¾, 3 is the numerator and 4 the denominator.

$$H_2N(CH_2)_n \left[\begin{array}{c} O \\ \parallel \\ C \end{array} \begin{array}{c} H \\ \mid \\ N \end{array} (CH_2)_n \right] \begin{array}{c} O \\ \parallel \\ C \end{array} \begin{array}{c} H \\ \mid \\ N \end{array} (CH_2)_n \begin{array}{c} O \\ \parallel \\ C \end{array} OH$$

Nylon polymer

nylon. Any of a class of synthetic polymeric materials in which the monomers are linked by -CO.NH- groups. Nylon is a *polyamide made either by copolymerization of a diamine with a dicarboxylic acid or by polymerization of a compound containing both amine and carboxyl groups in its molecules (see illustration). It is a translucent creamy-white material used in many moulded and machined articles and parts and in clothing fibres, bristles, rope, etc.

O

Oberon. *See* Uranus.

objective (object lens). The lens or lens system that is nearest the object in a telescope, microscope, or other optical instrument. It forms an image that is viewed through the *eyepiece.

object lens. *See* objective.

oblate. *See* ellipsoid.

obtuse. Denoting an angle that is greater than $90°$ and less than $180°$.

occlusion. The trapping of one substance in small pockets in crystals of another substance. The occluded material may be solid, liquid, or gas—small amounts of liquid are often occluded when crystals are formed from solution. The term is also used for the absorption of a gas by a solid, in which case the gas molecules or atoms occupy *interstitial positions in the lattice and the substance formed is homogeneous.

occultation. The process in which one celestial object cuts off the light or radio emission from another by moving between it and the earth.

ochre. Any of various naturally occurring forms of iron III oxide, Fe_2O_3, used as red or yellow pigments. *Burnt ochre*, which has a brownish-red colour, is a pigment made by calcining natural yellow ochre.

octadecanoic acid. *See* stearic acid.

octagon. A polygon that has eight sides.

octahedron. A *polyhedron with eight faces. A *regular octahedron* has congruent equilateral triangles as its faces.

octahydrate. A solid compound with eight molecules of *water of crystallization per molecule of compound, as in Iron II fluoride octahydrate, $FeF_2.8H_2O$.

octal notation. A system of representing numbers using the *base 8, thus using eight integers, 0-7. For example, the number represented by 125 in decimal notation is $(1 \times 8^2) + (7 \times 8) + 5$, or 175 in octal notation. Each digit in an octal number can be represented by a three digit *binary number; e.g. 7 is 111, 5 is 101, etc. For this reason octal is used in computers, each digit corresponding to one byte.

octane. Any of eighteen isomeric alkanes with formula C_8H_{18}. The octanes are colourless liquids obtained from petroleum: the most important is *iso-octane*, $(CH_3)_2CCH_2C(CH_3)_2$, which is used as a standard for the *octane rating of fuels.

octane rating. A number measuring the ability of petrol to burn in an internal-combustion engine without *knocking. It is the volume percentage of iso-octane (C_8H_{18}) in a mixture of iso-octane and (normal) heptane (C_7H_{16}) with the same properties as the fuel when the two are compared under standard conditions. A similar measure, the *cetane rating, is used for diesel fuel.

octant. A segment of a circle equal in area to one eighth of the area of the circle: i.e. having an angle of 45°.

octave. An interval between two notes such that the frequencies of the two are in the ratio 2:1. An octave covers eight notes of a scale in music.

octet. A group of eight electrons in the shell of an atom. This configuration has a special stability: it is found in the inert-gas atoms (except helium) and in atoms in compounds. *See* covalent bond, ionic bond.

odd-even nucleus. An atomic nucleus that has an odd number of protons and an even number of neutrons.

odd-odd nucleus. An atomic nucleus that has odd numbers of neutrons and of protons.

oersted. A CGS-electromagnetic unit of magnetic field strength equal to the field strength that produces a force of one dyne on unit magnetic pole. It is equal to $10^3/4\pi$ amperes per metre. [After Hans Christian Oersted (1777–1851), Danish physicist.]

ohm. Symbol: Ω. The *SI unit of electrical resistance, equal to the resistance between two points of a conductor when a potential difference of one volt between the points produces a current flow of one ampere. In 1948 this definition replaced the *international ohm*, which was defined as the resistance at 0°C of a column of mercury of length 106.300 centimetres, mass 14.4521 grams, and uniform cross section. 1 Ω_{int} = 1.000 49 Ω. [After Georg Simon Ohm (1787–1854), German physicist.]

ohmic. Denoting a resistance or other electrical device that obeys Ohm's law.

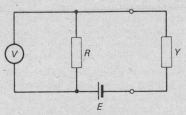

Ohmmeter circuit

ohmmeter. An instrument for measuring electrical resistance, usually a modified voltmeter containing a small dry battery (e.m.f. E). In the circuit (see illustration) Y is the unknown resistance connected to the meter: the voltage reading is $RE/(R + Y)$. The resistance R is adjusted so that the reading is full scale when $Y = 0$ and other values of Y give lower deflections. The scale is calibrated so that Y is read directly.

Ohm's law. The principle that at constant temperature the current (I) flowing through a conductor is directly proportional to the potential difference (V) applied across it. The constant of proportionality is the *resistance (R) of the material: i.e. $V = IR$.

oil-immersion. A method of increasing the resolving power of a microscope by placing a few drops of cedar-wood oil on the sample slide and lowering the objective so that it is immersed in the oil. The oil has the same refractive index as glass and serves to direct more light from the object into the lens.

oil of vitriol. *See* sulphuric acid.

Olbers' paradox. An astronomical paradox concerning the appearance of the night sky. The amount of light reaching earth from distant stars is inversely proportional to the square of the star's distance. On the other hand, the number of stars within a given solid angle is directly proportional to the square of the distance. It follows that the amount of light reaching the earth from stars is the same from all distances and that, if the universe contains an infinite number of uniformerly distributed stars, the night sky should be uniformly bright. It is thought that this is not the case because the recession of distant galaxies results in a lower brightness. [After Heinrich Wilhelm Olbers (1758–1840), German astronomer.]

oleate. Any salt or ester of oleic acid.

olefine (olefin). See alkene.

oleic acid (cis-9-octadecenoic acid). A yellow oily liquid, $CH_3(CH_2)_7CH:CH\cdot(CH_2)_7COOH$, found, in the form of its glycerides, in natural fats and oils. It is used in the manufacture of soaps, ointments, polishes, and lubricants. M.pt. 13°C; b.pt. 286°C (100 mmHg); r.d. 0.9.

olein. See triolein.

oleoresin. See balsam.

oleum. See sulphuric acid.

omega-minus. Symbol: Ω^-. An *elementary particle with negative charge and a mass 3276 times that of the electron. It is classified as a *hyperon.

onium ion. A positive ion formed by addition of a proton to a neutral molecule as in the formation of an ammonium, phosphonium, or oxonium ion.

opacity. The reciprocal of the *transmittance of a medium.

opaque. See translucent.

open-chain. Denoting a chemical compound with a straight or branched chain of atoms: i.e. a compound that does not contain a ring.

open-hearth process (Siemens-Martin process). A process for making steel by heating a mixture of pig iron and iron

oxide with producer gas on an open hearth.

operator. A mathematical symbol representing a particular operation. Thus, the differential operator d/dx represents differentiation with respect to the variable x.

opposition. The position in which two celestial objects, in practice the sun and any of the outer planets, have a difference of celestial *longitude of 180°. In this position, the outer planet in question is best seen, crossing the observer's meridian at midnight. The moon is at opposition when it becomes full.

optical activity. The property of certain substances of rotating the plane of *polarization of plane-polarized light. The effect is observed in passing a polarized beam through crystals, liquids, solutions, or vapour. It is caused by asymmetry of the molecules: a solution of one optical isomer will rotate the light in one direction and the other optical isomer causes equal rotation in the opposite sense (see optical isomerism).

The extent to which the plane of polarization is rotated depends on the wavelength of the light, the path length, and the concentration (of a solution). The optical activity of a substance is expressed by its *specific rotation* $[\alpha]_\lambda^t = \alpha V/lm$, where α is the rotation, V the volume, m the mass of dissolved or pure substance, and l the path length. In the notation, t is the temperature and λ the wavelength used (often D is used to indicate the D line of sodium). Optical rotation is measured in a *polarimeter.

A substance that rotates the plane in a clockwise sense (looking towards the oncoming light) is *dextrorotatory*: one that causes anticlockwise rotation is *laevorotatory*. Often prefixes *d*- (or (+)) or *l*- (or (-)) are used in the names of optically active isomers. These should not be confused with the *D*- and *L*- prefixed to the names of sugars and other natural compounds. *D*- compounds are named as having similar configurations to an arbitrarily chosen isomer of glyceraldehyde and *L-* compounds resemble the other isomer. Thus *D-* and *L-* refer to the configuration of the molecules, not the optical rotation, and it is possible for a *D-* isomer to be laevorotatory. *DL-* and *dl-* denote racemates.

optical axis. The axis of symmetry of a *lens, *mirror, or other optical system.

optical bench. A table, track, etc., on which sources, lenses, mirrors, and other optical components can be mounted and moved: used in experiments in optics.

optical centre. *See* lens.

optical character recognition (OCR). A method of feeding information into a computer in the form of conventional printed characters, which are identified by the input device (the *optical character reader*). A special typeface is usually necessary.

optical crown. Any of various types of optical glass with low dispersion of light. *Compare* optical flint.

optical flat. A flat surface on which the irregularities are not greater than the wavelength of light. Optically flat surfaces are necessary in many optical devices and experiments.

optical flint. Any of various types of optical glass with high dispersion of light. *Compare* optical crown.

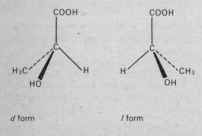

d form   *l* form

Fig. 1: Lactic acid

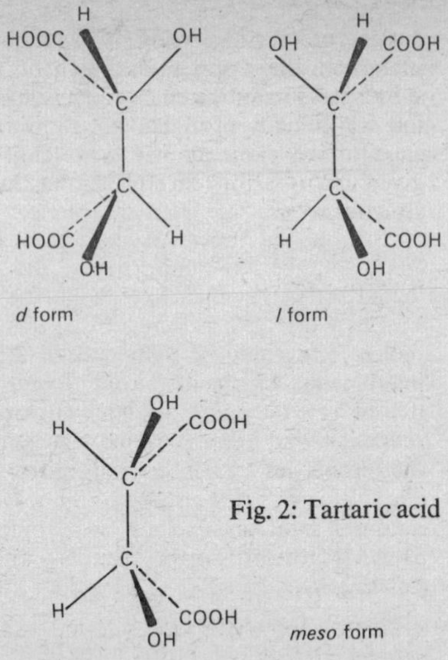

Fig. 2: Tartaric acid

meso form

optical isomerism (enantiomorphism). A type of *stereoisomerism in which a compound can have two structures, one of which is a mirror image of the other. Optical isomerism can occur in many substances, particularly those that contain an *asymmetric carbon atom*, i.e. a carbon atom to which is attached four different groups arranged tetrahedrally. This occurs in lactic acid, in which the central carbon atom is asymmetric and one structure is a mirror image of the other (see fig. 1). The two forms are said to be *optical isomers* or *enantiomers*. They have very similar physical properties but differ in their *optical activity. Tartaric acid is more complicated because it has two asymmetric carbon atoms in its molecules (see fig. 2). In addition to the *l-* and *d-* forms it has a third form—the *meso* form—which is not a mirror image of either. Stereoisomers that are not mirror images are called *diastereoisomers*. The *meso* form of tartaric acid is optically inactive because one half of the molecule is a mirror image of the other.

optical pyrometer. *See* pyrometer.

optical rotatory dispersion. A technique in which the *optical activity of a substance is investigated as a function of the wavelength of light. A graph of specific rotation against wavelength gives information on the molecular structure.

optic axis. The direction in a bire-fringent crystal in which no *double refraction occurs.

optics. The study of light and optical instruments. *Geometrical optics* is the use of ray diagrams and the laws of reflection and refraction to determine the properties of lenses and mirrors. *Physical optics* is the study of the wave nature of light, and includes such topics as diffraction, interference, and polarization.

orbit. 1. The path of one celestial body around another.
2. The path of an electron around the nucleus of an *atom, often visualized as a circle or ellipse.

orbital. A wave function in an atom or molecule. *See* atomic orbital, molecular orbital.

order. 1. The power to which a concentration is raised in an expression for the rate of a chemical reaction. Thus, if the rate of reaction between A and B is proportional to $[A][B]^2$, the reaction is said to be first order in A and second order in B. The overall order is three. The order of reaction is distinct from the molecularity.
2. *See* differentiation.
3. *See* differential equation.

order of magnitude. An approximate magnitude to within a factor of ten. For instance, values of 0.01 and 0.03 are the same order of magnitude: 10 and 1000 differ by two orders of magnitude.

ordinary ray. *See* double refraction.

ordinate. The *y* coordinate of a point on a two-dimensional Cartesian graph: i.e. the distance of the point from the *x* axis,

measured along the *y* axis from the origin. *Compare* abscissa.

ore. A mineral used as a source of a metal or other solid element.

organic. Relating to carbon compounds and their chemistry. The term was originally applied to compounds produced by living matter but now applies to any carbon compound with the exception of simple compounds such as oxides, carbon disulphide, cyanides, cyanates, and carbonates, which are considered to be *inorganic. The study and preparation of organic compounds is the province of *organic chemistry*.

organo-. Prefix indicating that compounds contain organic groups. *Organo-mercury* compounds, for example, are organometallic compounds of mercury.

organometallic. Denoting an organic metal compound. Organometallic compounds have direct bonds between the metal atom and carbon atom and include alkyls, aryls, Grignard reagents, and sandwich compounds. Metal carbonates, cyanides, cyanates, and carbides are not classed as organometallic.

origin. The point of intersection (0,0) of the coordinate axes in Cartesian coordinates; the point from which distances are measured.

orpiment. Arsenic trisulphide, As_2S_3, found naturally as a yellow solid and also synthesized for use as a pigment.

ortho. 1. Indicating that substituents in a benzene ring are in adjacent (1,2) positions. *See* structural isomerism.
2. Denoting the most hydrated acid of a series of acids formed from the same anhydride.

orthochromatic film. Photographic film that is sensitive to green light in addition to blue. *Compare* panchromatic film.

orthogonal. Indicating that lines, planes, curves, or other surfaces cross at right angles; perpendicular.

orthohydrogen. The form of molecular hydrogen in which the atomic nuclei have parallel *spins. It exists in equilibrium with *parahydrogen*, in which the spins are antiparallel. At room temperature the ratio of orthohydrogen to parahydrogen is about 3:1 and as the temperature is lowered the equilibrium proportion of parahydrogen increases. Pure parahydrogen can be obtained by using certain catalysts at liquid-air temperature. Chemically the two forms are the same although they differ slightly in their physical properties.

orthophosphate. See phosphate.

orthophosphite. See phosphite.

orthophosphoric acid. See phosphoric acid.

orthorhombic. See crystal system.

orthosilicate. See silicate.

orthosilicic acid. See silicic acid.

orthovanadate. See vanadate.

orthovanadic acid. See vanadate.

oscillating universe. See big-bang theory.

oscillator. An electronic circuit for producing an oscillating signal.

oscilloscope. See cathode-ray oscilloscope.

osmium. Symbol: Os. A hard brittle lustrous bluish-white *transition element found associated with platinum. It is used as a catalyst and in making platinum alloys—osmium-iridium alloys are used in pen nibs, compass bearings, and other applications requiring a hard corrosion-resistant material. Chemically, osmium resembles iridium. It forms a large number of complexes in a range of oxidation states. A.N. 76; A.W. 190.2; m.pt. 3045°C; b.pt. 5027°C; r.d. 22.57; valency 1–8.

osmium tetroxide. A yellow highly poisonous crystalline solid, OsO_4, with a pungent odour. It is prepared by heating osmium in air and is used as an oxidizing agent and catalyst. M.pt. 40°C; b.pt. 130°C; r.d. 4.9.

osmometer. Any apparatus for measuring osmotic pressure.

osmosis. Preferential flow of certain substances in solution through a *semipermeable membrane. If the membrane separates a solution from a pure solvent, the solvent will flow through the membrane into the solution. Osmosis can be stopped by applying a pressure to the solution—the value that prevents osmosis occurring is the *osmotic pressure* (symbol: Π). Osmosis is a *colligative property, depending on the number of molecules present. In dilute solutions the osmotic pressure of a solution is the pressure that the solute would exert if it were a gas occupying the same volume: i.e. $\Pi V = RT$, where V is the volume of solution, R is the gas constant, and T is the thermodynamic temperature.

osmotic pressure. See osmosis.

Ostwald's dilution law. The expression $K = \alpha^2/(1 - \alpha)V$, for dissociation of an acid in solution. K is a constant (the equilibrium constant), V the volume of water, and α the degree of dissociation. If α is small, the law can be written $\alpha = (KV)^{1/2}$. [After Wilhelm Ostwald (1853–1932), German chemist.]

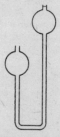

Ostwald viscometer

Ostwald viscometer. An apparatus for measuring the viscosity of liquids by comparing their rate of flow with that of a standard substance. The liquid is held in the upper bulb (see illustration) and the time (t) taken for it to flow to the lower bulb is measured. The density (ρ) is also found. The experiment is repeated with an equal volume of standard substance with viscosity η_o and density ρ_o. If this flows in a time t_o, the required viscosity η is given by $\eta/\eta_o = \rho t_o / \rho_o t$.

ounce. See Appendix.

output. 1. The signal, current, voltage, etc., delivered by an electrical circuit or device.
2. The processed data delivered by a computer. The term also refers to the part of the computer system that converts the data into usable form. Commonly this is some form of printer for producing conventional printed characters but may also be a device for producing punched cards, magnetic tape, or a visual display (see visual display unit). See also input.

overdamped. See damping.

overtone. A *partial of higher frequency than the fundamental in a musical note. See harmonic.

oxalate. Any salt or ester of oxalic acid.

oxalic acid (ethanedioic acid). A colourless poisonous crystalline solid, HOOC.COOH, present in many plants. It is prepared by passing carbon monoxide into sodium hydroxide solution under pressure and is used as a reducing agent and metal cleaner. M.pt. 101°C; r.d. 1.6.

oxidant. An oxidizing agent, particularly one that supplies oxygen. The term is often used for the liquid oxygen, hydrogen peroxide, or similar substance that oxidizes the fuel in a rocket.

oxidation. A process in which oxygen is combined with a substance or hydrogen is removed from a compound. Thus, the reaction $C + O_2 = CO_2$ is an oxidation of carbon and $H_2S + O = H_2O + S$, where O is supplied by an oxidizing agent, is an oxidation of hydrogen sulphide. The term has been extended to any reaction in which electrons are lost, as in the oxidation of ferrous ions to ferric ions: $Fe^{2+} - e = Fe^{3+}$. Compare reduction.

oxidation state or **number.** A measure of the electronic state of an atom in a particular compound, equal to the number of electrons it has more than or less than the number in the free atom. In ionic compounds the oxidation states are simply the charges of the ions, with due regard to sign. Thus, in calcium chloride, $CaCl_2$, the calcium has an oxidation state of $+2$ (Ca^{2+}) and the chlorine has a state of -1 (Cl^-). In covalently bonded compounds the electrons are assigned to the more electronegative of the pair of atoms. Thus in ammonia, NH_3, the nitrogen atom is in a -3 oxidation state and the hydrogen is in a $+1$ state (as if the compound was $3H^+N^{3-}$). A pure element has an oxidation state of 0. Coordinate bonds are considered to have no effect on the oxidation state. Thus the compound $Ni(CO)_4$ has nickel in the 0 oxidation state. The *complex ion $[Cu(NH_3)_4]^{2+}$ contains copper in its $+2$ oxidation state. It is considered to be a cupric ion (Cu^{2+}) with four ammonia molecules coordinated to it. The use of oxidation numbers is the basis of a system of nomenclature for inorganic compounds. Thus ferrous oxide, FeO, is called iron (II) oxide and ferric oxide, Fe_2O_3, is iron (III) oxide. See also valency.

oxide. Any compound of oxygen with another element. Oxides can be classified into two main types. Covalent oxides have simple covalently bonded molecules. Examples are carbon dioxide, sulphur trioxide, and nitrogen dioxide. Most covalent oxides are acidic, although some, such as carbon monoxide, are neutral. Ionic oxides are metal compounds containing the O^{2-} ion. They are all basic compounds: some are amphoteric. In addition, some metals can form *peroxides*, which contain the O_2^{2-} ion, *superoxides*, which contain the O_2^- ion, and *ozonides*, which contain the O_3^- ion.

oxidizing acid. An acid that acts as an oxidizing agent. The term is used for compounds, such as concentrated nitric acid, which dissolve metals that are lower than hydrogen in the *electrochemical series. Copper, for example, is dissolved according to the equations: $Cu + 2HNO_3 = CuO + H_2O + 2NO_2$; $CuO + 2HNO_3 = Cu(NO_3)_2 + 2H_2O$. In contrast, *nonoxidizing acids do not dissolve such metals.

oxidizing agent. A compound that causes *oxidation, either by supplying oxygen, abstracting hydrogen, or accepting electrons. Many compounds act as oxidizing agents in particular circumstances but the term is also used for compounds that are particularly effective oxidizing agents. Examples are oxygen, hydrogen peroxide, ferric salts, and concentrated nitric and sulphuric acids. See also oxidant.

oxonium ion. A positive ion of formula R_3O^+, where R is either hydrogen or an organic group. The simplest example is the hydroxonium ion, H_3O^+.

oxy. Designating a compound that contains oxygen.

oxygen. Symbol: O. A colourless odourless tasteless gaseous element forming 28% of the atmosphere by volume; the most abundant element in the earth's crust (49.2% by weight). Oxygen is essential for combustion and respiration of plants and animals. It is obtained by liquefying air and used in blast furnaces, welding, respirators, and the manufacture of organic chemicals. The element is diatomic, O_2, and highly reactive, combining with most other elements. It is one of the *chalconide elements. A highly active triatomic allotrope, *ozone, also exists. A.N. 8; A.W. 15.9994; m.pt. -218.4°C; b.pt. -183.0°C; r.d. 1.105 (air = 1).

ozone. A pale blue gaseous allotrope of oxygen, O_3, prepared by passing an electrical discharge through oxygen.

Ozone is a poisonous unstable compound, a powerful oxidizing agent, used for water purification. It reacts with unsaturated organic compounds to give *ozonides. M.pt. -192°C; b.pt. -112°C.

ozonide. 1. A type of compound prepared by addition of ozone to double bonds. Ozonides contain a three-membered ring of two carbon atoms and an oxygen atom in their molecules and are unstable, often explosive, substances. The process of forming ozonides (*ozonolysis*) is used as a tool in organic chemistry to detect the positions of double bonds in molecules.
2. An inorganic compound containing the O_3^- ion. An example is potassium ozonide, KO_3.

ozonizer. An apparatus for producing ozone by maintaining an electrical discharge in a stream of oxygen.

ozonolysis. See ozonide.

ozonosphere (ozone layer). A layer containing ozone in the earth's atmosphere. It lies between heights of 15 and 30 kilometres and absorbs the sun's higher-energy ultraviolet radiation.

P

packing fraction. The difference between the atomic mass of an isotope and its mass number, divided by its mass number: i.e. $f = (M - A)/A$. The fractions are sometimes multiplied by 10^4. They give an indication of the stability of atomic nuclei, being proportional to the energy binding the nucleons together.

pair production. Production of an elementary particle and its antiparticle from a photon in the field of an atomic nucleus. A gamma-ray photon, for example, can be converted into an electron and a positron. The total mass produced is equivalent to the energy lost by the photon according to the law of conservation of *mass-energy. See also annihilation.

palaeomagnetism. The study and measurement of the magnetization of rocks, used to investigate the way the earth's field has changed with time. The method is most useful for studying igneous rocks, which are formed at high temperatures and contain iron compounds. As they cool the rocks are magnetized by the earth's field and their intensity and direction of magnetization gives an indication of the intensity and direction of the field at the time the rocks were formed. Such studies show that the earth's polarity has changed direction many times in the past. The technique can also be used as a method of dating rocks.

palladic. See palladium.

palladium. Symbol: Pd. A steel-white soft ductile *transition element occurring in certain copper and nickel ores. The metal is used in jewellery and as a hydrogenation catalyst. It is fairly unreactive and does not tarnish in air, although an oxide forms at high temperatures. It dissolves slowly in cold nitric acid and hot hydrochloric and sulphuric acids and dissolves quickly in fused alkalis. Palladium is also capable of absorbing (occluding) up to 900 times its own volume of hydrogen. Unlike nickel it has very few simple ionic salts containing cations. Most of its compounds are complexes, being mainly palladium II (*palladous*) compounds with a few palladium IV (*palladic*) compounds. It also forms some complexes in other oxidation states. A.N. 46; A.W. 106.4; m.pt. 1552°C; b.pt. 3140°C; r.d. 12.02; valency 2, 3, or 4.

palladous. See palladium.

palmitic acid (hexadecanoic acid). A white crystalline solid, $CH_3(CH_2)_{14}$-COOH, occurring, in the form of its glyceride, in many oils and fats, from which it may be extracted by hydrolysis. One of the more common fatty acids, palmitic acid is used in soaps and lubricants and as a food additive. M.pt. 63°C; b.pt. 351°C; r.d. 0.85.

palmitin. See tripalmitin.

panchromatic film. Photographic film that is sensitive to all the colours of the visible spectrum. *Compare* orthochromatic film.

paper chromatography. See chromatography.

para. Indicating that substituents in a benzene ring are attached to carbon atoms on opposite sides of the ring: i.e. to the 1,4 positions. See structural isomerism.

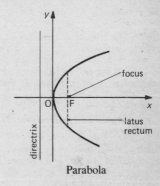

Parabola

parabola. A *conic with an eccentricity equal to 1. It has an equation of the type $y^2 = 4mx$, where $(m,0)$ is the focus and $y = -m$ is the directrix. In this form the vertex of the parabola is at the origin and the axis is the x axis. A line drawn from the focus to any point on the parabola makes an angle with the normal at that point equal to the angle made by a line parallel to the x axis, a property known as the *reflection property* of the parabola (*see* paraboloid).

parabolic. Having the shape of a parabola or a paraboloid.

paraboloid. A surface that has parabolic cross sections. There are two types. A *hyperbolic paraboloid* ($x^2/a^2 + y^2/b^2 = 2cz$) has parabolic cross sections in two planes and a hyperbolic cross section in the third. An *elliptic paraboloid* ($x^2/a^2 - y^2/b^2 = 2cz$) has an elliptical cross section in its third plane. A *paraboloid* *of revolution* is a special type of elliptic paraboloid produced by rotating a parabola about its x axis. This shape is used in devices for producing a parallel beam of radiation from a point source or for focusing a parallel beam to a point, as a result of the reflection property of the *parabola. Parabolic reflectors are used in searchlights, electric fires, radar antenna, radio telescopes, etc.

parachor. A property of pure liquids, $\gamma^{1/4}M(\rho_l - \rho_v)$, where γ is the surface tension, M the molecular weight, and ρ_l and ρ_v the densities of liquid and vapour. The value is related to the volume of the molecules and has been used in determining molecular structure.

paracyanogen. A white polymeric substance formed by heating cyanogen.

paraffin. 1. *See* alkane.
2. (kerosene). A mixture of hydrocarbons boiling in the range 150–300°C obtained by distillation of petroleum. It is used as a fuel, especially for domestic heaters.

paraffin wax. A white solid mixture of hydrocarbons melting in the range 50–60°C obtained from petroleum. It is used in candles, waxed paper and cloth, and polishes.

paraformaldehyde. A white solid polymer of *formaldehyde, $(HCHO)_n$, where n is 6 or more. It is made by dehydrating aqueous formaldehyde and is used as a disinfectant. M.pt. 120–170°C.

parahydrogen. *See* orthohydrogen.

Paraldehyde

paraldehyde. A colourless cyclic polymer of *acetaldehyde, $C_6H_{12}O_3$, made by treating acetaldehyde with sulphuric acid. It is less reactive than acetaldehyde and is used as a substitute for it. M.pt. 12°C; b.pt. 124°C; r.d. 0.99.

parallax. 1. The apparent change in the separation between two objects when they are viewed from different positions. If two objects are in line with the eye they appear to move apart as the observer moves to the left or right. "No parallax" only occurs when both objects occupy the same position: this is used as a test for finding the position of the image in simple experiments in optics. Parallax errors can occur in measuring instruments when the pointer is a small distance in front of the scale. In many instruments a mirror is placed behind the pointer, which is read from an angle at which it is in line with its reflection—the error is then eliminated.
2. annual parallax. The maximum angle subtended at a given star by the earth's mean orbital radius. The *parsec is a unit of distance based on parallax.

parallel. 1. Denoting lines or planes that are equally distant from each other at all points.
2. *in parallel.* Denoting electrical circuit elements joined so that each element is connected across the same two points of the circuit. If resistances R_1, R_2, R_3, --- are in parallel, the total resistance R is given by: $(1/R) = (1/R_1) + (1/R_2) + ...$ Capacitances in parallel have a total capacitance equal to their sum: i.e. $C = C_1 + C_2 + C_3 + ...$ *Compare* series.

parallelepiped. A solid whose faces are all parallelograms.

parallelogram. A quadrilateral that has both pairs of opposite sides parallel. The area of a quadrilateral is the length of the base multiplied by the perpendicular distance between the base and the opposite side.

parallelogram rule. A method of adding two *vectors by representing them by lines forming adjacent sides of a parallelogram, so that the vector sum (the *resultant*) is represented by the diagonal of the parallelogram. For example, in a parallelogram OACB, *OC* is the resultant of *OA* and *OB* and its length and direction can be found by solving the triangle OAC. A physical example of vector addition would be two forces acting simultaneously on a body, for which the resultant force could be found by a *parallelogram of forces*. The method can be extended to adding more than two vectors by simply placing them end to end to form a polygon. If vectors are represented in three dimensions in the form $a = ia_x + ja_y + ka_z$, then addition of vectors follows the rule: $a + b = i(a_x + b_x) + j(a_y + b_y) + k(a_z + b_z)$.

paramagnetism. A type of magnetic behaviour in which the material has a fairly low positive susceptibility that is inversely proportional to temperature. A paramagnetic sample will tend to move towards an applied magnetic field. The effect is caused by the spins of unpaired electrons in the atoms or molecules, which give the atoms a magnetic moment. The effect of paramagnetism always overwhelms the diamagnetic behaviour of the solid. Measurement of paramagnetic susceptibility is one method of finding the number of unpaired electrons in compounds of *transition metals. See also* magnetism.

parameter. An element of a mathematical expression that may, by varying, cause different values to be assigned to the whole. For example, if the equation of a circle is of the form $y^2 = r^2 - x^2$, then x and y are variables and r is a parameter, determining which circle centred on the origin is described.

Paraquat. Ⓡ A yellow water-soluble highly poisonous solid, $C_9H_{20}N_2(SO_4)_4$, widely used as a herbicide.

parent. *See* daughter.

parity. Symbol: *P*. A property of elementary particles depending on the symmetry of their wave function with respect to changes in sign of the coordinates. The wave function will generally be a mathematical function of coordinates x, y, and z. If these are replaced by $-x$, $-y$, and $-z$ the wave function either remains the same, in which case the particle has parity $+1$, or changes sign, in which case it has parity -1. In any system of particles interacting by strong or electromagnetic interactions the principle of *conservation of parity* applies: i.e. the total parity is unchanged in the interaction. This does not occur in processes that involve weak interactions. In beta decay it is found that the emitted electrons always have a spin in the opposite direction to their direction of motion—a violation of the principle of parity invariance.

parsec. A unit of distance used in astronomy, equal to the distance corresponding to a *parallax of one second of arc. It is equivalent to 3.26 light-years or 3.086×10^{16} metres.

partial. A tone that is an integral multiple of the fundamental frequency in a musical note (*see* quality). The partials of a note are the same as its *harmonics. The fundamental is the first partial. Higher frequencies are also called overtones.

partial derivative. A *derivative taken with respect to only one variable, other variables being treated as constants. The symbol ∂ is used. For example, if $y = 3x^2z^3$, $\partial y/ \partial x = 6xz^3$ and $\partial y/ \partial z = 9x^2z^2$.

partial pressure. The pressure that a component of a mixture of gases would exert if it were present alone in the same volume as the mixture. *See* Dalton's law of partial pressures.

particle physics. The branch of physics concerned with the properties, interactions, and structure of elementary particles.

partition coefficient. The ratio of the concentrations of a given substance in two different phases of a system. Thus, if a substance is added to a system of two immiscible liquids the ratio of its concentration in one to its concentration in the other is its partition coefficient for that system. The value depends on the temperature.

parton. An ultimate fundamental particle postulated as a basic unit of other elementary particles. In the simplest theories the partons are *quarks.

pascal. Symbol: Pa. The *SI unit of pressure, equal to a pressure of one newton per square metre. [After Blaise Pascal (1623–62), French scientist.]

Paschen series. A *spectral series in the infrared emission of hydrogen with wavelengths given by $1/\lambda = R(1/3^2 - 1/m^2)$, where R is the *Rydberg constant and m is 4, 5, 6, etc. [After Friedrich Paschen (1865–1947), German physicist.]

passive. 1. Denoting a solid substance that is chemically unreactive as a result of the formation of a thin protective layer on its surface. Aluminium, for instance, is highly reactive and quickly forms a transparent oxide, which coats the surface and prevents further attack. Similarly, iron is rendered passive by concentrated nitric acid, again by the formation of an oxide.
2. Denoting an electronic component that does not amplify the signal. Resistors, capacitors, and inductors are passive components.

patronite. A naturally occurring sulphide of vanadium, VS_4, used as a vanadium ore.

Pauli exclusion principle. The principle that no two electrons in an atom can have the same set of quantum numbers. [After Wolfgang Pauli (1900–58), Austrian physicist.]

pearl ash. See potassium carbonate.

pearlite. A constituent of many steels consisting of alternate layers of *ferrite and *cementite. It is formed from *austenite by slow quenching from high temperatures.

Peltier effect. A *thermoelectric effect in which an electric current passed through a junction between two different solids causes heat to be produced or absorbed, depending on the direction of the current. The phenomenon is the converse of the Seebeck effect. Semiconductor junctions can be used as cooling elements. [After Jean Peltier (1785–1845), French scientist.]

pendulum. 1. A body suspended from a point above its centre of gravity and allowed to swing freely under the action of gravity. The *simple pendulum* consists of a weight hung from a fixed support by a string of negligible weight. For small displacements the bob undergoes *simple harmonic motion with a period given by $2\pi(l/g)^{1/2}$, where l is the length and g the acceleration of free fall. A *compound pendulum* is any rigid body swinging on a pivot. Its period is $2\pi[(k^2 + h^2)^{1/2}/hg]^{1/2}$, where h is the distance between the pivot and the centre of gravity and k is the *radius of gyration. Pendulums can be used for measuring g, the *acceleration of free fall, and are used in prospecting for oil and minerals (see also gravimeter). *Kater's pendulum is an example of a compound pendulum designed for determining g. Many other specialized types of pendulum exist, including compensated pendulums for regulating clocks, the *ballistic pendulum for measuring velocities, and *Foucault's pendulum for demonstrating the rotation of the earth.
2. Any of various analogous devices in which a mass is suspended on a spring or torsion wire and allowed to rotate with a backwards and forwards twisting action. Such devices are called *torsion pendulums*.

penicillin. Any of a class of related

antibiotics produced by the *Penicillium* mould or made synthetically.

pennyweight. *See Appendix.*

pentaerythritol. A white crystalline powder, $C(CH_2OH)_4$, prepared by reacting acetaldehyde with excess formaldehyde. It is used in making synthetic resins. M.pt. 262°C; r.d. 1.4.

pentagon. A polygon that has five sides.

pentahydrate. A solid compound with five molecules of *water of crystallization per molecule of compound, as in copper sulphate pentahydrate, $CuSO_4.5H_2O$.

pentane. A colourless flammable liquid straight-chain alkane, C_5H_{12}, obtained from petroleum. Pentane is also called *n-pentane* to distinguish it from *iso-pentane and *neopentane. M.pt. -129.7°C; b.pt. 36.1°C; r.d. 0.63.

pentanoic acid (valeric acid). A colourless liquid, $CH_3(CH_2)_3COOH$, with a penetrating odour, made by oxidation of pentanol. It is used as an intermediate for perfumes. M.pt. -34°C; b.pt. 185°C; r.d. 0.9.

pentavalent (quinquevalent). Having a valency of five.

pentode. A *thermionic valve with five electrodes. It is a *tetrode modified by the inclusion of a third grid, the *suppressor grid*, between the anode and the screen grid to reduce the effects of *secondary emission from the anode. It is held at a negative potential.

pentose. A simple *sugar that has five carbon atoms in its molecules.

pentyl group. A monovalent group C_5H_{11}-, derived from pentane by removing one atom of hydrogen. The term is often used for the group with a straight chain of atoms: i.e. for the *n-pentyl group*. Other isomers exist including the *tert-pentyl group*, $CH_3CH_2C(CH_3)_2$-, the *isopentyl group*, $(CH_3)_2CHCH_2CH_2$-, and the *neopentyl group*, $(CH_3)_3CCH_2$-. The pentyl group was formerly called the *amyl group*.

penumbra. **1.** A region of partial *shadow surrounding the umbra. **2.** The outer, lighter portion of a *sunspot.

peptide. A compound formed by two or more amino acids linked together. The amino group ($-NH_2$) of one acid can be thought of as reacting with the carboxyl group ($-COOH$) of another to give the group $-NH-CO-$. This is known as the *peptide linkage*. *See also* polypeptide, protein.

peptization. Dispersion of a substance to form a colloid by addition of a chemical to stabilize the sol.

perboric acid. A hypothetical acid, HBO_3, known in the form of *perborate* salts.

perchlorate. Any salt or ester of perchloric acid.

perchloric acid. A colourless hygroscopic explosive liquid, $HClO_4$, made by distilling potassium perchlorate with 96% sulphuric acid under reduced pressure. It is a strong oxidizing agent, used in analytical chemistry, electroplating of metals, and explosives. M.pt. -112°C; b.pt. 16°C (8 mmHg); r.d. 1.8.

perfect gas. *See* ideal gas.

periclase. A natural form of magnesium oxide, MgO.

perigee. The point at which the moon or any other orbiting satellite is closest to the earth. *Compare* apogee.

perihelion. The point of closest approach to the sun attained by a planet, asteroid, comet, or other orbiting body. *Compare* aphelion.

perimeter. The distance around a plane figure. Thus, the perimeter of a polygon is the sum of the lengths of its sides and the perimeter of a circle is its circumference.

period. The constant interval between identical states of a system whose properties vary periodically. For a system in which a quantity varies with time, as in a *wave motion or *simple harmonic motion, the period is the time taken to complete a *cycle. The period of a crystal lattice in a particular direction is the distance between similar lattice points in that direction.

periodate. Any salt of periodic acid.

periodic acid. A white crystalline acid, HIO_4, made by the action of perchloric acid on iodine. Sublimes at $110°C$.

periodic table. A tabular arrangement of the elements in order of increasing atomic number such that similarities are displayed between groups of elements. Early attempts at classifying the elements include *Dobereiner's triads and *Newlands's law of octaves. In the modern form of the table (see illustration) the columns are called *groups*. All the elements in a group have similar electron configurations and a consequent similarity in chemical properties. Thus, in group IA the alkali metals all have a single electron in their outer shell and are electropositive metals. There is a tendency for electropositive behaviour to increase down the group because the atoms are larger. Thus potassium has a lower ionization potential than sodium and is more reactive. This behaviour is very marked in group IV, where there is a transition from the typical nonmetal, carbon, to the metal, lead. The rows across the table are called *periods*. Those from lithium to neon and sodium to argon are *short periods*, the others are *long periods*. Across a period there is a change from electropositive metallic behaviour to electronegative nonmetallic behaviour. The table also shows the transition series (see transition metal) and the *lanthanides and *actinides.

periscope. An apparatus for viewing objects when there is no direct line of sight to the eye. Essentially, it contains a mirror at 45° to the incident light, thus reflecting the light at right angles along the periscope tube onto a second mirror that reflects the light out of the apparatus, so that the reflected light is parallel to the original incident light but is displaced. Total internal reflection by right-angled prisms is sometimes used instead of reflection by mirrors.

permanent gas. A gas that cannot be liquefied by pressure alone: i.e. a gas with a *critical temperature below normal temperature, so that some cooling is necessary for the gas to be liquefied.

permanent hardness. See hard water.

permanent magnetism. See magnetism.

permanganate. A salt containing the ion MnO_4^-. Permanganates are powerful oxidizing agents derived from the hypothetical *permanganic acid*, $HMnO_4$. The permanganate ion is an intense purple colour. In strongly basic solutions it changes to the *manganate ion.

permanganic acid. See permanganate.

permeability. See relative permeability.

permittivity. See relative permittivity.

permutation. An ordered selection of a number of entities from a set of entities. The number of permutations of r elements from a set of n is denoted ^nP_r and is given by $n!/(n - r)!$. A permutation differs from a *combination in that the order is considered. There are six permutations of two letters from the letters A, B, and C: namely AB, BA, AC, CA, BC, and CB. There are only three combinations because pairs like AB and BA are the same combination of letters.

peroxide. 1. Any inorganic compound containing the O_2^{2-} ion. Peroxides can be hydrolysed to hydrogen peroxide.

peroxodisulphuric acid (persulphuric acid). An acid, $HOSO_2OOSO_2OH$, produced in solution by the electrolysis of potassium sulphate in dilute sulphuric acid at high current density.

peroxosulphuric acid (Caro's acid). A white crystalline strongly oxidizing acid, $HOSO_2OOH$, formed by the action of hydrogen peroxide on sulphuric acid.

Periodic Table of the Elements

IA	IIA	IIIB	IVB	VB	VIB	VIIB	VIII	VIII	VIII	IB	IIB	IIIA	IVA	VA	VIA	VIIA	O
1 H																	2 He
3 Li	4 Be											5 B	6 C	7 N	8 O	9 F	10 Ne
11 Na	12 Mg											13 Al	14 Si	15 P	16 S	17 Cl	18 Ar
19 K	20 Ca	21 Sc	22 Ti	23 V	24 Cr	25 Mn	26 Fe	27 Co	28 Ni	29 Cu	30 Zn	31 Ga	32 Ge	33 As	34 Se	35 Br	36 Kr
37 Rb	38 Sr	39 Y	40 Zr	41 Nb	42 Mo	43 Tc	44 Ru	45 Rh	46 Pd	47 Ag	48 Cd	49 In	50 Sn	51 Sb	52 Te	53 I	54 Xe
55 Cs	56 Ba	57* La	72 Hf	73 Ta	74 W	75 Re	76 Os	77 Ir	78 Pt	79 Au	80 Hg	81 Tl	82 Pb	83 Bi	84 Po	85 At	86 Rn
87 Fr	88 Ra	89† Ac															

Transition metals

*Lanthanides	57 La	58 Ce	59 Pr	60 Nd	61 Pm	62 Sm	63 Eu	64 Gd	65 Tb	66 Dy	67 Ho	68 Er	69 Tm	70 Yb	71 Lu
†Actinides	89 Ac	90 Th	91 Pa	92 U	93 Np	94 Pu	95 Am	96 Cm	97 Bk	98 Cf	99 Es	100 Fm	101 Md	102 No	103 Lr

perpendicular. Indicating that a line or plane is at right angles to another line or plane: a line or plane that is at right angles to another. A line is perpendicular to a curve or surface if it is perpendicular to its tangent line or plane.

perpetual motion. Continuous motion without any supply of energy. A machine or device that, once set in motion, would continue moving for ever would only be possible in the absence of forces due to friction, viscosity, etc. It would be impossible to construct such a machine that performed useful work as its operation would contravene the first law of thermodynamics.

Perspex. ® *See* polymethyl methacrylate.

persulphate. Any salt of persulphuric acid.

persulphuric acid. *See* peroxodisulphuric acid.

petrochemicals. Chemicals manufactured or obtained from petroleum or natural gas.

petrol. A mixture of octane, heptane, and other hydrocarbons used as a fuel in internal-combustion engines. It is obtained from petroleum and usually contains additives to prevent knocking, retard corrosion, etc.

petrolatum (petroleum jelly). A translucent semisolid amorphous mixture of hydrocarbons obtained from petroleum. It is used in medical dressings and as a lubricant.

petroleum. A mixture of hydrocarbons and other organic compounds formed originally from marine sediments. The crude oil is refined by distillation to give petrol, paraffin oil, lubricating oils, waxes, and tar. Petroleum is the raw material for the production of many organic chemicals.

petroleum ether (ligroine, benzine). A colourless flammable liquid, a mixture of hydrocarbons incorrectly named an ether. It is prepared by distillation of petroleum and used as a laboratory solvent.

pewter. An alloy of tin (about 80%) and lead with small amounts of antimony (up to 3%).

pH. A measure of the acidity or alkalinity of a solution, equal to the logarithm to base 10 of the reciprocal of the concentration of hydrogen ions: i.e. pH $= \log_{10}(1/[H^+])$. Neutral solutions have a pH of 7, acid solutions a pH less than 7, and alkaline solutions a pH greater than 7.

phase. 1. The stage that a periodically varying system has reached in its cycle at a given time measured from a specified reference point. Two alternating currents, waves, etc., of the same frequency are said to be *in phase* if they have their maximum and minimum values at the same time. Otherwise they are *out of phase*. A varying quantity can be represented by a line (vector) rotating about a point (*see* simple harmonic motion) and the difference in phase of two such quantities can be expressed by the angle between their rotating vectors. This is the *phase angle*: it is zero when the two systems are in phase.
2. Any of the apparent changes in shape that the moon cyclically passes through on its orbit around the earth. As more and then less of the moon's face is illuminated by the sun it is seen to grow from a slender crescent to a fully rounded disc and back again during the course of its synodic period (*see* period). When the moon lies between the sun and the earth (inferior *conjunction) it cannot be seen at all, except possibly as a black disc during a total solar eclipse. This phase is referred to as *new*. With increasing *elongation a crescent appears, popularly but not astronomically also called *new moon*. At *quadrature, half the moon is illuminated, this being *first quarter*. As more of the surface becomes visible the moon

reaches a stage where it is called *gibbous*. It continues onward until it reaches superior conjunction, with the earth lying between the sun and the moon. The whole of the lunar disc is now illuminated and the phase is called *full*. As the elongation between the sun and the moon decreases the phases are repeated in reverse order, the phase at quadrature this time being called *last quarter*, until new moon is once again reached.

3. A homogeneous part of a system, divided from other parts of the system by definite boundaries. Thus, a mixture of ice crystals in water contains two phases as does a suspension of solid particles in a gas or an alloy consisting of small inclusions of one metal in a matrix of another. A solution of salt in water, a solid solution, and a mixture of gases are all single phases.

phase modulation. *See* modulation.

phase rule. The principle that at equilibrium $P + F = C + 2$, where P is the number of phases in the system, C the number of *components, and F the number of degrees of freedom.

phenetole (phenyl ethyl ether). A colourless liquid, $C_6H_5OC_2H_5$. B.pt. 172°C; r.d. 0.97.

phenol. 1. (carbolic acid). A white hygroscopic crystalline solid, C_6H_5OH, which turns pink if impure or in light. An important aromatic hydroxy compound, it is manufactured by preparing sodium benzene sulphonate and treating with acid. Phenol is used to make resins, polymers, weed-killers, and pharmaceuticals. M.pt. 42°C; b.pt. 182°C; r.d. 1.1.

2. Any of a class of compounds that contain a hydroxyl (-OH) group bound to an aromatic ring. The phenols are acidic compounds. *Compare* alcohol.

phenol-formaldehyde resin. Any of a class of synthetic resins made by *condensation of formaldehyde with phenol or with a substituted phenol. The reaction is catalysed by acid or alkali and a

powder is produced, which is mixed with filler and used for moulding thermosetting plastics. Phenol-formaldehyde resins, such as Bakelite, are the most widely used *phenolic resins. They are nonflammable, moisture resistant, and thermally stable and are used for many moulded articles, laminates, and adhesives.

phenolic resin. Any of a class of synthetic thermosetting resins made by copolymerizing phenol, or a substituted phenol, with an aldehyde such as formaldehyde or furfural. The *phenolformaldehyde resins are the most widely used phenolics.

phenolphthalein. A colourless crystalline compound, $C_{20}H_{14}O_4$, used as an indicator. It is colourless in acid solutions and red in alkaline solutions, changing in the pH range 8.3-10.4.

phenyl ethyl ether. *See* phenetole.

phenyl group. The monovalent group C_6H_5-, derived from benzene.

phenyl hydrazine. A yellow oil or crystalline solid, $C_6H_5NHNH_2$. It forms *phenylhydrazones with aldehydes and ketones. M.pt. 190°C; b.pt. 243°C; r.d. 1.1.

phenylhydrazone. Any of a class of compounds with the general formula $R_2C = NNH.C_6H_5$, where R is an alkyl or aryl group or a hydrogen atom. The phenylhydrazones, which are crystalline solids, are made by reaction between phenylhydrazine and an aldehyde or ketone.

phenyl methyl ketone. *See* acetophenone.

Philosopher's stone. *See* alchemy.

philosopher's wool. *See* zinc oxide.

phlogiston. A substance formerly supposed to be present in all flammable substances and to be released during combustion, leaving ash. The phlogiston theory was disproved in the eighteenth century by Lavoisier.

Phobos. *See* Mars.

phon. A unit of loudness of sound defined with reference to a standard tone of frequency 1000 hertz. The intensity of this standard is varied until it is judged to be as loud as the sound to be measured. The loudness of the sound is then *n* phons, where the reference is *n* decibels above a standard datum level.

phonon. A quantized lattice vibration in a solid.

phosgene (carbonyl chloride). A colourless poisonous gas, $COCl_2$, with an odour of freshly mown hay, made by passing carbon monoxide and chlorine over activated carbon. It is used as a war gas and as a reagent in the manufacture of many organic chemicals. M.pt. $-118\,°C$; b.pt. $8\,°C$.

phosphate. Any salt or ester of a phosphoric acid. Phosphates include *orthophosphates*, M_3PO_4, *diphosphates* (*pyrophosphates*), M_4PO_7, and various *polyphosphates*. M is a monovalent metal.

phosphide. A compound of phosphorus with a more electropositive element.

phosphine. A colourless poisonous spontaneously flammable gas, PH_3, with the odour of rotting fish. It is prepared by dropping potassium hydroxide on to phosphorus and is used as a doping agent for semiconductors. M.pt. $-133\,°C$; b.pt. $-85\,°C$; r.d. 1.2.

phosphite. Any salt or ester of a phosphorous acid. The phosphites include the *orthophosphites*, M_2HPO_3, the *diphosphites* (*pyrophospites*), $M_2H_2P_2O_5$, and the *hypophosphites*, MH_2PO_2. M is a monovalent metal.

phosphonium ion. The monovalent ion PH_4^+ produced by protonation of phosphine. Phosphonium salts contain these ions and are analogous to *ammonium compounds.

phosphor. Any substance that exhibits fluorescence or phosphorescence.

phosphor bronze. *See* bronze.

phosphorescence. *Luminescence that persists after the source of excitation has been removed. It is distinguished from *fluorescence—in which emission ceases immediately. In scientific usage the term can either denote a luminescence that persists long enough to be recognized with the eye, or can denote one that lasts longer than 10^{-8} seconds. Phosphorescence was originally the "cold light" emitted from slowly oxidizing phosphorus. In general non-scientific usage the term is often used for cold light of the type emitted by seaweed or other organic matter (*see* bioluminescence, chemiluminescence).

phosphoric acid. Any of a number of oxyacids of phosphorus. The term is often applied to *orthophosphoric acid*, H_3PO_4, which is a syrupy colourless liquid made by the action of sulphuric acid on a phosphate. It is used in the manufacture of fertilizers and soaps. M.pt. $42.35\,°C$; r.d. 1.8. *Diphosphoric acid* (*pyrophosphoric acid*), $H_4P_2O_7$, is a glassy solid produced by heating orthophosphoric acid. It is the first member of a series of *polyphosphoric acids* with the general formula $H_{n+2}P_nO_{3n+1}$. The series includes *metaphosphoric acid*, a highly polymeric deliquescent glassy solid, $(HPO_3)_x$, which is used as a dehydrating agent.

phosphorous acid. A white crystalline hygroscopic dibasic acid, H_3PO_3, obtained by the action of water on phosphorous III oxide. M.pt. $73.6\,°C$; r.d. 1.7. This compound is also called *orthophosphorous acid*. Other acids include *diphosphorous acid* (*pyrophosphorous acid*), $H_4P_2O_5$, which is dibasic; *hypophosphorous acid*, H_3PO_2, which is monobasic; and the polymeric *metaphosphorous acid*, HPO_2.

phosphorus. Symbol: P. A nonmetallic element occurring chiefly in phosphate rocks. It exists in three allotropic forms that differ in chemical activity. White phosphorus is manufactured by reducing phosphates with carbon and silica in an electric furnace. It is the most reactive form of phosphorus, burning spontaneously in air, and is insoluble in water and soluble in organic solvents. It contains P_4 molecules, in which the four atoms are arranged at the corners of a tetrahedron. When heated at about 400°C for a few hours it converts into red phosphorus, a more stable allotrope of complex structure. Black phosphorus is obtained by heating white phosphorus at 200–300°C using a mercury catalyst. It is stable in air. The element is used in the manufacture of phosphoric acid, alloys, and semiconductors. Red phosphorus is used in matches. Phosphorus forms covalently bonded compounds with a valency of three, as in PCl_3, and with a valency of 5, as in PCl_5. It also forms binary phosphides with many other elements. A.N. 15; A.W. 30.9738; m.pt. 44.1°C (white); b.pt. 280°C (white); red phosphorus sublimes at 416°C; r.d. 1.82 (white), 2.20 (red), 2.25–2.69 (black).

phosphorus chloride. Either of two chlorides of phosphorus made by reaction between the elements. *Phosphorus trichloride*, PCl_3, is a colourless fuming liquid. M.pt. −112°C; b.pt. 76°C; r.d. 1.6. *Phosphorus pentachloride*, PCl_5, is a yellow volatile solid. Sublimes at 160°C; r.d. 3.6. Both compounds are used as chlorinating agents.

phosphorus oxide. Any of three oxides of phosphorus, P_4O_6, P_4O_8, and P_4O_{10}. The most important is *phosphorus pentoxide*, P_4O_{10}, which is a white hygroscopic powder made by burning phosphorus in dry air. It is the anhydride of phosphoric acid and is used as a dehydrating agent and catalyst. The formula is often written P_2O_5.

phosphorus oxychloride (phosphoryl chloride). A colourless fuming liquid, $POCl_3$, with a pungent odour. It is made by reacting phosphorus pentoxide and chlorine, and is used to manufacture organic phosphates and as a chlorinating agent. M.pt. 125°C; b.pt. 107°C; r.d. 1.7.

phosphorus pentoxide. *See* phosphorus oxide.

phosphoryl chloride. *See* phosphorus oxychloride.

phot. A unit of intensity of illumination equal to one lumen per square centimetre.

photocathode. A cathode from which electrons are emitted as a result of the *photoelectric effect.

photocell (photoelectric cell). Any device for converting light or other electromagnetic radiation directly into an electric current. Photocells depend for their action on the *photoelectric effect, *photoconductivity, or the *photovoltaic effect.

photochemistry. The branch of chemistry concerned with chemical reactions produced by light or ultraviolet radiation.

photochromism. Change of colour occurring when certain substances are exposed to light. Sometimes the change is reversible. The effect is displayed by some dyes.

photoconductivity. The increase of the electrical conductivity of certain solids, usually semiconductors, as a result of the impact of photons. It occurs when the photons have sufficient energy to increase the energy of an electron in a filled band, so that it is promoted into the conduction band.

photodisintegration. Disintegration of an atomic nucleus when it is hit by a gamma-ray photon. The photon may displace a nucleon or may cause fission (*photofission*) of the nucleus.

photoelectric cell. *See* photocell.

photoelectric effect. The ejection of electrons from a solid as a result of irradiation by light or other electromagnetic radiation. The number of electrons emitted depends on the intensity of the light and not on its frequency: the energy of the electrons is proportional to the frequency. This is explained by the theory that each electron is ejected by a single photon of radiation with energy $h\nu$, where h is the Planck constant and ν the frequency of the radiation. The energy E of the electrons is given by the *Einstein photoelectric equation*: $E = h\nu - \phi$, where ϕ is the *work function of the solid, i.e. the minimum energy required to remove an electron. Some metals, such as caesium and rubidium, have a low enough work function for electrons to be ejected by visible light but most solids only emit in the ultraviolet region of the spectrum.

photoelectron. An electron emitted by the photoelectric effect or by photoionization.

photoelectron spectroscopy. A form of electron spectroscopy in which the sample, usually a gas or solid, is irradiated with a beam of monochromatic ultraviolet radiation or X-rays. Electrons are ejected by photoemission or the photoelectric effect and measurement of their energy gives information on the ionization potentials of molecules and on the energy levels in the sample.

photoemission. The emission of electrons by the photoelectric effect or by photoionization.

photofission. *See* photodisintegration.

photoionization. Ionization of atoms or molecules by light or other electromagnetic radiation. The process can be thought of as a collision of a photon, with energy $h\nu$ where h is the Planck constant and ν the frequency, with an atom. If the photon energy exceeds the *ionization potential an ion is formed: $M + h\nu = M^+ + e$.

photoluminescence. *Luminescence occurring as the result of irradiation by electromagnetic radiation. Incident radiation is absorbed and produces atoms or molecules in excited states. They then decay, either directly back to the ground state or via an intermediate excited state, with the emission of light or other radiation. The wavelength of the emitted radiation is always longer than that of the absorbed radiation. Photoluminescence is the mechanism causing the brightness of fluorescent paints and materials. It is also used in detergent whiteners, which are compounds that absorb ultraviolet radiation and then emit blue light over a long period of time—thus giving a blue cast to a white fabric and counteracting any yellowing.

photolysis. The dissociation of a chemical compound into other compounds, atoms, and free radicals by irradiation with electromagnetic radiation, usually light or ultraviolet radiation. *See also* flash photolysis.

photometer. An instrument for measuring luminous intensity, usually by comparing a light source with a standard source. Typical photometers are the *Lummer-Brodhun, *flicker, and *grease-spot photometers.

photometry. The measurement of intensities of light sources and of illumination by use of photometers.

photomultiplier. A device for detecting photons, consisting of an *electron multiplier with a photocathode on the front. Photons hitting this electrode produce electrons by the photoelectric effect and these are amplified and detected by the electron multiplier.

photon. A quantum of electromagnetic radiation. The photon is usually thought of as an elementary particle with zero rest mass and an energy $h\nu$, where h is the Planck constant and ν the frequency of radiation.

photopic vision. Vision by the *eye when the cones of the retina are the main receptors. This is the type of vision occurring at normal luminance levels. The cones can distinguish different colours. *Compare* scotopic vision.

photosphere. The surface region of the *sun.

photovoltaic effect. The production of an e.m.f. between two layers of different materials when the surface is irradiated by light or other electromagnetic radiation. The phenomenon is shown by a thin layer of copper oxide on copper and by gold on selenium (*see* selenium cell).

phthalic acid. A colourless crystalline dibasic carboxylic acid, $C_6H_4(COOH)_2$. The carboxyl groups are in *ortho* positions. M.pt. 207°C; r.d. 1.6.

phthalic anhydride. A white crystalline compound, $C_8H_4O_3$, made by dehydrating phthalic acid or by oxidation of naphthalene. It is used as a drying agent and a raw material for making dyes. M.pt. 131°C; sublimes at 285°C; r.d. 1.5. *See also* anhydride.

physical change. Any change that does not involve the production of different chemical compounds.

physical chemistry. The branch of chemistry concerned with such topics as the physical properties of chemical compounds and general investigations of their structure, chemical bonding, and mechanism and rates of reaction.

physical optics. *See* optics.

physics. The study of matter and energy without reference to chemical changes occurring. "Traditional" physics includes such topics as heat, light, sound, mechanics, electricity, and magnetism. "Modern" physics includes atomic, nuclear, and particle physics, relativity, quantum mechanics, and astrophysics.

physisorption. *Adsorption in which the adsorbed substance is held by van der Waals forces.

pi (π). The ratio of the circumference of a circle to its radius: 3.141 592 653 ....

picrate. Any salt or ester of picric acid.

picric acid (2,4,6-trinitrophenol). A poisonous yellow explosive crystalline solid, $C_6H_2(NO_2)OH$, used in explosives

and the manufacture of dyes. M.pt. 122°C; r.d. 1.8.

piezoelectric effect. An effect observed in certain crystals whereby they develop a potential difference across a pair of opposite faces when subjected to a stress. The phenomenon is observed in Rochelle salt and quartz and is used in transducers. *Compare* electrostriction.

pig iron. Impure iron obtained by smelting iron in a *blast furnace: used for making steel.

pigment. An insoluble *colourant. Pigments are used in dispersion, as in paints, or can be melted together with the substance to be coloured, as in many synthetic fibres and plastics. They may be organic compounds or insoluble inorganic solids.

pi-meson. *See* pion.

pincushion distortion. *See* distortion.

pint. *See Appendix.*

pion (pi-meson). A type of *meson having either zero charge and a mass 264.2 times that of the electron, or a positive or negative charge and a mass 273.2 times that of the electron.

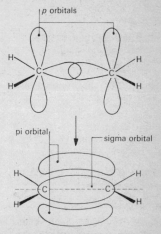

Molecular orbitals in ethylene

pi orbital (π-orbital). A type of *molecular orbital in which there are two regions on each side of a line between the atoms. Pi orbitals form one of the bonds in a double bond between atoms. In ethylene, for example, each carbon atom is sp^2 hybridized and two of these *hybrid orbitals form a *sigma orbital between the carbon atoms. The remaining p orbitals, one on each carbon atom, overlap laterally and form a pi orbital, which has lobes each side of the sigma orbital. It is occupied by a pair of electrons with opposite spins. The double bond between the carbon atoms thus consists of a sigma bond and *pi bond*.

piperazine. A colourless crystalline heterocyclic solid, $C_4H_{10}N_2$, used as a corrosion inhibitor and insecticide. M.pt. 104°C; b.pt. 145°C.

pipette. A graduated glass tube for measuring out a known fixed quantity of a liquid.

pitch. The frequency of a sound as judged by comparison with the frequency of a pure tone.

pitchblende. A heavy black radioactive mineral consisting mainly of uranium oxides. It is the most important ore of uranium.

pK. The logarithm to base 10 of the reciprocal of the equilibrium constant of a given chemical reaction: i.e. pK $=$ $\log_{10}(1/K)$. It is usually used as an indication of the strengths of acids, K being the dissociation constant. Weak acids have high pK values.

Planck constant. Symbol: h. The universal constant $6.626\ 196 \times 10^{-34}$ joule-second. A vibration of frequency f has energy quantized in units of hf. [After Max Planck (1858–1947), German physicist.]

plane-polarized light. *See* polarization.

planet. Any of the nonluminous bodies that orbit the sun. Strictly the term is applied to the group of major bodies of which nine are known. In order of succession outwards from the sun they are: Mercury, Venus, the earth, Mars, Jupiter, Saturn, Uranus, Neptune, and Pluto. Technically, the *asteroids are also planets. The planets in the solar system fall into two main categories: namely, the inner or terrestrial planets, Mercury, Venus, the earth, and Mars, and the outer or giant planets, Jupiter, Saturn, Uranus, and Neptune. The terrestrial planets are relatively small and complicated in chemical composition, the earth being the largest; the giant planets are much larger in size and possess atmospheres of simple composition, usually methane and ammonia, with cores of highly condensed frozen hydrogen. The outermost planet, Pluto, is probably comparable to the earth in size but similar to the giant planets in composition.

planetoid. *See* asteroid.

plano-concave. *See* lens.

plano-convex. *See* lens.

plasma. A mixture of ions and electrons. The plasma in an arc, spark, or other electrical discharge is composed of simple ions and electrons together with unionized gas. The plasma occurring in thermonuclear reactions, as in the sun, is highly ionized and consists of a mixture of atomic nuclei and electrons.

plaster of Paris. A hemihydrate of calcium sulphate, $CaSO_4.\frac{1}{2}H_2O$, made by partial dehydration of gypsum by heating. When mixed with water it sets to a hard mass.

plastic. 1. *See* elasticity.
2. Any of various synthetic materials that contain or consist of *polymers and are moulded at some stage in the manufacturing process. *Thermosetting* plastics are polymeric materials that harden on heating to give a rigid product that cannot then be softened. *Thermoplastic* materials soften when heated and harden again when cooled. The terms *plastic* and *resin* are often used synonymously, although in strict technical usage *resin* is used for the

original polymer whereas *plastic* refers to the final material, which may contain pigments, fillers, plasticizer, stabilizer, and antioxidant.

plasticizer. A material mixed with a synthetic resin to improve the flexibility or other properties of the plastic. Camphor, for example, is used in celluloid.

platinic. *See* platinum.

platinous. *See* platinum.

platinum. Symbol: Pt. A silvery-white malleable and ductile *transition element occurring native and associated with certain copper and nickel ores. It is used as a catalyst, in jewellery, and in thermocouples, electrodes, and other pieces of scientific equipment. Platinum is often alloyed with iridium and rhodium. With cobalt it forms an alloy used in permanent magnets.

The metal is rather less reactive than palladium: it does not oxidize and is inert to all mineral acids. Like palladium it absorbs hydrogen, combines with many nonmetals at high temperatures, and dissolves in fused alkalis. Unlike nickel, it forms no simple ionic salts: its compounds are mainly platinum II (*platinous*) and platinum IV (*platinic*) complexes. A.N. 78; A.W. 195.09; m.pt. 1772°C; b.pt. 3820°C; r.d. 21.45; valency 1, 2, 3, 4, or 6.

platinum chloride. Either of two chlorides of platinum. *Platinum IV chloride* (platinic chloride), $PtCl_4$, is a brown solid when anhydrous, made by evaporating a solution of platinum in aqua regia. It also exists as *chloroplatinic acid*, $H_2PtCl_6.6H_2O$. Platinum IV chloride is used in platinum plating and photography. R.d. 4.3. *Platinum II chloride* (platinous chloride) is a green-grey powder, $PtCl_2$, made by heating platinum with chlorine. It is used to make platinum salts. R.d. 5.9.

platinum metals. The group of related transition metals ruthenium, osmium, rhodium, iridium, palladium, and platinum.

pleochroism. The phenomenon in which certain crystals have different colours when viewed from different directions.

plumbago. *See* graphite.

plumbic. *See* lead.

plumbous. *See* lead.

Pluto. The ninth planet in order of succession outwards from the sun and the furthest of all the planets currently known in the entire solar system. Lying at a mean distance of 5907 million kilometres from the sun, Pluto orbits it at an average velocity of less than 5 km per second, taking 248.5 years to complete one revolution. Pluto is one of the most obscure objects in the night sky. No reliable facts are available for the planet, but it seems certain that it is radically different from the other outer planets. Its diameter is probably about 5900 km (less than half that of earth) and its mass must be assumed to be of the order of one tenth to one third that of earth. It is thought that the temperature of the sunlit side of Pluto probably does not rise much above −214°C; as a result many astronomers believe that there can be no atmosphere except possibly for traces of helium and neon. Pluto is therefore believed to be a very small planet and it has been suggested that it was originally a satellite of Neptune that escaped. In fact at a perihelion of 4505 million km Pluto's highly eccentric orbit falls within that of Neptune. Of all the planets, Pluto's path is the most deviant, having an eccentricity of 0.25.

plutonium. Symbol: Pu. A silvery metallic transuranic *actinide element made in nuclear *reactors by neutron bombardment of uranium. It is extensively used in nuclear reactors and nuclear weapons. The fissile isotope, ^{239}Pu, has a half-life of 24 360 years. A.N. 94; m.pt. 639.5°C; b.pt. 3235°C; r.d. 19.84; valency 3, 4, 5, or 6.

pneumatic. Operating by air pressure.

p-n-p transistor. *See* transistor.

point-contact transistor. *See* transistor.

point defect. *See* defect.

poise. Symbol: P. A CGS unit of dynamic *viscosity equal to the dynamic viscosity of a fluid for which a tangential force of one dyne per square centimetre is necessary to create a velocity gradient of one centimetre per second per centimetre. [After Jean Poiseuille (1799–1869), French physicist.]

Poiseuille's equation. The equation $Q = \pi(p_1 - p_2)r^4/8\eta l$, giving the rate of fluid flow Q through a circular pipe of radius r and length l. The dynamic viscosity is η and $(p_1 - p_2)$ is the pressure difference between the ends of the tube. The equation can be used in a method of determining viscosity by measuring the rate of flow under steady conditions.

poison. A substance that destroys the activity of a catalyst.

Poisson's ratio. Symbol: ν. The ratio of the lateral *strain in a body to the longitudinal strain. Thus if a rod of material of length l and diameter d undergoes tensile elongation to produce an increase in length Δl and a decrease in diameter Δd, Poisson's ratio is $(\Delta d/d)/(\Delta l/l)$. If the volume does not change, the ratio is 0.5. [After Siméon Denis Poisson (1781–1840), French mathematician.]

polar. Denoting a substance whose molecules have a dipole moment. Water, for example, is a polar compound because of unequal sharing of electrons between the oxygen and hydrogen atoms.
Polar solvents are capable of dissolving ionic solids because part of the energy required to separate the ions in the crystal is recovered in *solvation of the ions. They are not usually capable of dissolving *nonpolar compounds.

polar bond. *See* electrovalent bond.

polar coordinates. *Coordinates used to locate the position of a point by its

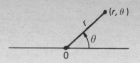

planar polar coordinates

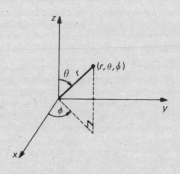

spherical polar coordinates

Polar coordinates

distance from a fixed point (the *pole*) and its angular displacement from a line. In two dimensions the coordinates are (r,θ), r being the *radius vector* and θ the *azimuth* or *amplitude* of the point. In three dimensions *spherical coordinates* can be used, in which the point is located by a radius vector, a *colatitude* θ measured from a vertical line, and a *longitude* ϕ measured from a horizontal line (see illustration). *See also* cylindrical coordinates.

polarimeter. An instrument for determining the specific rotation of optically active substances (*see* optical activity). It consists of a *polarizer*, which is a Nicol prism or other device to polarize light, a transparent cell to contain the sample, and an *analyser*, similar to the polarizer but capable of being rotated and fitted with an angular scale. A beam of monochromatic light passes through the polarizer, cell, and analyser and is observed through an eyepiece. The analyser is rotated until the light has maximum intensity—the angular rotation can then be read on the scale.

polariscope. Any instrument for viewing objects in polarized light. Strains in glass or transparent plastics are observed as coloured interference lines in the material.

polarization. 1. Restriction of the direction of vibration in light or other *electromagnetic radiation. Normal light consists of transverse vibrations of electric and magnetic fields: the fields vibrate at right angles to the direction of light in all possible planes. Under certain conditions, the vibration can be restricted to one particular plane—the light is then said to be *plane-polarized*. Plane polarization can occur by reflection (*see* Brewster's law), or by *double refraction. Other types of polarized light can also be produced. *Circular polarization* occurs when the electric vector describes a circle around the direction of the light beam. *Elliptical polarization* occurs when the vector describes an ellipse. *See also* optical activity.
2. Separation of charges in a *dielectric material. When an electric field is applied across an insulator current does not flow through it, but the molecules are polarized so that they act as dipoles. In this way charge is stored in a capacitor.
3. A reduction in the e.m.f. of a cell as a result of the formation of a layer of small bubbles of hydrogen on the cathode. These decrease the effective area of the electrode and also produce a back e.m.f. The effect is reduced by use of a depolarizer.

polarography. An analytical technique for identifying and determining the concentrations of ions by electrolysis of a solution. The cathode is a dropping-mercury electrode: i.e. one in which mercury is continually forming at and dropping away from the end of a vertical glass tube. In this way the electrode is kept clean and has a low surface area. A slowly increasing voltage is applied between this electrode and an anode

and different positive ions in the solution are discharged at different values of the voltage. A graph of current against voltage (a *polarogram*) shows a series of steps, each corresponding to a particular type of ion.

Polaroid. ® A type of film containing many minute doubly refracting crystals aligned with their axes parallel. The film polarizes transmitted light and is used in sunglasses and other devices to reduce glare.

pole. 1. *See* magnetic pole.
2. *See* mirror.

polonium. A radioactive element: one of the decay products of radium. Polonium is one of the *chalconides. A.N. 84; A.W. 210; m.pt. 225°C; b.pt. 950°C.

polyamide. Any polymer in which the monomers are linked by amide groups, -CO.NH-. The various forms of *nylon are the most common examples.

polyatomic. Having molecules composed of several atoms. The term is often used to distinguish compounds from diatomic and monatomic compounds.

polycarbonate resin. Any of a class of synthetic resins that are polyesters of carbonic acids and dihydric phenols. Usually, polycarbonate resin is made by reacting phosgene and bisphenol. It is a strong transparent thermoplastic material used for many moulded articles.

polychromatic. Denoting light or other electromagnetic radiation that has a mixture of wavelengths; not monochromatic.

polycyclic. Denoting a chemical compound that contains two or more rings of atoms in its molecular structure.

polyene. An organic compound containing three or more double bonds in its molecules.

polyester resin. Any of a class of synthetic resins made by reacting polyhydric alcohols with polybasic acids or

$$n \ \text{HO(CH}_2)_2\text{OH} + n \ \text{HO.CO} \langle \bigcirc \rangle \text{CO.OH}$$

ethanediol   terephthalic acid

$$\text{HO(CH}_2)_2-\text{O}-\left[\overset{\overset{\text{O}}{\|}}{\text{C}}-\langle \bigcirc \rangle-\overset{\overset{\text{O}}{\|}}{\text{C}}-\text{O(CH}_2)_2\,\text{O}-\right]_n + n\,\text{H}_2\text{O}$$

Polyester polymer

their anhydrides. Saturated polyesters can be made by condensation of simple diols, such as ethane diol, with unsaturated acids, such as phthalic acid (see illustration). Compounds of this type are widely used in synthetic fibres like Terylene and Dacron. If unsaturated acids, such as maleic and fumaric acids, are used, then the polyester itself is unsaturated. This resin can then be further polymerized to a thermosetting plastic; often styrene is added at this stage and the material is reinforced with glass fibre. The result is very strong, light, and hard wearing making it suitable for use in car bodies, boat hulls, and similar applications. *Alkyd* (or *glyptal*) resins are made from phthalic acid and glycerol, often with inclusion of styrene. They are used in paints and as moulding materials.

polyether. Any polymer in which the monomers are linked by ether groups (C-O-C). An example is the *epoxy resins.

polyethylene. See polythene.

polygon. Any plane figure with three or more sides.

polyhedron. A geometric solid whose plane faces are polygons. For any simple polyhedron, the sum of the number of vertices and faces less the number of edges is equal to two. This is known as *Euler's theorem*. A *regular* polyhedron is one that has all its faces congruent. Only five types of polyhedron may be regular, namely the tetrahedron, hexahedron, octahedron, dodecahedron, and icosahedron.

polyhydric. Denoting an alcohol that contains two or more hydroxyl groups in its molecules.

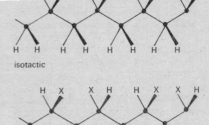

isotactic

syndiotactic

Tactic polymers

polymer. A compound whose molecules consist of repeated molecular units. Originally the term was applied to molecules of any size produced by addition of identical molecules (*monomers*), so that the polymer had the same empirical formula as the monomer. Now, it is usually applied to compounds of high molecular weight formed from numbers of simpler units. Many natural polymers exist, including rubber, proteins, starch, and cellulose. Synthetic polymers are usually organic and are extensively used in *plastics, textiles, paints, and adhesives.

Most polymers do not have a simple formula. Instead, they are composed of a mixture of large molecules with a distribution of molecular weights. The properties of polymeric materials also depend on the structure of the molecules, which is classified in various ways. The simplest molecules have straight chains: more complicated polymers contain side chains. Polymers are also characterized by the arrangement of monomers in the chain. *Homopolymers* are formed of the same type of monomer. A polymer consisting of two or more different units is a *copolymer*, in which the types of monomer may have a random order or may have some type of regular structure. A *block copolymer* has short chains of one monomer alternating with chains of another (see illustration). A *graft copolymer* has a main chain of one type of monomer with side chains of the other. If the groups attached to atoms along the chain of a polymer have a regular arrangement in space the polymer is said to be *tactic*: otherwise it is *atactic*. Atactic polymers are of two types. In *isotactic* polymers the same configuration is repeated at each point along the chain. *Syndiotactic* polymers have *asymmetric atoms in their chain and the arrangement is alternately inverted along the chain. Tactic polymers are *stereospecific.

The properties of polymeric materials depend not only on the structure of the chains but also on the shape of the chains and the way they are arranged in the bulk material. Thus, if the chains are tangled together or otherwise orientated at random the material will be amorphous. If they are parallel the polymer will have some degree of crystallinity. In some materials polymer chains are linked together by side chains, leading to an extended three-dimensional structure, which again may be random or regular.

polymerization. A chemical reaction leading to the formation of a polymer. Synthetic polymers may be produced by two types of process. In *addition polymerization* the monomers simply add together with no other compound formed. This happens for monomers that contain double bonds, as in the production of *polythene, or rings, as in the production of *epoxy resins. In *condensation polymerization* water, an alcohol, or other simple molecule is eliminated. This occurs when the monomer contains two reactive groups, as in the production of *nylon. Addition and condensation copolymers can also be made (*copolymerization*) by reaction of two or more different compounds.

polymethylmethacrylate. A colourless transparent thermoplastic material made by polymerization of methyl methacrylate. It is widely used as a substitute for glass. Polymethylmethacrylate is produced under the trade name *Perspex*.

polymolybdate. *See* molybdate.

polymolybdenic acid. *See* molybdate.

polymorphism. Existence of more than one solid (crystalline) form of the same chemical substance. If two forms exist the phenomenon is *dimorphism.

polynomial. A mathematical expression of the form $A_o + A_1x + A_2x^2 + A_3x^3 + \ldots$

polypeptide. A chain of *amino acids held together by *peptide linkages. Polypeptides are found in proteins.

polyphosphate. *See* phosphate.

polyphosphoric acid. *See* phosphoric acid.

polypropylene. A thermoplastic material made by polymerization of propylene. It is similar to high-density *polythene but is stronger, lighter, and more rigid.

polysaccharide. A *sugar or other *carbohydrate with molecules containing more than one monosaccharide unit.

Polystyrene polymer

polystyrene. A clear colourless thermoplastic material made by polymerization of styrene (see illustration). It is a clear rather brittle flammable substance used in electrical insulation and in making moulded articles. *Expanded polystyrene* is a light rigid foam extensively used in packing and in thermal insulation.

polysulphide. See sulphide.

polytetrafluoroethylene (PTFE). A synthetic material produced by polymerization of tetrafluoroethylene. It has several useful properties, notably its ability to withstand temperatures up to above 400°C and its low coefficient of friction (equal to that of wet ice). PTFE is used in coating nonstick cooking utensils and in gaskets, bearings, and electrical insulation. It is sold under the trade names *Teflon* and *Fluon.*

polythene (polyethylene). A white translucent waxy thermoplastic material made by polymerizing ethylene at high pressure. Polythene is made in the gas phase and is produced in two forms. Low-density polythene is a soft material made at high pressures. High-density polythene is made at lower pressures with the use of solid catalysts. It is more crystalline and more rigid than the low-density form and softens at a higher temperature. Both types are widely used: the high-density material being used in large numbers of moulded articles and the low-density form being also used for flexible tubes and sheets.

polytungstate. See tungstate.

polytungstic acid. See tungstate.

polyurethane. Any of a class of synthetic resins formed by copolymerization of diisocyanates and polyhydric alcohols. All contain the urethane group, -NH.CO.O-. Many types of thermosetting and thermoplastic polyurethanes exist. They are used in fibres, moulded articles, adhesives, rubbers, paints, lacquers, and foams.

polyvanadate. See vanadate.

polyvanadic acid. See vanadate.

polyvinyl acetate (PVA). A *vinyl resin or plastic produced by polymerization of vinyl acetate. It is a soft material used in emulsion paints and adhesives.

polyvinyl chloride (PVC). A *vinyl resin or plastic produced by polymerization of vinyl chloride. The plastic is used in two forms: rigid PVC, which is moulded into many products, and flexible PVC, produced by addition of a plasticizer, which has an immense range of applications in moulded articles, sheets, tubes, electrical insulation, clothing, etc. PVC is tough, nonflammable, resistant to moisture, and a good electrical insulator. After polythene, it is the second most widely used synthetic plastic.

polywater. A supposedly polymeric form of water made by condensing water in very narrow glass or quartz tubes. It is now known to be impure water, containing ions from the glass or quartz.

population inversion. See laser.

population type. See star.

positive column. A large luminous column near the anode in a *discharge tube.

positive rays. Streams of positive ions, as produced in a discharge tube. See also canal rays.

positron. The *antiparticle of the electron: an *elementary particle with a positive charge equal in magnitude to the electron's charge and a mass equal to the mass of the electron. Positrons are produced during *pair production and in one form of *beta decay.

positronium. A short-lived entity consisting of an electron and a positron revolving around a common centre of mass, analogous to a hydrogen atom. Positronium has a mean life of the order of 10^{-7} second, undergoing *annihilation to photons.

potash. Any of various compounds of potassium, including potassium carbonate, potassium hydroxide (*caustic potash*), and potassium aluminium sulphate (*potash alum*).

potassium. Symbol: K. A soft silvery metallic element that is widely distributed in nature, large amounts being found in salt deposits such as carnallite and kainite. Potassium can be obtained by electrolysis. The element has few uses but it is essential to plant growth and its salts are extensively used in fertilizers. It is one of the *alkali metals. A.N. 19; A.W. 39.102; m.pt. 63°C; b.pt. 770°C; r.d. 0.86; valency 1.

potassium-argon dating. A method of *dating geological samples by using the decay of potassium-40, a radioisotope occurring in small amounts in stable potassium. It decays slowly with a half-life of 1.28×10^9 years to give argon-40. Measurement of the ratio of the two isotopes present gives an estimate of the sample's age. The method is useful for ages of up to about 10^{10} years.

potassium bicarbonate. *See* potassium hydrogen carbonate.

potassium bromide. A white crystalline solid, KBr, made by heating a mixture of iron bromide and potassium carbonate. It is used in photography. M.pt. 730°C; b.pt. 1435°C; r.d. 2.7.

potassium carbonate (potash). A white deliquescent alkaline powder, K_2CO_3, extracted from natural sources of potassium chloride by treatment of a solution with carbon dioxide. It is used in glass manufacture and the preparation of other potassium salts. M.pt. 891°C; r.d. 2.4.

potassium chlorate. A colourless or white crystalline substance, $KClO_3$, made from potassium chloride and calcium chlorate. It is used in making explosives and matches. M.pt. 356°C; r.d. 2.3.

potassium chloride. A white crystalline solid, KCl, occurring naturally in carnallite and in certain brines. It is used in fertilizers and as a source of other potassium salts. M.pt. 790°C; sublimes at 1500°C; r.d. 1.9.

potassium chromate. A yellow crystalline solid, K_2CrO_4, made by roasting chromite, potassium carbonate, and limestone, then treating the product with aqueous potassium sulphate. It is used as an analytical reagent, mordant, and pigment. M.pt. 971°C; r.d. 2.7.

potassium cyanide. An extremely poisonous white deliquescent crystalline solid, KCN, with a faint odour of bitter almonds. It is made by heating potassium carbonate with carbon in a current of ammonia and is used in case-hardening steel and in the extraction of *gold and silver. M.pt. 634°C; r.d. 1.5.

potassium dichromate. A poisonous red crystalline solid, $K_2Cr_2O_7$, made by adding sulphuric acid to aqueous potassium chromate. It is a strong oxidizing agent and is used in electroplating, dyeing, tanning, and the production of many chemicals. M.pt. 396°C; r.d. 2.7.

potassium hydrogen carbonate (potassium bicarbonate). A colourless crystalline slightly alkaline powder, $KHCO_3$, made by passing carbon dioxide into aqueous potassium carbonate. It is used in soft drinks and fire extinguishers. R.d. 2.17.

potassium hydrogen tartrate (cream of tartar). A colourless crystalline compound, $C_4H_5O_6K$, used as an ingredient of baking powders. It is deposited as argol in wine vats.

potassium hydroxide (caustic potash). A white deliquescent strongly alkaline solid, KOH, manufactured by electrolysis of concentrated potassium chloride solution and used in making soft soaps. M.pt. 360°C; r.d. 2.0.

potassium iodide. A white crystalline solid, KI, prepared by the reaction of hydrogen iodide and potassium hydrogen carbonate. It readily liberates iodine and is used in medicine, photography, and as a fuel additive. M.pt. 686°C; b.pt. 1330°C; r.d. 3.1.

potassium nitrate (saltpetre, nitre). A white crystalline solid, KNO_3, used as a source of oxygen in fireworks, a preservative for food, and a nitrogen fertilizer. M.pt. 33°C; r.d. 2.1.

potassium permanganate. A purple crystalline solid, $KMnO_4$, with a metallic sheen. It is prepared by oxidation of potassium manganate and is used as an oxidizing agent and disinfectant. Decomposes at 240°C; r.d. 2.7.

potassium sodium tartrate (Rochelle salt). A colourless efflorescent crystalline substance, $KNaC_4H_4O_6.4H_2O$, used in baking powders. Crystals of potassium sodium tartrate show the piezoelectric effect. R.d. 1.77.

potassium sulphate. A colourless or white crystalline substance, K_2SO_4, used in fertilizers and the manufacture of glass. M.pt. 1072°C; r.d. 2.7.

potential. Symbol: V. The work done in bringing unit electric charge from infinity to a specified point in an electric field. The electric *potential difference* between two points is the work done in moving the charge from one point to the other. It is independent of the path taken. Electric potential and potential difference is measured in volts.
The potential can also be defined for magnetic and gravitational fields by the work necessary to bring unit pole or unit mass to the point.

potential divider (voltage divider). A resistor or chain of resistors with a

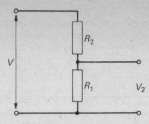

Potential divider

voltage applied across it and a tapping connection so that a fraction of this voltage can be drawn. In the diagram, $V_2 = R_1V/(R_1 + R_2)$. The connection may be fixed to give a constant fraction of the voltage or may be variable as in a *potentiometer.

potential energy. Symbol: E_p. The energy that a system has by virtue of its position or state, determined by the work necessary to change the system from a reference position to its present state. Thus, a stretched spring has potential energy equal to the work done in stretching it from its equilibrium position, which is taken as the reference state. An object a height h above the ground has a potential energy of mgh, where m is its mass and g is the acceleration of free fall. *See also* kinetic energy.

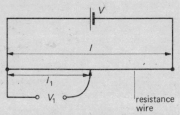

Potentiometer

potentiometer. 1. An apparatus for measuring potential difference by use of a *potential divider. A simple form has a straight length of uniform resistance wire with a constant voltage connected across it. The potential difference to be measured is connected between one end of the resistance wire and a sliding contact, which is moved along the wire

until no current flows as indicated by a galvanometer (see illustration). The unknown voltage V_1 is given by $V_1 = l_1 V / l$. Usually a standard cell with voltage V_2 is introduced to find a new balance, l_2: then $V_1 / V_2 = l_1 / l_2$. The device gives the true potential difference or e.m.f. because at balance no current is drawn.
2. A circuit component consisting of a resistance with a sliding contact, suitable for use in giving a controlled variable voltage.

potentiometric. Denoting an experimental technique that depends on measurements of potential. A *potentiometric titration* is one in which the end point is determined by monitoring the electrode potential of an electrode in the mixture.

pound. 1. Symbol: lb. An imperial unit of mass equal to 0.453 592 37 kilogram. This is the unit in *avoirdupois weight: it was formerly defined as the mass of a block of platinum, the change to a definition based on the kilogram occurring in 1963.
2. Units of mass in troy and apothecaries weight. *See Appendix.*
3. *See* pound-weight.

poundal. Symbol: pdl. An FPS unit of force equal to the force that would give a mass of one pound an acceleration of one foot per second per second. It is equivalent to 0.138 255 newton.

pound-weight. Symbol: lbwt. A unit of force equal to the force that would give a body of mass one pound an acceleration equal to the local *acceleration of free fall. The force in poundals is the force in pounds-weight multiplied by the local value of g in feet per second. The unit, which is sometimes loosely called the *pound*, varies slightly from place to place. The *pound-force* (symbol: lbf) is the force that would give a mass of one pound an acceleration equal to the standard acceleration of free fall (32.174 ft s^{-2}): 1 lbf = 32.174 pdl.

power. 1. The rate of performing *work, measured in units such as *watts and *horsepower.
2. A measure of the magnifying ability of an optical system, equal to the ratio of the size of the image produced to the size of the object.
3. The number of times a number or expression is multiplied by itself. For example a to the power 4 is $a \times a \times a \times a$, written a^4, where 4 is the *exponent.

praseodymium. Symbol: Pr. A soft silvery metallic element belonging to the *lanthanide series. A.N. 59; A.W. 140.907; m.pt. 931°C; b.pt. 3212°C; r.d. 6.7; valency 3 or 4.

precession. 1. The motion of a body that is spinning on an axis (OA) in which one end of the axis (A) moves around a second axis passing through O. The precessional motion results from an applied torque. It is seen in tops and *gyroscopes. *See also* Larmor precession.
2. This phenomenon as it occurs in planets, particularly the earth. Because of precession the vernal and autumnal *equinoxes are moving westward along the *ecliptic; during the 26 000-year cycle over which the earth's axis describes its small circle as a result of precession, the Pole Star changes. At present the Pole Star is represented by Polaris, the brightest star in Ursa Minor. 5000 years ago the Pole Star was Alpha Draconis. In general, precessional effects are caused by the gravitational influence of the moon and sun, but the influence of the other planets also causes precession of the earth.

precipitate. A visible suspension of solid particles produced in a solution by chemical reaction. For example, if sodium chloride and silver nitrate solutions are mixed together, the insoluble silver chloride appears as a white precipitate, which eventually settles to the bottom of the vessel.

precipitation. 1. The process of forming a chemical precipitate.

2. A process occurring in certain alloys that are supersaturated solid solutions, in which a separate phase is formed. The process is the basis of a method of hardening alloys (*precipitation hardening*).

presbyopia. A defect of the eye in which near objects cannot be focused clearly. It is caused by hardening of the lens so that the ciliary muscles cannot produce enough accommodation and the near point is further away than the normal 33 centimetres. Presbyopia can be corrected with convex spectacle lenses worn for close work. It is sometimes called *far sight*, which should not be confused with long sight (*hypermetropia). *See* eye.

pressure. Symbol: p. The force acting per unit surface area. The pressure at a point below the surface of a fluid acts in all directions and is given by the height of the fluid multiplied by its density. Pressure is measured in pascals. Other units are the bar, the atmosphere, and the mmHg.

pressurized-water reactor. A type of *fission reactor in which water is used as a coolant, maintained at a high pressure to prevent it from boiling.

primary. 1. The body around which a satellite orbits. The earth is the moon's primary in the earth-moon system; the sun is the earth's primary in the solar system.
2. *See* transformer.

primary cell. *See* cell.

primary colours. A set of three colours that can be mixed in suitable proportions to produce any other colour. Red, green, and blue are an example: another set of primary colours is cyan (greenish-blue), magenta, and yellow.
If red, green, and blue lights are mixed in equal proportions, white light is produced by an additive process. Similarly, three pigments of primary colours mixed in equal proportions

combine by a subtractive process to give black.

primary standard. *See* unit.

prime number. A number that has no factors other than itself and one.

principal axis. *See* mirror, lens.

principal focus. *See* mirror, lens.

principal quantum number. *See* atom.

printed circuit. An electrical circuit in which the connections between components are made by thin channels of metal coated onto an insulating board.

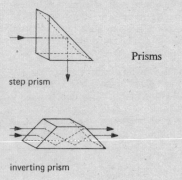

Prisms

step prism

inverting prism

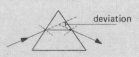

deviation

deviation

prism. 1. A solid geometrical figure with two identical parallel faces, consisting of congruent polygons with their corresponding sides parallel, the other faces being formed by joining corresponding vertices of these polygons.
2. A piece of glass, quartz, rock salt, or other transparent material cut in the shape of a prism and used for dispersing or changing the direction of light or ultraviolet or infrared radiation. The commonest type are triangular prisms used for dispersing light. Prisms are also

used for inverting images or for internal reflection (see illustration).

prismatic. Denoting optical instruments that contain prisms. *Prismatic binoculars*, for example, are *binoculars in which the path length is increased and the image erected by pairs of prisms.

prismatic colours. The colours produced when white light is split up by dispersion. The prismatic colours are the colours of the rainbow: violet, indigo, blue, green, yellow, orange, and red.

probability. A numerical expression of the likelihood that an event will occur. Probabilities are usually written as fractions. If b is the total number of ways in which an event can occur and of these, a events can occur in a specified way, then a/b is the probability of one of these specified events occurring. For example, in drawing one card from a pack of playing cards there are 52 cards that may be drawn. Of these, 4 cards are aces: the probability of drawing an ace is $4/52 = 1/13$. In this definition of probability the possible events are all considered as independent and equally likely. Probabilities are combined by multiplication: the probability of a coin turning up "heads" three times is $\frac{1}{2} \times \frac{1}{2} \times \frac{1}{2} = \frac{1}{8}$.

producer gas. A mixture of carbon monoxide and nitrogen made by passing air over hot coke: $2C + O_2 + 4N_2 = 2CO + 4N_2$. The reaction is exothermic and no further external heating is required once the process has started. Producer gas is used as a fuel: its calorific value is less than that of *water gas and it is often made at the same time.

product. 1. The result of multiplying two or more numbers or quantities together.
2. A compound produced in a chemical reaction, as distinguished from the reactants.

program. A set of instructions fed into a computer.

prolate. *See* ellipsoid.

promethium. Symbol: Pr. A radioactive element belonging to the *lanthanide series. It does not occur naturally, being obtained from nuclear reactors. The most stable isotope, ^{145}Pr, has a half-life of 17.7 years. A.N. 61; valency 3.

prominence. A towering eruption of gas in the upper chromosphere of the *sun that throws matter out into space in vast streamers. Prominences, which are composed of incandescent hydrogen, helium, and calcium, climb high above the sun's surface and appear to stand there motionless over long periods. They are associated with *sunspots and are subject to the 11-year cycle.

proof. A measure of the strength of alcoholic drinks. *Proof spirit* contains 49.28% by weight of pure alcohol (52.1% by volume at 16°C). Alcoholic strengths can then be specified as being *under proof* or *over proof*. Each degree under or over proof corresponds to 0.5 per cent of alcohol by volume. Usually liquors are described as having *degrees of proof*. The number of degrees of proof is equal to the percentage of proof spirit in a mixture of proof spirit and water: 70° proof contains 70 parts of proof spirit and 30 parts of water. The percentage of alcohol by volume is converted into degrees proof by multiplying by 1.7535. Thus pure alcohol is 175.35 degrees proof. In America a similar system is used with the difference that proof spirit is defined as containing 50% of alcohol by volume.

propanal. *See* propionaldehyde.

propane. A colourless flammable gaseous alkane, C_3H_8, obtained from petroleum and used as a fuel. M.pt. -190°C; b.pt. -42°C; r.d. 1.56 (air = 1).

propanoic acid. *See* propionic acid.

propanol. One of two isomeric alcohols, C_3H_7OH, which are colourless flammable liquids used as solvents and flavourings. *Propan-1-ol* (n-propyl alcohol), $CH_3CH_2CH_2OH$, is made by oxidizing natural-gas hydrocarbons. M.pt. $-127°C$; b.pt. $97°C$; r.d. 0.8. *Propan-2-ol* (iso-propyl alcohol), $(CH_3)_2CHOH$, is made by treating propene with sulphuric acid followed by hydrolysis. It is the principal source of acetone and its derivatives. M.pt. $-86°C$; b.pt. $82°C$; r.d. 0.8.

propellant. 1. The fuel, including the oxidant, used in a rocket.
2. An explosive charge used to fire a bullet or shell.
3. A volatile inert gaseous substance, liquefied under pressure, used to expel the contents of an aerosol can.

propene (propylene). A colourless slightly soluble gaseous alkene, $CH_3CH:CH_2$, made by cracking petroleum hydrocarbons and used in making polypropylene. M.pt. $-185.2°C$; b.pt. $-48°C$; r.d. 1.46 (air = 1).

proper fraction. *See* fraction.

propionaldehyde (propanal). A colourless liquid aldehyde, C_2H_5CHO. M.pt. $-81°C$; b.pt. $48.8°C$; r.d. 0.81.

propionic acid (propanoic acid). A colourless liquid carboxylic acid, C_2H_5COOH. M.pt. $-20.8°C$; b.pt. $141°C$; r.d. 0.99

propyl group. The group, C_3H_7-. The term is often used for the *n-propyl group*, $H_3CCH_2CH_2$-. The other isomer is the iso-propyl group, $(CH_3)_2CH$-.

protactinium. Symbol: Pa. An actinide element found in uranium ores. Its most stable isotope, ^{231}Pa, has a half-life of 33 000 years. The element was formerly called *protoactinium*. A.N. 91; m.pt. about $1200°C$; b.pt. about $4000°C$; r.d. 15.4 (calc.).

proteins. Any of a large number of naturally occurring organic compounds found in all living matter. Proteins consist of chains of amino acids joined by *peptide linkages. They have high molecular weights (up to 10 million) and complex structures. The particular nature of a protein depends on the amino acids present and the sequence in which they are bound. It also depends on the way the chains are linked and folded.

protium. The isotope of hydrogen with mass number one.

protoactinium. *See* protactinium.

proton. An *elementary particle with a positive charge equal in magnitude to the charge of the electron and a rest mass of $1.672\,62 × 10^{-27}$ kg (about 1836 times the mass of the electron). Protons are stable particles. They form the *nuclei of hydrogen atoms and together with neutrons, are present in all other atomic nuclei. Positive hydrogen ions are protons although in solution these are solvated by water molecules (*see* solvation).

proton number. *See* atomic number.

Prussian blue. Any of a number of blue pigments containing ferrocyanide or ferricyanide ions.

prussic acid. *See* hydrogen cyanide.

pseudoaromatic. Denoting a chemical compound that has molecules containing a ring of conjugated double bonds but is not an *aromatic compound. *Cyclooctatetraene*, C_8H_8, for instance, has an eight-membered ring of alternating double and single bonds yet behaves like an alkene. It does not have a planar ring of carbon atoms.

pseudohalogen. Any of a small group of compounds that resemble the halogens in their reactions. Pseudohalogens have a formula of the type R_2, where R is a simple group of atoms, and they form R^- ions. Cyanogen, $(CN)_2$, and thiocyanogen, $(SCN)_2$, are the commonest examples.

psi particle. Any of three heavy short-lived elementary particles. The psi particles have masses of 3.095 GeV, 3.684 GeV, and 4.1 GeV and have zero charge. It is thought that they may be formed from charmed quarks (*see also* charm).

psychrometer. *See* wet- and dry-bulb hygrometer.

psychrometry. The measurement of atmospheric *humidity.

PTFE. *See* polytetrafluoroethylene.

Ptolemaic system. The system of celestial mechanics developed by the Alexandrian astronomer, Ptolemy (Claudius Ptolemaeus; *c.* A.D. 127–151), and based upon the concept that the earth lay at the centre of the universe while the sun, moon, planets, and stars orbited around it within a set of concentric spherical shells. The paths of the celestial bodies in these spheres were considered to be circular and the more complex motions of the planets were accounted for by an elaborate system of epicycles (small circles around which the planets moved and whose centres orbited around the earth). This theory, which was the crystallization of contemporary cosmological thinking, was later superseded by the revolutionary *Copernican system.

p-type. *See* semiconductor.

puddling. The process of heating cast iron with iron oxide to reduce the carbon content and produce *wrought iron.

pulsar. A star that acts as a source of regularly fluctuating electromagnetic radiation, the period of the pulses usually being very rapid, occasionally even as high as 50 times a second. These are probably *neutron stars, which, as they rotate, emit radiation in directions that are strictly controlled by the directions of their magnetic fields. The energy output is only detectable when the "beam" is turned towards the observer. *See also* black hole.

pulsating star. A star that is subject to variations in the light emitted by it owing to periodic and regular expansion and contraction of its surface atmosphere.

pulse. A sudden increase in magnitude of a physical quantity shortly followed by a decrease.

pulse-height analyser. An electronic circuit for sorting voltage pulses according to their amplitude. The pulses obtained from certain types of counter have an amplitude proportional to the energy of the orginal particle, and the use of a pulse-height analyser can yield an energy spectrum of the radiation. *See* scintillation counter.

purple of Cassius. A purple pigment made by mixing tin IV oxide and colloidal gold. It is used for colouring glass.

PVA. *See* polyvinyl acetate.

PVC. *See* polyvinyl chloride.

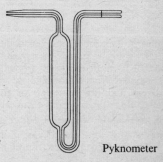

Pyknometer

pyknometer. A bulb with two attached capillary tubes (see illustration) used for measuring the density of liquids. It is filled by sucking the liquid into the bulb until it reaches the graduation on one tube, thus giving a known volume of liquid. *See* relative-density bottle.

pyramid. A solid geometrical figure whose base is a polyhedron and whose other faces are triangles with a common apex. If this apex lies above the centroid of the base, the solid is a *right pyramid.*

pyranose. *See* sugar.

pyrene. A colourless hydrocarbon, $C_{16}H_{10}$, obtained from coal tar. M.pt. 156°C; b.pt. 404°C; r.d. 1.3.

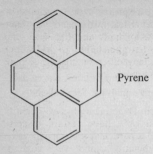

Pyrene

Pyrex. ®A type of heat-resistant borosilicate glass widely used in laboratory apparatus, cooking utensils, etc.

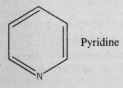

Pyridine

pyridine. A flammable colourless liquid, C_5H_5N, with a strong disagreeable odour. It is a heterocyclic aromatic base, extracted from coal tar and used as a solvent and raw material for the manufacture of other organic chemicals. M.pt. $-42°C$; b.pt. $115°C$; r.d. 0.9.

pyrites. Any of several naturally occurring metal sulphides, especially iron pyrites, FeS_2.

pyrocatechol. See catechol.

pyrochlore. A brown or black naturally occurring oxide, $NaCaNb_2O_5F$, used as an ore of niobium.

pyroelectricity. The production of opposite electric charges on opposite faces of certain crystals when they are heated. Tourmaline is an example of a pyroelectric substance.

pyrogallic acid. See pyrogallol.

pyrogallol (pyrogallic acid). A white crystalline solid, $C_6H_3(OH)_3$, made by heating gallic acid (3,4,5-trihydroxybenzoic acid) with water in an autoclave. It is used in photography and in

the absorption of oxygen for gas analysis. M.pt. $132°C$; b.pt. $309°C$; r.d. 1.5.

pyrolusite. A soft black, blue, or greyish naturally occurring oxide, MnO_2, important as an ore of manganese.

pyrolysis. The decomposition or dissociation of a chemical compound as a result of heat.

pyrometer. Any of several instruments for measuring high temperatures by means of the heat radiation emitted. The *optical pyrometer* consists of a tube with an objective lens, which produces an image of the source at a point inside the tube. A tungsten filament is mounted in the focal plane of this lens and is viewed by an eyepiece. The filament is heated by an electric current, which is varied until it has the same brightness as the source, i.e. until it "disappears". The size of the current gives a measure of the temperature: the instrument is precalibrated. In the *total-radiation pyrometer*, all the radiation is focused onto a bolometer, thermopile, or similar detector.

pyrometry. The science of measurement of high temperatures. The term is usually understood to mean measurement of high temperatures by means of the radiation emitted.

pyrophoric. Denoting materials that emit sparks when struck. Pyrophoric alloys, such as Misch metal, are used in lighter flints.

pyrophosphate. See phosphate.

pyrophosphite. See phosphite.

pyrophosphoric acid. See phosphoric acid.

pyrosilicate. See silicate.

pyrosilicic acid. See silicate.

pyrosulphuric acid. See sulphuric acid.

pyrovanadate. See vanadate.

pyrovanadic acid. See vanadate.

Pythagoras's theorem. The theorem that in a right-angled triangle the square of the length of the hypoteneuse is equal to the sum of the squares of the lengths of the other two sides: i.e. $BC^2 = AB^2 + AC^2$, with the right angle at A. [After Pythagoras of Samos, c. 580–500 B.C., Greek philosopher and mathematician.]

Q

quadrant. A segment of a circle equal in area to one quarter of the circle: i.e. having an angle of 90°.

quadrant electrometer. A type of *electrometer in which a light metal vane on a torsion wire can move inside four metal quadrants. Pairs of opposite quadrants are connected together. A high steady potential (about 120 volts) is applied to the vane. A low potential difference applied across the pairs of quadrants causes a deflection which is directly proportional to the potential difference. The instrument draws no current. It is now hardly ever used.

quadratic equation. An equation whose highest term has a power of two: i.e. an equation of the form $ax^2 + bx + c = 0$. The roots are given by $x = [-b \pm \sqrt{(b^2 - 4ac)}]/2a$.

quadrature. The relative positions of two celestial objects, in practice the sun and the moon or a planet, having a difference in celestial *longitude of 90°. *Eastern* and *western quadrature* coincide roughly with the first and last quarters of the moon respectively.

quadrilateral. Any plane figure bounded by four straight lines.

quadrivalent (tetravalent). Having a valency of four.

qualitative analysis. *See* analysis.

quality (timbre). A property of a sound arising from the fact that it is a combination of a fundamental frequency and a number of *partials (or overtones)

of higher frequency. Notes produced by different musical instruments may have the same loudness and frequency and yet be distinguishable to a listener because they differ in quality. A pure tone has a single frequency.

quantitative analysis. *See* analysis.

quantized. Denoting a physical quantity that can only have certain discrete values in a particular system. The value of the quantity cannot vary continuously but must change in steps.

quantum. A discrete amount of energy: the smallest amount of energy by which a system can change. The photon, for example, is the quantum of electromagnetic radiation, a phonon is a quantum of vibrational energy in a crystal, etc.

quantum electrodynamics. The quantum mechanical theory of electromagnetic interactions between particles and between particles and electromagnetic radiation.

quantum electronics. The application of quantum mechanics to the study of the generation of microwave power in solid crystals.

quantum mechanics. A mathematical theory that developed from quantum theory and is used to explain the behaviour of atoms, molecules, and elementary particles. In early quantum theory, the basic idea that energy is quantized was introduced in an *ad hoc* manner to explain particular phenomena (black-body radiation, the photoelectric effect, atomic spectra, etc.). De Broglie's theory that a particle can also behave like a wave was used by Schrödinger in his formulation of *wave mechanics, in which the wave-like nature of matter leads directly to the idea that only certain energy levels are possible, corresponding to the existence of standing waves. At about the same time an alternative formulation of quantum mechanics was developed based on the *Heisenberg uncertainty principle.

quantum number. An integer or half integer giving the possible values of a quantized property of a system. For example, an electron in the *Bohr atom can have angular momenta of $nh/2\pi$, where n can be 1, 2, 3, etc.: n is the quantum number which determines the possible values of the angular momentum.

Elementary particles also have quantum numbers to describe their properties. The *charge of a particle can be zero or can be equal to the charge of the electron and of equal or opposite sign. These states can be described by quantum numbers 0, -1, or 1, characterizing the charge. Other properties, such as spin, baryon number, hypercharge, and parity, are similarly described by quantum numbers.

quantum theory. A mathematical theory involving the idea that the energy of a system can change only in discrete amounts (quanta), not continuously. The concept was first introduced by Max Planck in explaining the form of the curves of intensity against wavelength for the radiation emitted from a *black body. He assumed that the radiant energy could only be produced in discrete units. Other early applications of quantum theory were Einstein's explanation of the *photoelectric effect and the theory of the *Bohr atom. See also quantum mechanics.

quark. Any of three hypothetical particles that are postulated, together with their antiparticles, as constituents of the known hadronic particles. Any particle can be explained as a combination of quarks and antiquarks. Despite many searches, no quark has as yet been discovered. See also charm.

quart. See Appendix.

quartz. A colourless or white natural crystalline form of silica, SiO_2. Quartz is one of the commonest minerals. It is used in high temperature laboratory apparatus and in optical devices for its transparency to ultraviolet radiation.

quartz clock. A type of *clock in which the frequency of an oscillator circuit is stabilized by coupling it to a crystal of quartz using the piezoelectric effect.

quasar (quasi-stellar object). Any of several hundred recently discovered extragalactic sources of very strong electromagnetic radiation. These objects are starlike in appearance and have enormous *red shifts. If these are interpreted as Doppler effects, the velocity of recession of quasars must be well over half the speed of light and their distances must be several thousand million light-years. Consequently their energy output appears to be in the region of three or four hundred times that of a galaxy, perhaps even more. How such energy outputs occur is so far unexplained.

quaternary ammonium compound. Any compound with the general formula $R_4N^+X^-$, where X is an acid radical and R is either an organic group or a hydrogen atom.

quenching. 1. Fast cooling of a hot metal, used to harden steel, etc.
2. Sudden termination of the discharge in a Geiger counter.
3. A process in which the lifetime of excited atoms, ions, etc., is reduced by addition of a substance that deactivates them.

quicklime. See calcium oxide.

quicksilver. See mercury.

quiet sun. The sun when it has no sunspots, flares, or prominences.

quinhydrone. A reddish solid that is a mixture of hydroquinone and quinone held by intermolecular forces, $C_6H_4(OH)_2.C_6H_4O_2$. In solution it gives a mixture of the two compounds. A half-cell consisting of a platinum electrode in such a mixture is called a *quinhydrone electrode*. The redox reaction $C_6H_4(OH)_2 = C_6H_4O_2 + 2H^+ + 2e$, occurs with electron transfer between the solution and the platinum and the electrode can be used to measure pH.

quinine. A white bitter alkaloid, $C_{20}H_{24}N_2O_2.3H_2O$, extracted from cinchona bark and used as an antimalarial drug, often in the form of a soluble salt. M.pt. $57\,^{\circ}C$.

quinol. *See* hydroquinone.

Quinone

quinone. 1. (benzoquinone). A yellow crystalline solid, $C_6H_4O_2$, prepared by the oxidation of aniline with chromic acid and used in manufacturing dyes and quinol. M.pt. $116\,^{\circ}C$; r.d. 1.3. *See also* quinhydrone.
2. Any of a class of compounds in which a CH group in a benzene ring has been replaced by a CO group.

quinquevalent. *See* pentavalent.

quotient. The result of dividing one number by another; a ratio.

Q-value. The amount of energy released in a specified nuclear reaction.

R

racemic. Denoting a substance composed of equal amounts of its optically active isomers (*see* optical isomerism). A racemic mixture is not optically active. Racemic compounds are denoted by *dl-*, as in *dl-*tartaric acid.

racemic acid. The racemic form of tartaric acid.

rad. A unit of dose of ionizing radiation equal to the dose equivalent to an absorbed energy of 0.01 joule per kilogram of material.

radar. A system for detecting distant objects by reflecting electromagnetic waves off them. A beam, consisting of regular pulses of microwaves, is transmitted from a movable aerial. Pulses hitting a distant object are reflected back to the aerial and the distance of the object is measured by the time taken between transmission of a pulse and reception of the reflected pulse. In a typical radar system, the aerial moves round and the detected signals are fed to a cathode-ray screen which gives a display of the surrounding area.
If the object is moving the difference between the frequency of the transmitted pulse and that of the reflected pulse (caused by the Doppler effect) can be used to distinguish motion of the object and to determine its velocity (*Doppler radar*). The term is an acronym for *radar detection and ranging*.

radian. The angle subtended at the centre of a circle by an arc whose length is equal to the radius of the circle. A complete revolution is 2π radians. π radians is equivalent to 180 degrees.

radiant heat. *See* infrared radiation.

radiation. Particles or waves emitted from a source, in particular electromagnetic radiation, sound waves, or streams of particles thrown out by radioactive materials.

radiation belts. *See* Van Allen belts.

radiation pressure. A pressure exerted on a surface by the impact of electromagnetic radiation.

radiative. Involving the production of electromagnetic radiation. Thus a *radiative collision* is a collision resulting in the emission of a photon.

radical. 1. *See* free radical, group.
2. A quantity that is a root of another quantity. Thus $\sqrt{x}$ is a radical: $\sqrt{}$ is a radical sign.

radio. The communication of signals by means of *radiofrequency electromagnetic waves. The term embraces all forms of communication, including television, but is sometimes used specifically for sound broadcasting. A radio transmitter generates a *carrier wave of fixed frequency. The information to be transmitted is converted into an electrical signal (as by a microphone or television camera) which modulates this carrier wave (see modulation). The modulated carrier wave is projected from an aerial and carried, either as a *ground wave or *sky wave, to the aerial of the receiver where the signal is demodulated.

radio-. Prefix indicating a radioactive isotope of an element.

radioactive. Exhibiting radioactivity.

radioactive series. A series of radioisotopes such that each member of the series is produced by alpha or beta decay from the preceding members. Thus, in the uranium series the first member is uranium-238, which undergoes alpha decay to thorium-234. This undergoes beta decay to protactinium-234, which in turn undergoes beta decay to uranium-234, and so on. Three other series exist: the thorium series, the actinium series, and the neptunium series. All end with stable isotopes of lead.

radioactivity. Spontaneous disintegration of the nuclei of certain isotopes with emission of beta rays (electrons), alpha rays (helium nuclei), or gamma rays. Electrons are produced by *beta decay (in some cases positrons are emitted). They have kinetic energies up to the MeV range and are not very penetrating. Alpha particles are produced by *alpha decay. They produce more ionization than beta rays and have energies up to about 10 MeV. Like beta rays, they are also easily absorbed by matter but are less penetrating. Gamma rays are photons of electromagnetic radiation. They are formed when beta or alpha decay leads to the production of a nucleus in an excited state, which decays to its ground state with emission of a gamma-ray photon. They are not efficient at ionizing matter and, in consequence, are highly penetrating, being capable of passing through several centimetres of lead.

In any sample of radioactive material the number of disintegrations per unit time, and thus the rate of decomposi-

tion, is proportional to the amount of material present. Thus the rate follows an equation of the form $-dN/dt = \lambda N$, where N is the number of atoms and λ is a constant for the particular substance—its decay constant. The number of atoms present follows the equation $N = N_0\exp(-\lambda t)$, where N_0 is the initial number: the amount of radioactivity shows an exponential decay. The reciprocal of the decay constant is the mean life of the isotope (τ), which is equal to $T_{1/2}/0.693$, where $T_{1/2}$ is the half-life—the time for half the sample to decay. The activity of an isotope is measured in *curies.

Natural radioactivity is exhibited by natural isotopes of heavy elements: these are arranged in *radioactive series. Artificial radioisotopes may also be produced by bombardment with particles.

radio astronomy. The branch of astronomy involving the use of *radio telescopes.

radiocarbon dating (carbon-14 dating). A method of *dating archaeological specimens that are made of wood, cotton, or other once-living matter. The carbon dioxide in the air contains a small amount of carbon-14, a radioisotope produced by bombardment of nitrogen with cosmic rays. This is incorporated into living matter during the lifetime of the organism. Once the matter dies, the amount of carbon-14 present begins to decrease due to beta decay to give nitrogen-14. Thus, measurement of the amount of radiocarbon present, by using a counter, gives an estimate of the object's age. The half-life is 5730 years making the technique suitable for dating objects up to about 10 000 years old.

radiochemistry. The branch of chemistry concerned with compounds containing radioisotopes, their use in separating isotopes, chemical reactions, application to *tracer studies, etc.

radiofrequency. A frequency between 3 kilohertz and 300 gigahertz.

radiofrequency radiation. *Electromagnetic radiation with frequencies lying in the range 3 kilohertz to 300 gigahertz. Radiofrequency radiation is the portion of the electromagnetic spectrum with wavelengths longer than those of infrared radiation. It includes *microwaves and is divided into frequency bands.

radiogenic. Resulting from or produced by radioactive disintegration.

radiography. The use of radiation, usually X-rays, to investigate internal structure. The X-rays are passed through the object investigated and form a shadow image (*radiograph*) on a fluorescent screen or photographic plate.

radio interferometer. *See* radio telescope.

radioisotope. An isotope that is radioactive.

radiology. The science of X-rays, particularly their effects and applications.

radiolucent. Partially transmitting radiation. The term is usually applied to materials that partially transmit X-rays or gamma rays, distinguishing them from *radiotransparent* materials, which allow almost all the radiation through, and *radio-opaque* materials, which block most of the radiation.

radioluminescence. *Fluorescence caused by radioactive decay. The radiation—electrons, alpha particles, or gamma rays—causes excitation of the atoms and light emission results. Radioluminescence is the process responsible for the glow of self-luminous paints.

radiolysis. The process of decomposing a substance by irradiation with high-energy particles or X-rays. *See also* photolysis.

radionuclide. A nuclide that is radioactive.

radio-opaque. *See* radiolucent.

radio source. A source of radio waves in space. Radio sources include supernovae, pulsars, quasars, and some stars, including the sun.

radio telescope. An instrument for detecting radio waves from a particular direction in space. One type has a large parabolic aerial that can be moved to point to different areas of the sky. The *radio interferometer* has two or more separate aerials connected to the same radio receiver. The position of the radio source is determined by the interference of radio waves reaching the different aerials.

Radio telescopes are used in studying the intensity and nature of radio waves from inside and outside the Galaxy. Many celestial objects have been discovered that are not visible with optical telescopes.

radiotransparent. *See* radiolucent.

radium. Symbol: Ra. A white radioactive metallic element occurring as a product of radioactive decay of heavier elements. Its most stable isotope, ^{226}Ra, has a half-life of 1622 years. The element is used as a radioactive source in research and medicine. A.N. 88; A.W. 226.05; m.pt. 700°C; b.pt. 1140°C; valency 2.

radius. The line or distance between the centre of a circle or sphere and the circumference or surface.

radius of gyration. Symbol: k. A distance associated with a rotating body, such that $I = mk^2$, where m is the mass of the body and I is its moment of inertia.

radius vector. *See* polar coordinates.

radix. The base of a number system. For example, 10 is the radix of decimal notation.

raffinate. A liquid that has been purified by solvent extraction.

raffinose (melitose). A white crystalline powder, $C_{18}H_{32}O_{16}.5H_2O$, with a sweet taste, extracted from sugar beet. It is a trisaccharide sugar used in medicine and in making other saccharides. M.pt. 118°C; r.d. 1.5.

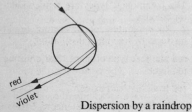

red

violet

Dispersion by a raindrop

rainbow. An arc of coloured bands seen in the sky when sunlight is refracted and dispersed by raindrops. The rainbow is composed of spectral colours with the violet band on the inside of the bow. It is produced by light that undergoes total internal reflection inside the drop (see illustration) and the light reaching the observer makes an angle of 40–42° with the direction of the observer's shadow. Sometimes a faint secondary bow is seen lying outside the primary bow: this is formed by two internal reflections and has its colours reversed.

Raman effect. The inelastic scattering of light or ultraviolet radiation by molecules. The scattered radiation and the incident radiation differ in frequency by discrete amounts, corresponding to vibrational and rotational changes in energy level of the molecules. The spectrum of the scattered radiation (*Raman spectrum*) can thus be used to give information on molecular structure. [After Chandrasekhara Raman (1880–1970), Indian physicist.]

Raney nickel. A form of nickel made by dissolving the aluminium from an aluminium-nickel alloy with sodium hydroxide. The nickel produced has a spongy texture and is an extremely effective hydrogenation catalyst.

Rankine scale. A scale of temperature based on the Fahrenheit scale. Temperatures are measured in degrees Rankine (°R) which are equivalent to Fahrenheit degrees. Absolute zero is 0°R (–459.67°F). °R = °F + 459.67. [After William John Rankine (1820–72), Scottish engineer.]

Raoult's law. The principle that the fractional decrease in vapour pressure occurring when a solute is dissolved in a solvent is equal to the ratio of the number of solute molecules to the total number of molecules: i.e. $(p - p_1)/p = N_1/(N + N_1)$, where p_1 is the vapour pressure of the solution, p that of the pure solvent, N_1 the number of solute molecules, and N the number of solvent molecules. [After Francois Raoult (1830–1901), French chemist.]

rare earth. *See* lanthanide.

rare gas. *See* inert gas.

rational number. A number that is an integer or can be expressed as a quotient of integers.

ray. A line used to represent the path of light or other radiation.

rayon. Either of two types of textile fibre made from cellulose. *Viscose rayon* is produced by dissolving wood pulp in a mixture of sodium hydroxide and carbon disulphide to form *viscose*, a thick brown solution of cellulose *xanthate. This is forced through fine holes into a bath containing acid, which decomposes the xanthate to regenerate cellulose fibres. *Acetate rayon* is made by treating wood pulp with a mixture of acetic anhydride with acetic and sulphuric acids to form cellulose acetate, which is dissolved in a solvent and forced through fine holes into the air. The solvent evaporates leaving the cellulose acetate fibre. Rayon was formerly called *artificial silk*.

reactance. Symbol: X. The imaginary part of the *impedance of an electrical circuit. The reactance of an inductance L is ωL and that of a capacitance C is

$1/\omega C$, where ω is the *angular frequency. Reactance is measured in ohms.

reactant. A substance taking part in a chemical reaction, as distinguished from a product of the reaction.

reaction. 1. *See* chemical reaction.
2. An equal and opposite force occurring when a force acts on a body. *See* Newton's laws of motion.

reactive dye. *See* dye.

reactor. 1. An apparatus for producing energy by nuclear fission or fusion. *See* fission reactor, fusion reactor.
2. A capacitance, inductor, or other element producing reactance in an electrical circuit.

reagent. A substance used for chemical analysis or synthesis.

realgar. A bright red mineral consisting of arsenic disulphide, As_2S_2.

real image. *See* image.

Réaumur scale. A temperature scale in which the melting point of ice is 0° and the boiling point of water is 80°. Temperatures are measured in *degrees Réaumur* (°R). [After René Antoine Réaumur (1683–1757), French scientist.]

reciprocal. One divided by a number or quantity: the reciprocal of 4 is ¼.

reciprocal ohm. *See* siemens.

reciprocal proportions, law of. The principle that if two elements combine individually with a third element then they combine with each other in the same relative proportions as they react with this third element. Oxygen and hydrogen, for example, both combine with carbon. The ratio C:O is 12:32 and the ratio C:H is 12:4, these being proportions by weight. In accordance with the law, hydrogen and oxygen combine together in the ratio 4:32 (or 2:16). The three compounds involved are CO_2, CH_4, and H_2O. The law involves the idea that elements have combining weights (or *equivalent weights) and is sometimes called the

law of equivalent proportions. It is one of three laws of *chemical combination.

rectangle. A parallelogram whose angles are all right angles.

rectifier. 1. An electronic component or circuit for converting an alternating current into a unidirectional current. The commonest types are semiconductor *diodes.
2. An apparatus for separating liquid mixtures by fractional distillation.

red giant. A type of giant *star emitting red light.

red lead. *See* lead oxide.

redox. Relating to simultaneous *oxidation and *reduction. A redox reaction is one in which one reactant is oxidized and the other is reduced.

red shift. The displacement of the spectral lines emitted by a moving body towards the red (longer wavelength) end of the visual spectrum. It is caused by the *Doppler effect and, when observed in the spectrum of distant stars and galaxies, it indicates that the body is receding from the earth. The greater the displacement, the higher is the velocity of recession. Displacement of lines to the red end of the spectrum can also result from the *Einstein shift, caused by the high gravitational field of a star. *See also* big-bang theory.

reduced equation. An equation of state of a gas in which the pressure (p), volume (V), and temperature (T) are replaced by the ratios p/p_c, V/V_c, and T/T_c, where p_c, V_c, and T_c are the critical pressure, volume, and temperature of the substance. The ratios are the reduced pressure, volume, and temperature.

reducing agent. A compound that causes *reduction, either by abstracting oxygen, supplying hydrogen, or donating electrons. Many compounds act as reducing agents in particular circumstances but the term is also used for compounds that are particularly effective reducing agents. Examples are hydrogen, carbon, carbon monoxide, and ferrous salts.

reducing sugar. See sugar.

reduction. A process in which oxygen is removed from or hydrogen is combined with a compound. Thus, the reaction $Fe_2O_3 + 3C = 2Fe + 3CO$ is a reduction of iron oxide and $3H_2 + N_2 = 2NH_3$ is a reduction of nitrogen. The term has been extended to any reaction in which electrons are gained, as in the reduction of ferric ions to ferrous ions: $Fe^{3+} + e = Fe^{2+}$. Compare oxidation.

refining. The process of purifying a substance.

reflecting telescope (reflector). A telescope in which the primary image is formed by a concave mirror. Reflecting telescopes are used in astronomy. Unlike refracting telescopes they do not suffer from chromatic aberration and spherical aberration is reduced by use of parabolic mirrors. Several types exist, including the *Newtonian, *Cassegrainian, *Gregorian, and *Schmidt telescopes.

reflection. The process in which light or other radiation that is incident on a surface bounces off the surface back into the original medium rather than passing into the second medium. There are two *laws of reflection*:
(1) the incident ray, the reflected ray, and the normal to the surface at the point of incidence all lie in the same plane;
(2) the angle of incidence equals the angle of reflection. The laws only apply for *specular reflection. See also total internal reflection.

reflector. See reflecting telescope.

reflex angle. An angle lying between 180° and 360°.

reflux. To boil a liquid in a vessel fitted with a condenser so that the vapour is continuously returned to the vessel.

reforming. The process of converting petroleum hydrocarbons into hydro-

carbons suitable for use in petrol by temperature, pressure, and use of catalysts.

refracting telescope (refractor). A telescope in which the image is formed by lenses rather than mirrors: examples are the *Kepler telescope and the *Galilean telescope. Refractors are used as *terrestrial telescopes. They are less suitable for use in astronomy than *reflecting telescopes because of their chromatic and spherical aberration and because of the difficulties involved in grinding large lenses.

refraction. The change in direction of a ray of light or other electromagnetic radiation as it passes from one medium to another medium in which it has a different velocity. The two *laws of refraction* are:
(1) the incident ray, the refracted ray, and the normal at the point of incidence all lie in the same plane;
(2) the sine of the angle that the incident ray makes with the normal divided by the sine of the angle that the refracted ray makes with the normal is a constant. Thus $\sin i / \sin r = \mu$ (this is *Snell's law*). See also refractive index.

refractive index. A measure of the extent to which a medium refracts light. When light passes from medium 1 to medium 2, the ratio $\sin i / \sin r$ is a constant for the two media (see refraction). It is called the *relative refractive index* and is given the symbol n_{12}. It is equal to the ratio of the velocities of light in the two media (i.e. c_1/c_2). It follows that $n_{12} = 1/n_{21}$ and $n_{12}.n_{23} = n_{13}$. The *absolute refractive index*, which has the symbol n, is the value for a medium when the first medium is free space. The absolute value for air is 1.000 29, and values of refractive index measured in air are thus close to absolute values. The refractive index depends on the wavelength of light used (see dispersion) and values are commonly given for the yellow D lines in the sodium emission spectrum ($\lambda = 589.3$ nm).

Refractive index can be found by a variety of methods. An object immersed in a liquid or viewed through a rectangular block of solid has an apparent depth d_a related to its true depth d_t by $d_t/d_a = n$. This is used in a simple method of measuring the refractive index of liquids or solids by using a travelling microscope to focus on the apparent position of an object. The refractive index of a solid in the form of a prism of known angle can be measured by using a spectrometer to determine its *deviation. This method can also be applied to a liquid contained in a hollow prism. Another technique is that of determining the angle at which total internal reflection occurs. Small samples of solid can be investigated by immersing them in a liquid mixture and changing the composition of the mixture until the sample "disappears". The liquid and solid then have the same refractive index. Other techniques depend on interference measurements.

refractometer. Any of various instruments or pieces of apparatus used to measure refractive index.

refractor. *See* refracting telescope.

refractory. Denoting a solid that has a high melting point and can withstand high temperatures.

refrigerant. A substance that is used as the working fluid in a refrigerator. Refrigerants are usually volatile liquids or easily liquefied gases.

Regnault's method. A method of measuring the density of gases by weighing a large bulb of known volume, first filled with the sample at a known temperature and pressure and then evacuated. [After Henri Victor Regnault (1810–78), French chemist.]

relative density (specific gravity). Abbreviation: r.d. The ratio of the density of a substance to the density of a reference substance under specified conditions. Relative densities of solids and liquids are measured with reference to water, the density of the solid or

liquid being taken at a specified temperature and the density of the water being taken at $4°C$, the temperature at which its density is a maximum. The temperature of the substance is usually understood to be $20°C$. The relative density measured in this way is numerically equal to its density in grams per cubic centimetre. Any substance with a relative density less then 1 will float on water. The relative density of gases is often measured with reference to air, both gases being at standard temperature and pressure.

relative humidity. *See* humidity.

relative permeability. Symbol: μ_r. A magnetic property of a substance that determines the way in which it changes the *magnetic flux density of an external field. It is given by $\mu_r = B_m/B_0$, where B_m and B_0 are flux densities in the medium and in free space respectively. An alternative definition is that it is the ratio of the self-inductance of a circuit when the circuit is surrounded by the medium to the self-inductance of the circuit in free space: i.e. $\mu_r = L_m/L_0$. Sometimes the permeability (μ) of the medium is defined by the ratio of the magnetic flux density (B) in the medium to the external magnetic field strength (H) that produces it: i.e. $\mu = B/H$. The relative permeability may then be expressed as $\mu_r = \mu/\mu_0$, where μ_0 is the *magnetic constant (or *permeability of free space*).

relative permittivity. Symbol: ϵ_r. The ratio of the capacitance of a system surrounded by a certain medium to the capacitance of the system in free space: i.e. $\epsilon_r = C_m/C_0$. The *permittivity* of a medium (ϵ) is the ratio of the electric displacement to the strength of the electric field. It is related to the relative permittivity by the equation: $\epsilon_r = \epsilon/\epsilon_0$, where ϵ_0 is the *electric constant (or *permittivity of free space*).

relativistic. Involving the theory of relativity or effects predicted by this theory.

relativistic mass. The mass of a body as predicted by the theory of relativity. The relativistic mass of a particle moving at velocity v is $m_0(1 - v^2/c^2)^{-1/2}$, where m_0 is the *rest mass.

relativistic velocity. A velocity sufficiently close to that of light for effects predicted by the theory of relativity to be significant.

relativity. A system of mechanics introduced by Einstein to replace Newtonian mechanics. The theory was developed in two parts.

The *special theory of relativity* deals with uniform motion. Up to the early part of the twentieth century the motion of bodies and the action of forces had been based on *Newton's laws of motion, involving such ideas as the constancy of mass and the simple method of obtaining relative velocities by vector addition of the velocities of the observer and the object observed. In particular, it was assumed that light was propagated through a stationary medium, the ether, at a fixed velocity c relative to the ether. At the beginning of this century several results were obtained that appeared to contradict the principles of Newtonian mechanics. The best known is the *Michelson-Morley experiment, which indicated that the velocity of light was the same when measured in the direction of the earth's rotation and perpendicular to this direction. The observer always appears to be at rest. Several attempts were made to explain this phenomenon and this experiment was the factor that inspired Einstein's theory, published in 1905. He assumed that the velocity of light is a constant for all observers, no matter how they move, and that there is no standard of absolute rest. The constant velocity of light is an experimental fact derived from the Michelson-Morley experiment.

The assumption leads to several non-classical results for observations made on a system that is moving at a constant velocity relative to the observer. An object of length, l_0 when at rest relative to an observer will appear to have a length $l_0(1 - v^2/c^2)^{1/2}$ when moving at velocity v. The apparent shortening is equivalent to a *Lorentz-Fitzgerald contraction. A similar change occurs in the mass: an object having a mass m_0 when at rest relative to an observer has a mass $m_0(1 - v^2/c^2)^{-1/2}$, when moving at a velocity v. This increase in mass is only significant at high velocities: it is the effect producing the limiting velocity of particles in the *cyclotron. The increase of mass with velocity led to the idea that mass and energy are interconvertible: i.e. that mass can be "destroyed" with the production of an equivalent amount of energy and that mass can be produced with loss of energy. The two are related by *Einstein's equation, $E = mc^2$. Another relativity effect is that of *time dilation*. If an observer A measures the passage of time on a clock and another observer B is moving at a velocity v relative to A, it appears to A that time as measured on B's clock is given by $t(1 - v^2/c^2)$. In other words, it appears to A that time is running more slowly for B. The effect is observed in the motion of some mesons in cosmic rays, which appear to have longer lifetimes because they are moving at relativistic velocities. One of the consequences of relativity theory is that there is no absolute value of time. Two observers moving relative to one another would assign different times to the occurrence of an event. In relativity theory, events are specified in a *space-time continuum.

The *general theory of relativity* deals with accelerated relative motion and is particularly concerned with gravitational effects. An observer who is travelling in a circular path experiences an acceleration directed towards the centre of his path and he appears to be subjected to an outwards, centrifugal, force. This force is proportional to his mass in the same way that the weight of an object resulting from gravitational attraction is proportional to the object's mass. Indeed, if the observer were in a sealed vehicle and were not aware of his motion, he might ascribe the force to a gravitational attraction by a body outside the vehicle. In other words, the effects of gravity thought of as action at a distance due to a body's gravitational field, can alternatively be described by a particular motion in space. This idea is used in the general theory to describe gravitational attraction as a property of space-time. Thus, the apparent force experienced by a body moving close to a large mass is explained by assigning *non-Euclidean geometrical properties to space-time, which is said to be "curved" by the presence of the mass. Tests of general relativity are usually based on astronomical measurements: an example is the *Einstein shift. *See also* unified-field theory.

relay. An electrical device in which an electrical current in one circuit can open or close a separate circuit. In the simplest type the first circuit contains a coil which produces a magnetic field to open or close a mechanical switch in the other circuit. Other types have electronic switching using semiconductors or valves.

remanence. The magnetization of a material remaining when the external field has been reduced to zero in a *hysteresis cycle. Substances such as steel, with high remanences, make good permanent magnets. Soft iron and other materials with low remanence are used in electromagnets.

resin. A solid or semisolid natural or synthetic polymeric substance. *See* plastic.

resistance. Symbol: *R*. A measure of the ability of a piece of material or an electrical component or circuit to resist the flow of current, equal to the potential difference across the circuit divided by the current it produces. In alternating-current circuits the resistance is the real part of the *impedance. The quantity is measured in ohms.

resistance thermometer. A type of *thermometer that depends for its action on changes in resistance with temperature. Usually a coil of fine platinum wire is used incorporated in a Wheatstone-bridge circuit. The resistance of the coil increases with temperature.

resistivity. Symbol: ρ. A measure of the ability of a material to resist the flow of an electric current, equal to the resistance per unit length of the material taken for unit cross-sectional area. Resistivity is given by the equation: $\rho = RA/l$, for a resistor of length l and constant cross-sectional area A. Its units are ohm-metres. Resistivity is sometimes incorrectly called *specific resistance*.

resistor. A component used in an electrical circuit to introduce resistance. Resistors are made of a coil of resistance wire or of carbon.

resolution. 1. The separation of a vector into components.
2. *See* resolving power.

resolving power (resolution). The ability of a microscope, telescope, or other optical instrument to produce distinguishable images of closely spaced objects. The resolving power depends on the wavelength of the light and on the amount of light collected by the instrument. The resolving power of a spectrometer is its ability to distinguish between closely spaced wavelengths, energies, etc.

resonance. 1. (in mechanics). The condition occurring when a vibrating system is subjected to a periodic force that has the same frequency as the natural vibrational frequency of the system. At resonance, the amplitude of vibration is a maximum.
2. (in chemistry). The phenomenon in which the molecular structure of a chemical compound is not described by any single simple formula but can be represented by a combination of two or more structures. For instance, the polar molecule HCl has a character between that of a covalent molecule H-Cl and an

ionic molecule H^+Cl^- and can be described as a *resonance hybrid* of the two. *Benzene, in which the bonds have a character between single bonds and double bonds, can be written as a resonance hybrid of two Kekulé structures. In chemistry, the separate resonance structures, which do not have any independent existence, are often written with a double-headed arrow (↔) between them.

3. (in electrical circuits). The condition occurring in an electrical circuit when its *impedance is a minimum and depends only on the resistance of the circuit, not on its reactance. Resonance occurs at the frequency f for which $2\pi fL = 1/2\pi fC$, where L is the inductance and C the capacitance. At this frequency the reactance is zero and the circuit is said to be *tuned*.

4. (in particle physics). Any of a large number of elementary particles with very short lifetimes. The resonances are excited states of more stable particles.

5. (in nuclear physics and spectroscopy). The absorption of a photon with the correct energy to excite a nucleus, atom, etc., from one energy level to a higher level.

resorcinol (1,3-dihydroxybenzene). A white crystalline solid phenol, $C_6H_4(OH)_2$, used in the manufacture of dyes and celluloid. M.pt. 111°C; b.pt. 281°C; r.d. 1.3.

restitution. Elasticity. The *coefficient of restitution* (symbol: e) for the impact of two spheres is the ratio of the relative velocity of the bodies before impact to that after impact, the velocities being taken along the line between the centres of the spheres at impact.

rest mass. The mass of a body when it is at rest relative to its observer, as distinguished from its *relativistic mass.

resultant. A *vector produced by the combination of two or more other vectors.

retina. *See* eye.

retort. 1. A piece of laboratory apparatus consisting of a glass bulb with a long narrow neck.
2. Any of various vessels for carrying out industrial chemical processes.

reverberatory furnace. A type of furnace in which the fuel is burnt in one section and the heat is directed down onto the contents by a curved roof made of refractory material. Reverberatory furnaces are used for smelting or refining metals in such a way that the fuel is not mixed with the material used.

reversible. 1. Denoting a chemical reaction that can proceed in either direction. An example is the hydrolysis of esters: $CH_3COOC_2H_5 + H_2O = CH_3COOH + C_2H_5OH$. The reverse reaction—esterification of an alcohol by an acid—also occurs. In general, such reactions reach a state of *chemical equilibrium which can be displaced in one direction or the other by changing the conditions or by continuously removing one compound from the system as it is formed. Reactions that proceed completely in one direction are said to be *irreversible*.
2. Denoting a change in the conditions of a system in which the system is always at thermodynamic equilibrium during the change. Reversible processes are theoretical ideals: they would have to be performed infinitely slowly and are never realized in practice. All real changes are *irreversible*.

Reynolds number. Symbol: Re. A number used in problems involving the flow of fluids, equal to $v\rho l/\eta$, where v is the fluid's velocity, ρ is the density, η the viscosity, and l is a chosen dimension of the system through which the fluid is moving. For example, l may be the radius of a pipe. At a certain value of the Reynolds number a change occurs from laminar flow to turbulent flow. [After Osborne Reynolds (1842–1912), British physicist and engineer.]

rhenium. Symbol: Re. A silvery-white lustrous *transition element obtained as a by-product in refining molybdenum. Alloys of rhenium with molybdenum are superconducting at very low temperatures. The element dissolves in concentrated nitric and sulphuric acids but not in hydrochloric acid. It reacts with oxygen at high temperatures. A.N. 75; A.W. 186.2; m.pt. 3180°C; b.pt. 5620°C; r.d. 21.0; valency 1-7.

rheology. The study of the flow properties of fluids. *See* viscosity.

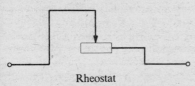

Rheostat

rheostat. A variable resistor, typically consisting of a coil of bare resistance wire with one end connected in the circuit, the other connection being made by a sliding contact at points along the coil (see illustration).

rhodium. Symbol: Rh. A silvery-white soft ductile *transition element occurring associated with platinum. It is used as a catalyst and a constituent of some platinum alloys. The element is not attacked by acids and only slowly dissolves in aqua regia. It reacts with oxygen, chlorine, and other nonmetals at red heat. The main rhodium compounds are complexes of rhodium III and, to a lesser extent, rhodium IV. A.N. 45; A.W. 102.9; m.pt. 1966°C; b.pt. 3727°C; r.d. 12.4; valency 3 or 4.

rhodochrosite. A red or pink naturally occurring carbonate, $MnCO_3$, used as a source of manganese.

rhombohedral. *See* crystal system.

rhombus. A parallelogram whose sides are all equal. A square is a rhombus that is also a rectangle.

rigidity modulus. *See* modulus.

ring. A closed circle of atoms in the structure of a molecule.

ring closure. *See* cyclization.

RMS. *See* root mean square.

Rochelle salt. *See* potassium sodium tartrate.

rock crystal. A transparent form of quartz, SiO_2.

rock salt. A transparent naturally occurring crystalline form of sodium chloride, NaCl.

rods. *See* eye.

röntgen. Symbol: R. A unit of exposure *dose equal to the dose that produces ions of one sign carrying a charge of 2.58×10^{-4} coulomb in air. [After Wilhelm Konrad Röntgen (1845–1923), German physicist.]

Röntgen rays. *See* X-rays.

root. 1. One of two or more identical factors of a given number. The *square root* of a number is a number that, multiplied by itself, gives the specified number. Thus the square root of 16, written 16 or $(16)^{1/2}$, is either $+4$ or -4, since $4 \times 4 = 16$ and $-4 \times -4 = 16$. The *cube root* is the number that gives the specified number when multiplied by itself twice. In general, the nth root of x is the number that multiplied by itself $n - 1$ times gives x.
2. A value that satisfies a given equation. For example, the equation $x^2 + x - 2 = 0$ is satisfied by $x = 1$ and $x = -2$: these are its roots. In general, a quadratic equation has two roots, a cubic equation three roots, etc.

root mean square (RMS). The square root of the average of the squares of a number of values. For instance, if gas molecules have velocities v_1, v_2, v_3, etc., their RMS velocity is given by

$$\left(\frac{n_1 v_1^2 + n_2 v_2^2 + n_3 c_3^2 +}{n} \right)$$

Here n_1 molecules have velocity v_1, n_2 have velocity v_2, etc., and n is the total number of molecules. The RMS value of an alternating current or voltage is obtained by taking the square root of the average of the squares of all the values of current (or voltage) over a complete cycle. The RMS value of a quantity is useful because it measures the effect in a circuit: for example the heat produced by an alternating current in a resistance (R) is I^2Rt, where I is the RMS current and t is the time for which it flows. For this reason, RMS values are also called *effective values*. If the alternating quantity is represented by a sine wave, its RMS value is $A/\sqrt{2}$, where A is the amplitude (i.e. the peak value).

Rose's metal. An alloy of bismuth (about 50%), lead (about 25%), and tin. It has a low melting point (94°C) and is used for bending metal pipes, in constant temperature baths, and similar applications.

rotary converter. A device for converting an alternating current into a direct current, consisting essentially of an a.c. motor driving a d.c. generator.

rotor. The moving part of a *generator or *electric motor. It can be either the field winding or the armature.

rubber. 1. A natural polymeric material obtained from the sap of the tree *Hevea braziliensis*. The sap (*latex*) is first coagulated by acetic acid and either dried in air (*crepe rubber*) or in wood smoke (*smoked rubber*). This material is moulded and vulcanized (*see* vulcanization) with addition of a filler, such as carbon black. Natural rubber is a polymeric material made up of isoprene units.
2. Any of various synthetic elastomers made by polymerizing or copolymerizing styrene, butadiene, neoprene, and similar compounds.

rubidium. Symbol: Rb. A soft silvery-white metallic element occurring in some minerals and brines. It is made by electrolysis of the fused chloride. Rubidium is an *alkali metal. A.N. 37; A.W. 85.47; m.pt. 38.4°C; b.pt. 688°C; r.d. 1.53; valency 1.

rubidium-strontium dating. A method of *dating geological samples by using the decay of rubidium-87, a radioisotope occurring in some rocks. It decays with a half-life of 5×10^{11} years to give strontium-87. Measurement of the ratio of the two isotopes present gives an estimate of the sample's age. The method is useful for ages of over 10^9 years.

ruby. A naturally occurring red form of aluminium oxide, Al_2O_3, containing small amounts of chromium, used in lasers.

rust. A hydrated form of iron III oxide, $Fe_2O_3.H_2O$, formed by the action of moist air on iron or steel.

ruthenium. Symbol: Ru. A hard white *transition element found associated with platinum. It is used as a catalyst and a constituent of some platinum alloys. The metal is not attacked by acids, dissolves in fused alkalis, and combines with oxygen and halogens at high temperatures. It forms large numbers of complexes in several oxidation states. A.N. 44; A.W. 101.07; m.pt. 2310°C; b.pt. 3900°C; r.d. 12.41; valency 1-8.

rutile. A reddish-brown to black mineral consisting of titanium dioxide, TiO_2, used as an ore of titanium.

Rydberg constant. Symbol: R. The constant with a value $me^4/8h^3\epsilon_0^2c$ appearing in equations for the wavelengths of lines in the *Rydberg series and similar *spectral series (m and e are the mass and charge of the electron, h is the Planck constant, and ϵ_0 is the electric constant). For hydrogen, its value is $1.096\ 77 \times 10^7$ m^{-1}. [After Johannes Robert Rydberg (1854–1919), Swedish physicist.]

Rydberg series. A series of lines appearing in the absorption spectrum of a gas in visible and ultraviolet radiation. The lines get progressively closer together as the wavelength decreases until they merge into a continuum. The series is produced by excitation from the *ground state to various excited states and the onset of the continuum corresponds to complete removal of the electron to form an ion. *Rydberg spectra* can thus be used to measure *ionization potentials. *See also* spectral series.

S

saccharide. A *carbohydrate; a monosaccharide, disaccharide, polysaccharide, etc.

saccharimeter. A form of *polarimeter designed for measuring the strength of sugar solutions. *Compare* saccharometer.

saccharin. A white crystalline powder, $C_7H_5ONHSO_4$, with a taste five hundred times as sweet as cane sugar. It is used, often in the form of the sodium salt, as a sweetener. M.pt. 228°C.

saccharometer. A *hydrometer that is designed and calibrated for measuring the strengths of sugar solutions. *Compare* saccharimeter.

safety lamp. *See* Davy lamp.

sal ammoniac. *See* ammonium chloride.

salicylate. Any salt or ester of salicylic acid.

salicylic acid (1-hydroxybenzoic acid). A white crystalline powder, $C_6H_4(OH)$-$(COOH)$, that occurs naturally in several plants. It is used to make aspirin and dyes and as a food preservative. M.pt. 157°C; b.pt. 211°C (20 mmHg); r.d. 1.4.

saline. Denoting a solution that contains a salt or salts, especially one containing sodium chloride or another alkali-metal halide.

salt. 1. Any of a class of compounds produced, together with water, by reaction between an acid and a base. Salts are derived from acids by replacement of one or more hydrogen atoms with metal atoms or ions or with other positive ions. Typically, they are crystalline water-soluble solids consisting of mixtures of positive and negative ions, as in sodium chloride, Na^+Cl^-, or ammonium nitrate, NH_4^+ NO_3^-. Some metal salts are not purely ionic. Aluminium chloride, for example, is a low-melting solid containing Al_2Cl_6 molecules, in which the atoms are held together by covalent bonds. Salts such as $NaHCO_3$, in which acidic hydrogens are present, are *acid salts*. Similarly, salts such as $Al(OH)_2Cl$ are *basic salts*. **2. (common salt).** *See* sodium chloride.

salt bridge. A tube containing a solution of potassium chloride in a gel, used to obtain electrical contact between two half cells without mixing of the electrolytes.

saltcake. *See* sodium sulphate.

saltpetre. *See* potassium nitrate.

samarium. Symbol: Sm. A silvery metallic element belonging to the *lanthanide series. A.N. 62; A.W. 150.4; m.pt. 1072°C; b.pt. 1778°C; r.d. 7.5; valency 3.

Sandmeyer reaction. A chemical reaction in which a diazonium group is replaced by a halogen atom or cyanide group by reaction with the corresponding copper I salt. For example, $C_6H_5N_2^+X^- + CuCl = C_6H_5Cl + N_2$ $+ CuX$. In a variation of the method, the *Gatterman reaction*, the copper salt is replaced by a mixture of copper powder and a hydracid (HCl, HBr, etc.). [After Traugott Sandmeyer (1854–1922), German chemist.]

sandwich compound. A type of transition-metal complex in which a metal atom is "sandwiched" between two parallel hydrocarbon rings. The original example is *ferrocene which may be regarded as an iron atom coordinated to two cyclopentadienyl ions, $C_5H_5^-$. The coordination is to the pi-electrons of the whole ring, not to the carbon atoms, and in fact the compound obeys the Hückel rule and is aromatic. Other *metallocenes are known in addition to sandwich compounds formed with other ring compounds. Chromium, for example, forms a complex with benzene, $Cr(C_6H_6)_2$.

saponification. The hydrolysis of an *ester using an alkali. The reaction yields an alcohol and the salt of the carboxylic acid. If fats are saponified, a *soap is produced. The *saponification number* of a fat or oil is the number of milligrams of potassium hydroxide necessary to completely saponify one gram of fat.

sapphire. A naturally occurring blue form of aluminium oxide, Al_2O_3, containing small amounts of cobalt. It is used as a gemstone.

saros. *See* moon.

satellite. A smaller celestial body that orbits around a larger one. The earth is a satellite of the sun, whereas the moon is a satellite of the earth. In the solar system, of the planets lying within the asteroid belt, only the earth and Mars are known to have satellites. The *giant planets, by contrast, have numerous satellites, Jupiter having twelve, Saturn ten, Uranus five, and Neptune two.

saturated. 1. Denoting a vapour that is in equilibrium with its liquid or solid. In a saturated vapour, atoms or molecules are condensing from the vapour at the same rate as they are evaporating. The amount of substance in the vapour phase depends on the temperature and increases with rising temperature. *See* vapour pressure.
2. Denoting a solution in which the solute is in equilibrium with undissolved substance. Atoms or molecules are leaving a saturated solution at the same rate as they are dissolving. The concentration of dissolved substance depends on the temperature. It increases with temperature for liquids and solids and decreases with temperature for gases. *See also* solubility, supersaturated.
3. Denoting a magnetic material in which the amount of magnetization is a maximum: i.e. it can increase no further no matter how strong the magnetizing field. At saturation, all the domains are aligned in the direction of the external field. *See* ferromagnetism.
4. Denoting a chemical compound that contains no double or triple bonds, so that no further addition can occur.

saturation. The property of a *colour determined by its freedom from white. If the colour is a pure spectral colour (as caused by light of a single wavelength) it is said to be saturated. Colours that have the same hue but differ in saturation are called *tints*.

Saturn. The sixth planet in succession outwards from the sun and the second largest planet in the solar system. Lying at a mean distance of 1427 million kilometres from the sun, Saturn completes its orbit in 29.46 years. Like Jupiter it is a *giant planet, with a diameter of 119 300 km, a mass 95.14 times the earth's, and a mean relative density of 0.7. Like the other giant planets it has a dense atmosphere with latitudinal cloud belts consisting mainly of hydrogen, methane, and ammonia, and the planet's surface temperature has been estimated at about $-160°C$. Its rotation period is about 10 hours 14 minutes. Saturn has ten satellites, the largest being *Titan*, which is as large as Mercury. The most remarkable feature of Saturn is its unique system of rings, formed, it is believed, by solid particles that have failed to coalesce into a satellite because they are too close to the planet. Three rings have so far been definitely sighted, arranged concentrically and girdling the planet's equator. The outer diameter of the largest ring is about 272 000 km and the rings themselves are probably no more than 16 km thick, so that they almost disappear when seen edge-on.

sawtooth waveform. A waveform that repeatedly increases steadily to a peak value and drops suddenly to zero (see illustration).

Sawtooth waveform

scalar. A physical quantity that is characterized only by its magnitude and does not depend on directional measurement. *Compare* vector.

scalar product (dot product). The product of two *vectors yielding a scalar whose magnitude is the product of the magnitudes of the vectors and the cosine of the angle between them. The scalar product is written $a.b$ and read as "a dot b". Thus, if $a.b = c$, then c is a scalar of magnitude $|a| |b| \cos \theta$. A physical example of a scalar product is that of a force F making an angle θ with a given direction. The work W done by the force moving a distance d along this direction is $F.d = Fd \cos \theta$. If the vectors are represented in Cartesian coordinates, their scalar product has the form: $a.b = a_x b_x + a_y b_y + a_z b_z$. Two vectors can also be multiplied together to give a *vector product.

scaler. An electronic circuit or device for counting pulses, giving a single output pulse as a result of a certain number of input pulses. Scalers are used in counting particles of radiation.

scandium. Symbol: Sc. A silvery metallic element occurring as thortveitite (Sc_2SiO_7). Scandium belongs to group III of the periodic table but is usually classified with the *lanthanides. A.N. 21; A.W. 44.96; m.pt. 1539°C; b.pt. 2727°C; r.d. 3.02; valency 3.

scanning. The process of traversing an area, volume, range of values, etc. Examples are the repeated movement of a beam of electrons in a television camera or in a cathode-ray tube, the continuous movement of a radar aerial, and the continuous change in position of a spectrometer grating to cover the range of values of a spectrum.

scanning electron microscope. *See* electron microscope.

scattering. Deflection of incident particles (electrons, photons, etc.) by other particles in their path. Scattering is said to be *elastic or inelastic depending on the type of interaction. The type of interaction also determines the directions in which scattering occurs. For instance, charged particles can be scattered by interaction with the electrostatic field of atomic nuclei. High-energy neutrons, and other particles, may be scattered as a result of *strong interactions.

The scattering of electromagnetic radiation can occur in several ways. If the particles are relatively large, reflection occurs in random directions. This happens with particles of dust, etc. If the particles are small compared with the wavelength of the radiation, diffraction occurs and the amount of deflection depends on the wavelength (proportional to $1/\lambda^4$). This phenomenon is responsible for the blue of the sky—air molecules scatter blue light more than red light. Compton scattering (*see* Compton effect) involves the interaction of photons with electrons.

scheelite. A white, yellow, green, or brown naturally occurring tungstate of calcium, $CaWO_4$, used as an ore of tungsten.

Schiff's base. Any of a class of compounds with the general formula $ArN:CR_2$, where Ar is an aryl group and R is an aryl or alkyl group or a hydrogen atom. Schiff's bases are crystalline substances obtained by reacting an aromatic amine ($ArNH_2$) with an aldehyde or ketone (R_2CO). [After Hugo Schiff (1834–1915), German chemist.]

Schiff's reagent. A solution of the dye magenta that has been decolorized by sulphur dioxide. The colour is restored by aliphatic aldehydes, which oxidize the reduced form of the dye back to its original coloured form. Aromatic aldehydes and aliphatic ketones restore the colour more slowly: aromatic ketones do not react.

schlieren photography. The use of high-speed photography to demonstrate and study turbulent flow in fluids. Variations in the density of the fluid cause transient streaks, which are visible because of local variations in the refractive index. [From German, *Schliere*: a streak.]

Schmidt telescope. A type of reflecting telescope in which a curved photographic plate is used at the centre of curvature of the primary concave mirror, thus reducing coma, astigmatism, and spherical aberration of the image. [After Bernhard Schmidt (1879–1935), German astronomer.]

Schottky defect. See vacancy.

Schrödinger's equation. An equation used in *wave mechanics to describe the behaviour of a particle. It has the form:

$$\nabla^2 \psi + \frac{8\pi^2 m}{h^2}(E-U)\psi = 0,$$

where ∇ is the *Laplace operator, m is the mass of the particle, E is the total energy, U the potential energy, h the Planck constant, and ψ is the *wave function of the particle. [After Erwin Schrödinger (1887–1961), Austrian physicist.]

scintillation. The production of small flashes of light in certain substances (*scintillators*), caused by the impact of high-energy particles. The light is emitted by *fluorescence.

scintillation counter. A type of *counter in which particles of radiation are detected by the flashes of light that they produce in a scintillator. This is mounted on the front of a *photomultiplier, which produces an output pulse for every flash of light. The pulses are counted by a *scaler. The heights of the photomultiplier pulses depend on the

energies of the incident particles and a *pulse-height analyser can be used to study these energies. The device is then a *scintillation spectrometer.

scintillator. See scintillation.

sclera. See eye.

sclerometer. Any instrument for measuring the *hardness of materials. An example is the apparatus used in the *Brinell test. Another type of sclerometer, the *scleroscope*, determines hardness by dropping a standard ball from a prescribed height and measuring the height to which it rebounds.

scotopic vision. Vision by the *eye when the rods of the retina are the main receptors of light. This is the dominant process at very low luminance levels. The rods cannot distinguish different colours. *Compare* photopic vision.

screen grid. A grid placed between the anode and control grid of a thermionic *valve and held at constant potential in order to reduce the capacitance between the anode and the control grid.

screening. Shielding of a piece of apparatus in order to protect it from electric or magnetic fields. Electrical screening is effected by surrounding the apparatus with a conducting container. Magnetic screening uses a container of high permeability, such as Mumetal.

scruple. See Appendix.

search coil. A flat coil of wire used to measure magnetic flux. The coil is removed from the position in the magnetic field to a position outside the field or is turned through 180° in the field and the total induced current is measured by a *ballistic galvanometer.

Searle's method. A technique for measuring the thermal conductivity (λ) of good conductors. The specimen, in the form of a bar, is lagged and mounted horizontally with a steam bath connected to one end. Thermometers are placed to measure the temperatures (θ_1 and θ_2) at two positions a known distance (*d*) apart along the bar. Cooling water is passed through a copper spiral at the other end of the bar. The inlet and outlet temperatures of the water (θ_3 and θ_4) are also measured. In the steady state, the thermal conductivity is calculated from the equation:

$\lambda A \, (\theta_1 - \theta_2)/d = m \, (\theta_4 - \theta_3)$,

where *A* is the cross-sectional area of the bar and *m* the mass of water flowing per unit time.

secant. 1. *See* trigonometric function.
2. A line cutting off an arc of a curve.

second. 1. Symbol: s. The SI unit of time, equal to the interval of time occupied by 9 192 631 770 periods of vibration of the radiation produced by the transition between two hyperfine levels in the ground state of the caesium-133 atom. This standard of the second is obtained using a caesium clock. Formerly, the unit was defined by astronomical measurement. *See* time.
2. One sixtieth of a minute of angle. The symbol is " placed to the right of the number.

secondary. *See* transformer.

secondary cell. *See* accumulator.

secondary emission. Emission of electrons (*secondary electrons*) from a solid as a result of the impact of other electrons. If the incident primary electron has sufficient energy it can often eject several secondary electrons, an effect utilized in the electron multiplier.

secondary standard. *See* unit.

sector. A part of a circle bounded by two radii and the arc contained between them. A sector of a sphere is a sector of a circle rotated about a diameter.

sedimentation. The settling of particles suspended in a liquid. The rate at which this occurs, under gravity or centrifugal force (*see* ultracentrifuge), yields information on the average particle size.

Seebeck effect. The phenomenon in which an electromotive force is pro-duced in a circuit containing two conductors when both junctions are at different temperatures. *See* thermoelectric effect, thermocouple. [After Thomas Johann Seebeck (1770–1831), German physicist, inventor of the thermocouple.]

seed. A small crystal used to induce crystallization from a saturated or supersaturated solution or from a supercooled melt.

segment. 1. A part of a circle bounded by a chord and the arc it cuts off. A spherical segment is a part of a sphere contained between two parallel planes, one of which may be tangential to the sphere.
2. A part of a line or curve between two points on it.

seismograph. An instrument, effectively a large pendulum, for detecting and recording earthquakes, underground nuclear explosions, etc.

seismology. The scientific study of earthquakes.

selenate. Any salt or ester of selenic acid.

selenic acid. A strong oxidizing acid, H_2SeO_4. M.pt. 57–58°C; b.pt. 172°C; r.d. 2.95.

selenide. A compound of selenium with a more electropositive element or with an organic group.

selenium. Symbol: Se. A nonmetallic element existing in several allotropic forms, including a grey crystalline form, a red crystalline form, and a red or black amorphous form. It is obtained from the anode sludge produced in electrolytic copper refining and is used in semiconductors and particularly in photoelectric applications, such as photocells and xerography. Its chemistry is similar to that of sulphur: it is one of the *chalconide elements. A.N. 34; A.W. 78.96; m.pt. 217°C (grey); b.pt. 695°C (grey); r.d. 4.79 (grey); valency 2, 4, or 6.

selenium cell. A type of *photocell consisting of a metal support carrying a layer of selenium covered by a thin transparent layer of gold. Light falling on the cell produces a potential difference between the gold and the selenium by the *photovoltaic effect.

selenium rectifier. A type of *rectifier consisting of alternate layers of iron and selenium.

selenium sulphide. A brown powder, SeS_2. Decomposes at $100°C$.

selenology. The scientific study of the moon.

self-inductance. See electromagnetic induction.

semiconductor. A solid with an electrical conductivity that is intermediate between those of insulators and metals and that increases with increasing temperature. In contrast, the electrical conductivity of metals and other good conductors falls as the temperature rises. Semiconductors are usually metalloid elements or their compounds. Examples are germanium, silicon, lead telluride, and similar materials. They form covalent *crystals, in which the atoms are held together by covalent bonds. At low temperatures their electrons cannot move through the solid because they are held in position by the bonds. As the temperature is increased some electrons gain enough energy to escape from the covalent bonds and the "free" electrons can move through the crystal under the influence of an applied electric field, thus producing an electric current. In addition, the removal of electrons from the bonds between the atoms leaves a number of *holes*: i.e. missing electrons. These also contribute to the conductivity: an electron can migrate from one bond to occupy the hole in an adjacent bond, thus leaving another hole behind it. This is filled by another electron from the next bond, and so on. The net effect is of a positively charged hole moving in the opposite direction to the current. An alternative description of semiconductivity is given by the *band theory of solids. A semiconductor of this type is called an *intrinsic semiconductor*.

Extrinsic semiconductors are materials that owe their semiconductivity to the presence of small amounts of impurities. Often these are introduced in controlled amounts—a process known as *doping. *Donor* impurities are atoms that lose electrons and form positive ions, thus releasing electrons that can carry the current. The remaining electrons are strongly bound to the positive ions of the donor so hole conductivity does not occur. Thus the main part of the current is carried by electrons (known as the *majority carriers*). The remainder of the current is transported by *minority carriers*. A semiconductor of this type, in which most of the current is carried by negative electrons, is called an *n*-type semiconductor. An example is a silicon crystal doped with phosphorus atoms: the phosphorus has five outer electrons and can form four bonds to silicon atoms and release the extra electron.

Acceptor impurities are atoms that require extra electrons to form bonds. Boron, for example, has three outer electrons and in accepting an extra electron can form four bonds to neighbouring atoms. The ionization of the acceptor produces positive holes, which carry the current. A semiconductor in which the majority carriers are positive holes is called a *p-type* semiconductor.

Semiconductors are extensively used in a variety of electronic devices, including *thermistors, *transistors, phototransistors, and *integrated circuits.

semiconductor junction. A junction between two different layers of semiconductor. A *p–n* junction is a junction between a layer of p-type and a layer of n-type material.

semipermeable membrane. A barrier that permits the passage of some substances but is impermeable to others. See osmosis.

semipolar bond. *See* coordinate bond.

septivalent. *See* heptavalent.

sequence. A set of numbers or mathematical expressions in a fixed order, each term being determined by its position in the sequence. For example: 2, 6, 12, 20, 30, ... is a sequence for which the nth term is ($n^2 + n$). Sequences can be *convergent or *divergent. They are sometimes called *progressions*. *See also* series.

sequestration. The process of forming chemical complexes with metal ions to make them ineffective. Sequestration does not necessarily remove the ions but converts them into an inactive form. For example, sequestering agents, which are usually *chelating agents, can be used to counteract the effect of lime in soil by forming complexes with the calcium ions. Other applications include water softening and use in analytical chemistry.

series. 1. The summation of the terms of a *sequence. For example, the sequence 1, 1/4, 1/9, 1/16, ... $1/n^2$, ... has an associated series $1 + 1/4 + 1/9 + 1/16 + ...$ $1/n^2 +$ .... This series can be written in the form:

$$\sum_{n=1}^{n=\infty} \frac{1}{n^2}$$

A series, like a sequence, can be finite or infinite and an infinite series can be *convergent or *divergent. *See also* arithmetic progression, geometric progression.
2. *in series*. Denoting electrical circuit elements joined so that the same current flows through all of them. If resistances R_1, R_2, R_3, etc., are joined in series the total resistance is simply their sum, $R_1 + R_2 + R_3 + ...$. Capacitances C_1, C_2, C_3, etc., in series have a total capacitance, C, given by: $(1/C) = (1/C_1) + (1/C_2) + (1/C_3) + ...$. Cells connected in series have a total e.m.f., E, given by

the sum of the individual e.m.f.s: $E = E_1 + E_2 + E_3 + ---$. *Compare* parallel.

series-wound. *See* electric motor.

servomechanism. A mechanism in which mechanical motion requiring a large amount of power is controlled by a motion requiring a smaller amount of power. The system may involve feedback.

sesqui-. Prefix indicating a ratio of 2:3. A *sesquioxide*, for example, is one with the formula M_2O_3.

sesquiterpene. *See* terpene.

shade. *See* colour.

shadow. A dark region formed when an opaque body blocks part of the light falling onto a surface. If the light does not come from a point source, two regions are distinguished: the *umbra* in which all the light is obstructed and the *penumbra*, surrounding it, in which only part of the source's light is obstructed.

shadow bands. Parallel bands of shadow moving rapidly across the earth sometimes seen just before and after totality in a total eclipse. They are caused by a refraction effect in the earth's atmosphere.

shear. Deformation of a body in which planes within the solid move parallel to one another. A rectangular block, for example, would be deformed so that one pair of opposite faces became parallelograms. *See* stress, strain.

shell. A region around an atomic nucleus in which all the electrons have the same principal quantum number, n. In a simple model of the *atom, the electrons can only move in a certain number of circular orbits of fixed radius, each orbit having a fixed energy. Thus the electron shells may be thought of as a series of concentric spheres: in more advanced theories they are regions in space characterized by the quantum number and their energy. Shells with principal quantum numbers 1, 2, 3, ---

are designated by letters K, L, M, etc., so the K shell ($n = 1$) is the one nearest the nucleus. Each shell contains a maximum of $2n^2$ electrons.

shell model. A theoretical model of the *nucleus in which it is assumed that the nucleons move in a central field (in the same way that electrons in the atom move in the field of the nucleus). The application of quantum mechanics leads to the idea that the nucleons move in shells, analogous to the electron shells in the atom. The filled shells of particular stability correspond to nuclei that have *magic numbers.

sheradizing. A technique for coating steel with a layer of zinc, used to protect the steel from corrosion. The surface is covered with zinc powder and heated to just below the melting point of zinc, at which temperature a zinc-iron alloy forms at the interface and this is covered by a layer of pure zinc.

SHF. *See* super-high frequency.

shock wave. A very narrow region in which the flow of a fluid changes from subsonic to supersonic flow. Shock waves are caused when an object moves through a fluid at supersonic velocities. They are propagated as sound waves. *See* sonic boom.

shooting star. *See* meteor.

short circuit. A direct electrical connection of low resistance between two points in a circuit causing the current to take a reduced path and by-pass part of the circuit.

short sight. *See* myopia.

shower. A large number of particles originating from a single collision of a primary particle. *See* cosmic rays.

shunt. A resistor with a low value connected in parallel with an ammeter or similar measuring instrument so that a known fraction of the current is allowed to flow through the instrument. In this way, the instrument's operating range can be changed.

shunt-wound. *See* electric motor.

side chain. A group of two or more atoms attached at some point along a chain of atoms in a molecule. In isobutane, for example, the methyl group attached to the central carbon atom is regarded as a side chain attached to the main *straight chain of three carbon atoms. Isobutane thus has a *branched chain.

side reaction. A chemical reaction that takes place at the same time as the main chemical reaction and yields small amounts of other products.

sidereal. Denoting measurements taken with respect to the stars.

sidereal time. *Time measured with reference to the stars. The *sidereal day* (23 hours 56.06 minutes) is the time between successive transits of a chosen star across the observer's meridian. The *sidereal month* is the time between two successive conjunctions of the moon with a star. The *sidereal year* is the time for the sun to make one revolution of the celestial sphere, moving relative to the fixed stars. It is measured by the time between successive conjunctions with a chosen star. 1 sidereal year = 365.256 36 mean solar days; 1 sidereal month = 27.321 66 mean solar days. *See* solar time.

Siegbahn unit. *See* X-unit. [After Karl Siegbahn (b. 1886), Swedish chemist.]

siemens (mho, reciprocal ohm). Symbol: S. A unit of conductance equal to the conductance between two points of a conductor when a potential difference of one volt between these points causes a current of one ampere to flow. [After Sir William Siemens (1823–83), British engineer born in Germany.]

Siemens-Martin process. *See* open-hearth process.

sigma orbital (σ-orbital). A type of molecular orbital that is completely symmetrical about a line between the two atoms. Sigma orbitals are responsible for the single bonds (*sigma bonds*) between atoms in covalent compounds. *See also* pi orbital.

sigma particle. Symbol: Σ. An *elementary particle having either zero charge and a mass 2333 times that of the electron, a positive charge and a mass 2328 times that of the electron, or a negative charge and a mass 2343 times that of the electron. The three sigma particles are classified as *hyperons.

sigma pile. A source of neutrons together with a *moderator but no fissile materials. Such piles are used to study the properties of the moderator.

signal. A changing electric current, waveform, or other agency that carries information.

silane. Any of a class of hydrides of silicon with the general formula Si_n H_{2n+2}. They are named according to the number of silicon atoms: SiH_4 (silane), Si_2H_6 (disilane), Si_3H_8 (trisilane), etc. The silanes are formally similar to the alkanes but less stable due to the weakness of the Si-Si bonds: only the first six members of the series are known. A mixture of silanes is produced by the action of acid on magnesium silicide, Mg_2Si.

silica. Silicon dioxide, SiO_2. It is a hard crystalline colourless material that occurs in several natural crystalline forms, including quartz, flint, and sand. *Vitreous silica* is a glassy transparent material made by fusing silica. It has a low coefficient of thermal expansion and is used in heat-resistant apparatus.

silica gel. A form of silica produced by coagulation of a sodium silicate sol. This gives a gelatinous solid, which can be dehydrated to give a hard granular material. It is used as an absorbent, drying agent, and catalyst support. Cobalt salts absorbed in the gel give it a blue colour when dry and a pink colour when moist.

silicate. Any of a large number of oxy compounds of silicon, many of which occur naturally in rocks and minerals.

The simplest type are the *orthosilicates* with formula M_4SiO_4, where M is a monovalent metal. These contain the ion SiO_4^{4-}. An example of an orthosilicate is zircon, $ZrSiO_4$, although such compounds are probably not completely ionic.

More complex silicates contain SiO_4 tetrahedra linked together by sharing corners. Two such units are present in the *pyrosilicate* ion, $Si_2O_7^{6-}$, which is found in the scandium ore thortveitite, Sc_2SiO_7. The ions $Si_3O_9^{6-}$ and $Si_6O_{18}^{12-}$ both have cyclic structures in which the SiO_4 units are linked into a ring. The latter ion is found in beryl, $Be_3Al_2Si_6O_{18}$. The *pyroxenes* are a class of silicates containing infinite chains of SiO_4 units. They contain ions of the type $(SiO_3^{2-})_n$ together with sufficient positive ions to ensure neutrality. The *amphiboles* also have infinite chains with two strands held together by cross linkages. They contain ions of the type $(Si_4O_{11}^{6-})_n$. The various types of asbestos are amphiboles. The SiO_4 tetrahedra can also be joined together in infinite sheets, leading to the presence of $(Si_2O_5^{2-})_n$ ions. Such sheets are held together by the positive ions. This type of layer structure is found in micas. Finally the SiO_4 units characteristic of silicates can be linked together to form complex three-dimensional structures. In such lattices, some of the silicon atoms may be replaced by aluminium atoms, giving the *aluminosilicates*. The feldspars and the zeolites are examples of such compounds.

The compounds known as *metasilicates*, M_2SiO_3, do not contain SiO_3^{2-} ions. Instead they are either pyroxene compounds, having infinite chains, or compounds containing cyclic anions.

silicic acid. Any of a number of oxy acids of silicon. A mixture of silicic acids is produced by the action of acids on sodium silicates. It is best described as a mixture of hydrated forms of silica ($SiO_2.nH_2O$), from which no distinct acids can be isolated. The compounds *pyrosilicic acid*, H_6SiO_7, and *orthosilicic acid*, H_4SiO_4, are known in the form of their salts and esters. *Metasilicic acid,* H_2SiO_3, does not exist in this form: it is formally the parent acid of the polymeric metasilicates (*see* silicate).

silicide. Any compound of silicon with a more electropositive element.

silicon. Symbol: Si. A nonmetallic element occurring widely as silicates in many minerals and as the oxide in quartz and sand; the second most abundant element in the earth's crust (25.7% by weight). The element can be obtained as a brown amorphous powder or a brown crystalline electrically conducting material. It is made by reducing sand (SiO_2) with carbon in an electric furnace. High-purity silicon is obtained by the Van Arkel-de Boer process. The element is used in semiconductors and the production of some silicon compounds.

Chemically, it is fairly unreactive: the amorphous form will ignite in oxygen at high temperatures; both forms react with halogens and are soluble in alkali and hydrofluoric acid, although not in other acids. Unlike carbon, silicon does not form a large number of compounds with hydrogen (*see* silane) although it does have many oxygen compounds, such as the *silicates and *silicones. A.N. 14; A.W. 28.086; m.pt. 1410°C; b.pt. 2355°C; r.d. 2.33 (crystalline); valency 4.

silicon carbide (carborundum). A bluish-black crystalline solid, SiC, that is one of the hardest substances known. It is made by heating coke with sand in an electric furnace using salt as a flux. Silicon carbide is used as an abrasive, a refractory material, and a semiconductor. Sublimes at 2210°C; r.d. 3.20.

silicone. Any of a class of synthetic polymeric compounds of the general formula $[R_2SiO]_x$, where *R* is an organic group. They contain rings or chains of alternating silicon and oxygen atoms with the organic groups attached to silicon. Many different types exist: they are used as lubricating oils, water repellants, electrical insulators, and

constituents of polishes and lacquers. *Silicone rubbers* are synthetic polymers of dimethylsiloxane, $(CH_3)_2SiO$. They can be used at both high and low temperatures. *See also* siloxane.

silicon tetrachloride. A colourless fuming liquid, $SiCl_4$, obtained by passing chlorine over hot silicon carbide. It is used to make silicones and as a source of pure silicon. M.pt. -70°C; b.pt. 58°C; r.d. 1.50.

siloxane. Any of a class of silicon compounds containing Si-O-Si bonds with organic alkyl or aryl groups bound to the silicon atoms. A simple example is $R_3SiOSiR_3$, where R is the organic group. The *silicones are polymeric siloxanes.

silver. Symbol: Ag. A white lustrous ductile and malleable *transition element found native and as argentite (Ag_2S) and horn silver (AgCl). It is usually obtained during the refining of lead, copper, and nickel ores and a certain amount is recovered from the anode sludge produced in the electrolytic refining of copper. The metal is used for jewellery, coinage, decorative objects, and producing silver salts, especially those used in photography. It has the highest thermal and electrical conductivity of any element and is used in printed circuits. Silver is generally less reactive than copper, although it tarnishes in air because of attack by sulphur compounds. The most common compounds are those with silver in its +1 oxidation state: it forms silver I (*argentous*) ionic salts as well as numerous complexes. Silver II (*argentic*) compounds are nearly all complexes, although AgF_2 does contain Ag^{2+} ions. There are also a few silver III complexes. A.N. 47; A.W. 107.87; m.pt. 961.93°C; b.pt. 2212°C; r.d. 10.5; valency 1 or 2.

silver bromide. A pale yellow crystalline solid, AgBr, that darkens on exposure to light. It is used as a light-sensitive material in photographic film. M.pt. 432°C; r.d. 6.5.

silver chloride. A white powder, AgCl, that darkens on exposure to light. It is used in silver plating and photography. M.pt. 445°C; b.pt. 1550°C; r.d. 5.6.

silver glance. *See* argentite.

silver iodide. A pale yellow crystalline compound, AgI. M.pt. 556°C; b.pt. 1506°C; r.d. 6.01.

silver nitrate. A colourless poisonous crystalline solid, $AgNO_3$, made by dissolving silver in nitric acid. It is used in photography, in silvering mirrors, and as an analytical reagent. M.pt. 212°C; r.d. 4.3.

silver oxide. Either of two oxides of silver. *Silver I oxide*, Ag_2O, is a dark brown slightly soluble powder used as an oxidizing agent in chemical synthesis. Decomposes when heated; r.d. 7.14. It reacts with ozone to give *silver II oxide*, AgO.

silver point. *See* temperature scale.

silver sulphide. A black insoluble solid, Ag_2S, precipitated from solutions of silver salts by hydrogen sulphide. M.pt. 825°C; r.d. 7.32.

simple fraction. *See* fraction.

simple harmonic motion. Motion of a point moving along a path so that its acceleration is directed towards a fixed point on the path and is directly proportional to the displacement from this fixed point. The motion is thus described by the equation $d^2x/dt^2 = -k^2x$, x being the displacement from the point

with due regard to sign and k being a constant. The solution has the form $x = A \sin(kt + b)$, where A and b are constants.

A representation of simple harmonic motion is shown in the diagram. The point P moves around the circle with constant angular velocity ω. The foot of a perpendicular from P onto a diameter moves with simple harmonic motion. The equation describing this motion is $x = r \sin(\omega t + \alpha)$, where r, the radius of the circle, is the maximum displacement and α, the *epoch*, is the angular displacement at $t = 0$. The motion is a sinusoidal oscillation about O. The period of this oscillation is given by $T = 2\pi/\omega$.

It can be shown that simple harmonic motion occurs when a body or system moves under the influence of a restoring force F that is proportional to the displacement from its equilibrium position. Common examples of this are the motion of a pendulum (for small displacements) and the motion of a weight oscillating on the end of a vertical spring.

simple microscope. *See* microscope.

simple pendulum. *See* pendulum.

simultaneous equations. A set of two or more equations that together impose conditions on the same set of unknowns and whose solution is a value or set of values that satisfies all the equations. If solutions exist they are the points of

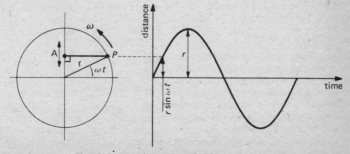

Simple harmonic motion of point A

intersection of the graphs of the individual equations.

sine. *See* trigonometric function.

sine wave. A waveform that has an equation in which one variable is proportional to the sine of another. In *simple harmonic motion, for example, the displacement is proportional to the sine of the time. The term is used with reference to the shape of the waveform; a waveform $y = A \cos x$ could also be described as a sine wave.

single bond. *See* covalent bond.

single crystal. A crystal that has a completely regular structure throughout. *See also* crystal.

sintered glass. Porous glass made by *sintering. It is used for filtration.

sintering. A process in which solid particles heated at a temperature below the melting point coalesce together to form a single mass. Glass and certain metals and ceramics can be sintered.

sinusoidal. Indicating a periodic quantity, motion, etc., that is described by a *sine wave.

siphon. A bent tube with one arm longer than the other, used for transferring liquid to a lower level over an intermediate barrier. The tube must be full of liquid for the siphoning action to begin. The siphon operates because of the excess weight of liquid in the longer arm.

SI units (Système International d'Unités). A system of units used, by international agreement, for all scientific purposes. It is based on the metre-kilogram-second (MKS) system and consists of seven base units and two supplementary units (see Table 1). Measurements of all other physical quantities are made in derived units, which consist of combinations of two or more base units. Fifteen of these derived units have special names (see Table 2). Base units and derived units with special names

have agreed symbols, which are used without a full stop.

Decimal multiples of both base and derived units are expressed using a set of standard prefixes with standard abbreviations (see Table 3). Where possible a prefix representing 10 raised to a power that is a multiple of three should be used (e.g. mm is preferred to cm).

sky wave (ionospheric wave). A radio wave transmitted between points by reflection from the *ionosphere rather than by direct transmission. *Compare* ground wave.

slag. A nonmetallic material produced on the surface of the liquid metal in smelting and refining metals. It is usually a mixture of oxides, silicates, sulphides, phosphates, etc., produced by combination of a *flux with the gangue of the ore or with impurities in the metal refined. *See also* basic slag.

slaked lime. *See* calcium hydroxide.

slaking. The process of adding water, as in the production of slaked lime.

slide rule. A mathematical instrument for simple calculations, consisting of fixed scales with a moving scale sliding between them. The scales are logarithmic: i.e. the distances along the scale are proportional to the logarithms of the index numbers. Numbers can be multiplied or divided by adding or subtracting distances using the moving scale.

slip. Motion of one plane of atoms in a crystal relative to another in a direction tangential to the planes. This sliding motion occurs in the plastic deformation of solids.

slip ring. *See* electric motor.

slow neutron. A *neutron with an energy of only a few electron volts.

slug. A unit of mass equal to the mass that would acquire an acceleration of one foot per second per second from a force of one pound-force. It is equivalent to 32.174 pounds (14.5939 kilograms).

Table 1. Base and Supplementary SI Units

physical quantity	name of SI unit	symbol for unit
length	metre	m
mass	kilogram(me)	kg
time	second	s
electric current	ampere	A
thermodynamic temperature	kelvin	K
luminous intensity	candela	cd
amount of substance	mole	mol
*plane angle	radian	rad
*solid angle	steradian	sr

*supplementary units

Table 2. Derived SI Units with Special Names

physical quantity	name of SI unit	symbol for SI unit
frequency	hertz	Hz
energy	joule	J
force	newton	N
power	watt	W
pressure	pascal	Pa
electric charge	coulomb	C
electric potential difference	volt	V
electric resistance	ohm	Ω
electric conductance	siemens	S
electric capacitance	farad	F
magnetic flux	weber	Wb
inductance	henry	H
magnetic flux density (magnetic induction)	tesla	T
luminous flux	lumen	lm
illuminance (illumination)	lux	lx

Table 3. Decimal Multiples and Submultiples to be used with SI Units

submultiple	prefix	symbol	multiple	prefix	symbol
10^{-1}	deci	d	10^{1}	deca	da
10^{-2}	centi	c	10^{2}	hecto	h
10^{-3}	milli	m	10^{3}	kilo	k
10^{-6}	micro	μ	10^{6}	mega	M
10^{-9}	nano	n	10^{9}	giga	G
10^{-12}	pico	p	10^{12}	tera	T
10^{-15}	femto	f			
10^{-18}	atto	a			

slurry. A thin suspension of a solid in a liquid.

smelting. The metallurgical process of extracting metals from their ores by heating them in a furnace with a flux and usually a reducing agent, such as carbon. A melt is produced composed of molten metal with a layer of molten *slag floating on top.

smetic. *See* liquid crystal.

smoke. A suspension of small solid particles in a gas.

smooth. Denoting contact between two bodies in which there is considered to be no frictional force.

Snell's law. The principle that when *refraction occurs the sine of the angle of incidence divided by the sine of the angle of refraction is a constant. [After Willebrord Snell (1591–1629), Dutch astronomer.]

soap. A salt of a fatty acid. Normal household soap (*hard soap*) is a mixture of sodium stearate, oleate, and palmitate. It is made by hydrolysing fats with caustic soda, thus converting the glycerides of stearic, oleic, and palmitic acids into the sodium salts and glycerol. *Soft soap* is a more liquid substance, produced by using potassium hydroxide instead of sodium hydroxide. Soaps owe their cleansing action to the structure of their anions, which consist of a long hydrocarbon chain attached to a carboxyl group. The hydrocarbon chain has an affinity for oil and grease and the charged carboxyl group has an affinity for water. Particles of grease or oil are thus emulsified in water. Insoluble salts of other metals with these acids are also called soaps. They are used as fillers and waterproofing agents.

soda. Any of various compounds of sodium, such as washing soda, caustic soda, and baking soda.

soda ash. *See* sodium carbonate.

soda lime. An off-white granular solid made by slaking calcium oxide with caustic-soda solution. It is a mixture of the hydroxides of sodium and calcium and is used as a drying agent and absorbent for carbon dioxide.

sodamide (sodium amide). A white crystalline solid, $NaNH_2$, with an odour of ammonia, made by passing ammonia over hot sodium. It is used in making sodium cyanide. M.pt. $210°C$; b.pt. $400°C$.

soda nitre. *See* sodium nitrate.

sodium. Symbol: Na. A soft silvery-white metallic element; the sixth most abundant element in the earth's crust, occurring principally as sodium chloride in sea water and rock salt. It is manufactured by electrolysis of a fused mixture of sodium chloride and calcium chloride. Sodium is used as a reducing agent in the manufacture of some chemicals and as a cooling agent in nuclear reactors. A.N. 11; A.W. 22.9898; m.pt. $97.5°C$; b.pt. $892°C$; r.d. 0.97; valency 1.

sodium alginate. *See* alginic acid.

sodium amide. *See* sodamide.

sodium bicarbonate (sodium hydrogen carbonate). A white soluble powder, $NaHCO_3$, obtained by passing carbon dioxide into a saturated solution of sodium carbonate. It is used in effervescent drinks, baking powders, fire extinguishers, and as a medical antacid. Decomposes at $270°C$; r.d. 2.16.

sodium bichromate. *See* sodium dichromate.

sodium bisulphate. *See* sodium hydrogen sulphate.

sodium bisulphite. *See* sodium hydrogen sulphite.

sodium carbonate. A white soluble salt obtained by the *Solvay process. The commercial form (*soda ash*) is a white anhydrous powder, Na_2CO_3, used in the manufacture of glass, soaps, paper, and other chemicals. The decahydrate (*washing soda*) is a white crystalline solid, $Na_2CO_3.10H_2O$, used as a domestic cleanser and water softener. M.pt. 851°C (anhydrous); r.d. 2.5 (anhydrous), 1.4 (decahydrate).

sodium chlorate. A colourless crystalline solid, $NaClO_3$, obtained by electrolysing a concentrated acidic solution of sodium chloride. It is used as a powerful oxidant and bleach in manufacturing paper, herbicides, and explosives. M.pt. 248°C; r.d. 2.5.

sodium chloride (salt, common salt). A colourless crystalline solid occurring naturally in rock salt and in sea water. It is used to make many chemicals including ceramics, glasses, soaps, and fertilizers, and in seasoning and preserving foods. M.pt. 804°C; b.pt. 1413°C; r.d. 2.2.

sodium cyanide. A white poisonous deliquescent crystalline solid, $NaCN$, made by reaction of hydrogen cyanide with caustic soda. It has a variety of uses, including the case hardening of steel, the extraction of gold, and the manufacture of other chemicals. M.pt. 564°C; b.pt. 1500°C.

sodium dichromate (sodium bichromate). A red-orange deliquescent crystalline solid, $Na_2Cr_2O_7.2H_2O$, manufactured by alkaline roasting of chromite ore, followed by leaching. It is used as an oxidizing agent and a source of chromium compounds. M.pt. 357°C (anhydrous); r.d. 2.5 (dihydrate).

sodium fluoride. A poisonous colourless crystalline solid, NaF, used in the fluoridation of water supplies and as an insecticide. M.pt. 988°C; b.pt. 1695°C; r.d. 2.6.

sodium glutamate. See sodium hydrogen glutamate.

sodium hydrogen carbonate. See sodium bicarbonate.

sodium hydrogen glutamate (sodium glutamate, monosodium glutamate). A white crystalline powder, $HOOC-(CH_2)_2CH(NH_2)COONa$, with a meaty taste, extracted from sugar-beet waste by alkaline hydrolysis or synthesized from acrylonitrile. It is widely used in flavouring meat.

sodium hydrogen sulphate (sodium bisulphate). An acidic white powder, $NaHSO_4$, obtained as a by-product in the manufacture of hydrochloric and nitric acids. It is used as a flux and an industrial cleaner. M.pt. 315°C; r.d. 2.7.

sodium hydrogen sulphite (sodium bisulphite). A white crystalline powder, $NaHSO_3$, with an odour of sulphur dioxide. It is prepared by saturating sodium carbonate solution with sulphur dioxide, and is used principally as a reducing agent, antiseptic, and bleach. R.d. 1.5.

sodium hydroxide (caustic soda). A white deliquescent solid, $NaOH$, that is strongly alkaline in solution and is very corrosive to organic tissue. It is manufactured by the electrolysis of sodium chloride solution and is used in making rayon, paper, detergents, and many other chemicals. M.pt. 318°C; b.pt. 1390°C; r.d. 2.1.

sodium hypochlorite. An unstable white crystalline solid, $NaOCl$, usually stored as a pale green aqueous solution. It is manufactured by electrolysis of cold dilute sodium chloride solution and is an oxidizing agent used in bleaching paper and textiles and as an antiseptic and fungicide. M.pt. 18°C.

sodium monoxide. See sodium oxide.

sodium nitrate (soda nitre). A colourless crystalline solid, $NaNO_3$, occurring naturally. It is used as an oxidizing agent and fertilizer and in making glass and pyrotechnics. The commercial salt is often called Chile saltpetre or caliche. M.pt. 308°C; b.pt. 380°C (decomposes); r.d. 2.3.

sodium nitrite. A white or yellowish crystalline solid or powder, $NaNO_2$, made by heating sodium nitrate. It is used for making azo dyes. M.pt. 271°C; r.d. 2.2.

sodium oxide. Any of three oxides of sodium. *Sodium monoxide*, Na_2O, is a white powder produced by direct combination of the element with a deficiency of oxygen. It reacts with water to give sodium hydroxide. Sublimes at 1300°C; r.d. 2.27. In excess oxygen *sodium peroxide*, Na_2O_2, is formed. This is a yellowish white powder used as an oxidizing and bleaching agent and as a source of hydrogen peroxide. M.pt. 460°C; r.d. 2.8. A *superoxide, NaO_2, also exists.

sodium peroxide. See sodium oxide.

sodium silicate. A mixture of *silicates obtained by fusing sodium carbonate and sand (SiO_2) in an electric furnace. The product contains $xNa_2O.ySiO_2$ with the ratio $x:y$ varying from 2:1 to about 1:4. The compounds form clear viscous colloidal solutions in water (*water glass*) and are used in making silica gel and detergents and in sizing textiles.

sodium sulphate. A white crystalline salt, Na_2SO_4, extracted from natural deposits and used in textile dyeing and detergents. The decahydrate, $Na_2SO_4.10$-H_2O, is known as *Glauber's salt*. M.pt. 888°C; r.d. 2.7.

sodium thiosulphate. A colourless crystalline solid, $Na_2S_2O_3.5H_2O$, made by reacting sulphur with sodium sulphite or as a by-product in the manufacture of sodium sulphide. It is a reducing agent, used as a fixer in photography (*hypo*) and as an antichlor. M.pt. 48°C; r.d. 1.7.

sodium-vapour lamp. A type of *fluorescent lamp in which an electrical discharge is passed through a low pressure of sodium vapour, causing it to emit a characteristic orange-yellow light.

soft. 1. Denoting electromagnetic or particle radiation with relatively low energy and, as a consequence, low penetrating or ionizing power. Soft X-rays are X-rays with long wavelength and low photon energy.
2. Denoting a *vacuum in which there

is a relatively high pressure of gas, such that ionization of gas plays an important role in any electrical phenomena occurring in the vacuum.

soft iron. Iron that contains very little carbon. Soft iron has a low remanence, losing most of its magnetization when the external magnetic field is removed. See magnet.

soft soap. See soap.

soft solder. See solder.

software. The programs used in a computer, as distinguished from the equipment itself—the *hardware*.

soft water. See hardness.

sol. A colloidal suspension of small particles of one substance in another. A *hydrosol* is a dispersion in water, an *aerosol* is a dispersion in air. A sol is distinguished from a *gel by the fact that the dispersed particles are independent. There are two main types.
Lyophobic (or *emulsoid*) sols are dilute suspensions that are easily precipitated. An example is a colloidal suspension of small metal particles produced by striking an arc under water. Lyophobic *colloids have little affinity for the dispersion medium and are irreversible. Smokes and mists are other examples.
Lyophilic (or *suspensoid*) sols are reversible systems in which there is affinity between the dispersed particles and the dispersion medium. An example is a solution of gelatin. This type of sol may be more concentrated and can be coagulated to form a gel.
When the dispersion medium is water the terms *hydrophobic* and *hydrophilic* are used.

solar cell. Any electrical device for converting solar energy directly into electrical energy. Solar cells are used to produce power in spacecraft. Essentially, they depend on the photovoltaic effect developed across a semiconductor junction. Another type of solar cell consists of an array of *thermocouples heated by the sun.

solar constant. The energy per unit area per unit time received from the sun at a point that is the earth's mean distance from the sun away. The solar constant gives the energy flux that would be received by the earth in the absence of its atmosphere. It has the value 1400 joules per square metre per second.

solar flare. An explosive and violent surge or outburst of radiation apparently caused by electrical storms in the chromosphere of the *sun. Flares are very bright and may last for up to half an hour, spreading horizontally throughout the chromosphere in a direction parallel to the surface of the photosphere. They emit large amounts of short-wave radiation and send out *cosmic rays that can affect the terrestrial ionosphere causing *magnetic storms. It is possible that they are associated with sunspots and faculae.

solar time. *Time measured by reference to the sun. The *solar day* is the time between successive transits of the sun across the meridian. Averaging this time out over a year gives the *mean solar day*. The *solar month* is the time taken for the moon to complete one revolution of the earth, returning to the same longitude. The *solar year* is the time between successive arrivals of the sun at the first point of Aries. 1 solar year = 365.242 19 mean solar days; 1 solar month = 27.321 58 mean solar days.

solar wind. Streams of electrons and protons emitted by the *sun. The solar wind is responsible for the formation of the *Van Allen belts and the *aurora. It also distorts the symmetry of the earth's magnetic field, causing it to extend further into space on the side of the earth that is distant from the sun. The intensity of the solar wind is greatest during the occurrence of solar flares and sunspots.

solder. Either of two alloys used for joining metals together. *Soft solder* is an alloy of tin (about 50%) and lead with a melting point in the range 200-300°C. The soft solder used in making electrical joints has a core of resin flux. *Hard solder* is an alloy of brass with a melting point above about 800°C. It is applied with a gas torch and a zinc chloride or similar flux—a process known as *brazing*.

solenoid. A long spiral coil of wire, usually cylindrical, through which an electric current can be passed to produce a magnetic field. Solenoids are used in electromagnets and other devices.

solid. 1. A state of matter characterized by incompressibility and resistance to shear stress. Unlike a gas, a solid has a fairly constant shape: unlike a liquid it will not flow to take the shape of its container.
The intermolecular forces involved are large enough to keep the atoms or molecules in fixed positions, about which they vibrate. Solids are classified into *amorphous solids, in which there is no regular structure, and crystalline solids, in which the atoms have a regular arrangement (*see* crystal).
2. A three-dimensional geometric figure, such as a cube or sphere.

solid angle. Symbol: ω or Ω. The region subtended by an area at a point, measured by the ratio of the area that the solid angle intercepts on the surface of a sphere to the square of the sphere's radius: $\omega = A/r^2$. Solid angles are measured in *steradians. The total solid angle around a point is 4π steradians.

solid solution. A homogeneous solid mixture of two or more substances, in which atoms, ions, or molecules of one substance occupy some of the lattice sites normally occupied by the other. The composition of solid solutions can vary within certain limits.

solid-state. Denoting an electronic device that contains transistors or other semiconductor devices, rather than thermionic valves.

solid-state physics. The experimental and theoretical study of the properties of the solid state, in particular the study of energy levels and the electrical and magnetic properties of metals and semiconductors.

solstice. One of two points at which the sun appears to be at its furthest points north or south. The ecliptic is then at its maximum distances from the celestial equator. In the northern hemisphere the *summer solstice* occurs on about June 21 and the *winter solstice* occurs on about December 22. These correspond to the longest and shortest days respectively. The terms are used both for the points of maximum displacement and the times at which these occur. *Compare* equinox.

solubility. The extent to which one substance will dissolve in another. Solubilities are expressed, under specified conditions of temperature and sometimes pressure, as the *concentration of the saturated solution.

solubility product. The equilibrium constant of dissolved ions in equilibrium with undissolved solid. The concentration of the undissolved solid is taken to be unity, so the solubility product is simply a product of ionic concentrations at saturation. Silver chloride, for instance, has a solubility product given by $K_s = [Ag^+][Cl^-]$. If the ionic product is exceeded, as by mixing silver nitrate with sodium chloride, silver chloride is precipitated until the ionic product falls to K_s. In general, the solubility product of a salt A^xB^y is $[A]^x[B]^y$.

soluble. Denoting a substance that dissolves. Unless the solvent is specified, it is understood to be water.

solute. *See* solution.

solution. A homogeneous mixture of two or more substances. The atoms or molecules of the components are completely intermingled and the solution thus consists of only one phase. When a solid or gas is dissolved in a liquid, the liquid is the *solvent* and the dissolved substance is the *solute*. When one liquid substance is dissolved in another, the one in excess is the solvent.

solvation. Association of molecules of a solvent with an ion in solution. In water, for example, a positive ion will be surrounded by a number of water molecules. In the case of transition metal ions, the closest water molecules are held by coordinate bonds. Any positive ion, however, will attract water molecules because of electrostatic forces between the ion and the *polar water molecules. The process is known as *hydration.

Solvay process (ammonia soda process). A process for the manufacture of sodium carbonate from sodium chloride and calcium carbonate. The calcium carbonate is heated to give calcium oxide and carbon dioxide. This is bubbled into a solution of sodium chloride that has been saturated with ammonia, thus precipitating sodium hydrogen carbonate and leaving ammonium chloride in solution. The sodium hydrogen carbonate is heated to yield sodium carbonate while the ammonium chloride is heated with calcium oxide, from the calcium carbonate, to regenerate the ammonia. [After Ernest Solvay (1833–1922), Belgian chemist.]

solvent. *See* solution.

solvent extraction. Extraction of one component from a mixture by use of a solvent that dissolves the required substance without dissolving other components. The method is often used for extracting natural products from the tissues of plants. Mixtures may also be separated by partition between two different mutually immiscible solvents. The mixture is dissolved in one solvent, this is shaken with the other, and the two are then left to separate into layers. With a suitable choice of solvents, one layer contains the required substance.

solvolysis. Reaction of a compound with its solvent. If the solvent is water the reaction is *hydrolysis.

sonar (asdic). A device for detecting objects under water by transmitting pulses of high-frequency sound and detecting the reflected pulses. The time difference between transmitting a pulse and collecting the reflected pulse gives a measure of the depth of the object. The system is similar to radar: the term is a shortened form of *sound navigation ranging.*

sonic boom. The loud bang produced by the *shock wave from an aircraft travelling at supersonic speed.

sorbite. A constituent produced in *steels by tempering martensite above about 500° C. It consists of cementite in a matrix of ferrite (alpha iron).

sorbitol. A white crystalline alcohol, $CH_2OH(CHOH)_4CH_2OH$, with a sweet taste. The *D-* compound is present in some fruits and can be made by reducing glucose. It is used as a sweetening agent and a raw material for the manufacture of vitamin C and polyurethane resins. M.pt. 110°C; b.pt. 295°C (3.5 mmHg); r.d. 1.5.

sorption. *Absorption or *adsorption by a solid.

sorption pump. A type of vacuum pump consisting of a vessel containing an absorbent material cooled in liquid nitrogen. Gas is absorbed on the material, usually activated charcoal or a molecular sieve, and removed from the system. Sorption pumps saturate in time and have to be regenerated by heating.

sound. Mechanical longitudinal vibration of air, water, or other material, carrying energy through the medium. Sound is transmitted in the form of longitudinal waves: alternate compressions and rarefactions of the medium travelling at a characteristic velocity. It is produced by a vibrating object which creates the disturbance and cannot be transmitted through a vacuum. The term is often used for those waves that can be detected by the human ear: i.e. waves with frequencies between about 20 and 20 000 hertz. However, any

analogous wave motion in any medium can also be designated as sound. Sound is characterized by the frequency of the wave motion (judged by its *pitch), the intensity of the sound (judged by its *loudness), and the *quality of the sound. The velocity in gases is independent of the pressure: in air it is 332 metres per second at 0°C.

Soxhlet apparatus. A laboratory apparatus for extracting soluble substances, consisting of a type of *reflux condenser with a special container, usually a thimble of porous paper, into which the material to be treated is placed. Solvent, heated in a flask, circulates continuously over the material and carries away its soluble constituents.

space charge. A region of net electric charge caused by a collection of electrons or ions in a semiconductor or in a thermionic valve or other electron tube.

space-time. The reference system of three dimensions of space and one of time used in the theory of *relativity to describe events. This system is sometimes called the *space-time continuum.*

spallation. A type of *nuclear reaction in which an atomic nucleus, hit by a high-energy photon or particle, emits several other particles or fragments.

spark. A *discharge of electricity occurring in air or other gases at normal pressures in the form of luminous wandering streamers between two points of opposite high electric potentials. A spark can only occur when the potential difference exceeds a certain value, the *sparking potential.* Usually the heating of the air by the spark causes a crackling noise. *See also* lightning.

spark chamber. An apparatus for detecting and studying the behaviour of high-energy ionizing particles. The chamber is filled with a gas such as neon and contains a stack of many parallel metal plates separated by narrow gaps. Alternate plates are given a high potential so that a particle passing through triggers a succession of short sparks along its track. The trace can be photographed from the side of the chamber.

spark gap. A gap between two electrodes designed so that a spark occurs when their potential difference exceeds a certain value.

spark photography. A photographic technique in which the camera is set up in the dark with the shutter open and the object is illuminated by a spark of very brief but known duration, thus allowing photography of a rapidly moving object.

special relativity. The special theory of *relativity.

specific. Per unit mass. The use of *specific* in the name of a physical quantity indicates that it is the property of unit mass, and therefore a property of a substance or material rather than of an object or system. *Specific volume*, for example, is volume per unit mass. In the names of some physical quantities *specific* does not have this meaning, in particular in *specific gravity and *specific resistance.

specific activity. Symbol: a. The *activity per unit mass of a radioisotope or radioactive material.

specific charge. The electric charge per unit mass. The term is usually used with reference to the charge-to-mass ratios of elementary particles.

specific gravity. See relative density.

specific heat capacity. Symbol: c. The *heat capacity of unit mass of a substance. It is the amount of heat required to raise the temperature of unit mass by unit temperature and is usually measured in joules per kilogram per kelvin. The heat required to raise the temperature of a gas depends on the way its pressure and volume change during the absorption of heat. Two *principal specific heat capacities* are used: one measured at constant pressure, c_p, and the other measured at

constant volume, c_v. The one at constant pressure is larger because some of the heat supplied is used in performing work. The ratio of these quantities, c_p/c_v, is given the symbol γ. Its value depends on the number of degrees of freedom of the gas molecules, being 1.66 for monatomic gases, 1.4 for diatomic gases, and approximately 1 for polyatomic gases. See also adiabatic equation.

specific resistance. See resistivity.

specific rotation. See optical activity.

spectral series. A series of lines in an absorption or emission spectrum, in which the lines all arise from transitions from or to the same energy level. In the absorption spectrum of hydrogen, for example, an electron in the first shell can absorb photons with frequencies that cause transitions to the second, third, fourth shells, etc., thus causing a series of dark lines. This is a *Rydberg spectrum. Similarly, if hydrogen is excited in some way it contains a mixture of hydrogen atoms with electrons in the second, third, fourth or higher orbits. These can decay with the emission of photons so that the electron falls back into the innermost shell giving a series of bright emission lines. This is the *Lyman series. Emission series can also occur when electrons move to the second, third, or fourth shells from higher levels, thus forming the *Balmer, *Paschen, and *Brackett series. Lines in such a spectral series have an equation of the type $1/\lambda = R(1/n^2 - 1/m^2)$, where R is the *Rydberg constant, n a constant integer, and m takes values of $n+1$, $n+2$, etc.

spectral type (spectral class). A classification of a *star according to the characteristics of its spectrum. The *Harvard classification* comprises several types of star. Type O, for example, have a blue colour and their spectrum is mainly absorption lines of helium ions. Type K have a red colour and their spectrum contains lines from many metals and some simple molecules.

spectrograph. A spectrometer or spectroscope designed to give a photographic record (the *spectrogram*) of a spectrum.

spectrographic analysis. Analysis of substances by the spectra they produce. A routine method of analysis involves photographing the emission spectrum produced by vaporizing a small amount of sample in an electric arc or a flame. The lines in the spectrum are characteristic of the elements and the intensities of the lines can be used to determine the amounts present. More detailed analysis of spectra can often show the nature of the molecules.

spectroheliograph. An instrument for photographing the sun using one of the wavelengths of light that it emits. It is essentially a monochromator for selecting light of one particular wavelength and focusing it onto a photographic plate.

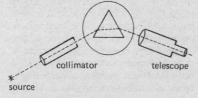

Prism spectrometer

spectrometer. Any of various instruments for producing and examining spectra. In a simple prism spectrometer (see illustration), the light is first passed through a *collimator onto the prism, which rests on a flat rotatable table. Refracted light is observed through the telescope, which can be moved round the prism. The instrument can be used for observations of simple emission and absorption spectra. It is also often used for measuring refractive indices and prism angles. Many other types of spectrometer exist for investigating *absorption and *emission spectra over almost the whole range of the electromagnetic spectrum. Spectrometers are often called *spectroscopes* or, when they

are designed for ultraviolet or visible radiation, *spectophotometers*. The term is also used for other instruments for analysing beams of electrons, alpha particles, ions, atoms, etc., in terms of their distribution of energy or mass. *See* mass spectrometer.

spectrophotometer. An instrument for producing a visible or ultraviolet spectrum and measuring the intensities of each wavelength. *See* spectrometer.

spectroscope. Any of various instruments for examining and measuring spectra. *See* spectrometer.

spectroscopic binary. *See* binary star.

spectroscopy. The production and analysis of spectra. Many different spectroscopic techniques exist depending on the type of radiation used. They give information on energy levels of atoms, molecules, ions, crystals, etc., and can be used to determine the structure and composition of molecules. Spectroscopy is also a valuable analytical technique (*see* spectrographic analysis). In astronomy it is used to investigate the composition and behaviour of stars, comets, and planets.

spectrum. A particular distribution of some property over the components of a system. Thus, a beam of particles has a spectrum of kinetic energies which can be displayed as a graph of number of particles (as ordinate) against particle energy. A complex sound wave has a spectrum of frequencies which can be represented by a graph of intensity against frequency. An electromagnetic radiation has a spectrum of intensity against wavelength (or frequency). The spectrum of the light or other electromagnetic radiation emitted or absorbed by a substance is characteristic of the substance (*see* emission spectrum, absorption spectrum). Such spectra are classified as *line, *band, or *continuous spectra according to their appearance. They are also designated by the type of electromagnetic radiation (*see* *visible, *ultraviolet, *infrared, *microwave, and *X-ray spectrum). *See also* electromagnetic spectrum.

specular reflection. Reflection in which the angle of reflection equals the angle of incidence. It is distinguished from *diffuse reflection and occurs when the surface is flat.

speculum. A mirror, especially one made of polished metal.

speculum metal. An alloy of copper and tin in the ratio 2:1, used for making metal mirrors.

speed. 1. The ratio of distance travelled to time taken. Speed has the same units as *velocity: unlike velocity it is not a vector quantity.
2. The extent to which a photographic film is sensitive to light.
3. *See* f-number.
4. The rate of operation of a pump, measured in litres per second, etc.

speltar. Commercial zinc containing about 3% lead and other impurities.

sphere. A surface or solid generated by rotating a circle about its diameter. The volume is $(4/3)\pi r^3$ and the surface area is $4\pi r^2$, where r is the radius.

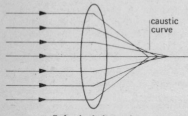

Spherical aberration

spherical aberration. An *aberration of lenses and mirrors with spherical surfaces, caused by the fact that rays striking the lens or mirror at points close to its edge are focused nearer than those close to the axis (see illustration). Thus a point object is not focused as a point image but as a small disc, the *circle of least confusion*. Spherical aberration is a geometrical effect—it arises because of the shape of the surface. In mirrors, it can be reduced by using a parabolic shape for distant objects or an ellipsoid for nearer objects. In lenses, it is reduced by making the lens in such a way that the deviation of the ray is shared equally at the two faces of the lens.

spherical coordinates. *See* polar coordinates.

spherical triangle. A triangle drawn on a spherical surface with sides formed by the arcs of great circles, i.e. circles whose centres are the centre of the sphere. The branch of mathematics concerned with the properties of spherical triangles is *spherical trigonometry*.

spheroid. *See* ellipsoid.

spherometer. An instrument for measuring the curvature of a lens, mirror, or other spherical surface. It has three legs and a central screw, which is adjusted to touch the surface. The radius of the surface is calculated from the reading on the screw.

spin. Symbol: s, J, or I. A property of *elementary particles corresponding to intrinsic angular momentum. The particle acts as if it were spinning on its axis and, according to quantum mechanics, only certain values of this angular momentum are possible. The angular momentum of a particle with spin can be represented by a vector along its axis. Such a particle also acts as a small magnet with a magnetic moment. When an external magnetic field is applied *precession occurs: the angular-momentum vector moves around the direction of the applied field. It can be demonstrated that only certain orientations are possible, namely those in which the components of the angular momentum along the field direction are $M_J h/2\pi$, where h is the Planck constant and M_J is called the *spin quantum number*. A particle with a spin J has M_J values of J, $J-1$, ... $-(J-1)$, $-J$. Thus the electron, with spin of ½, has spin quantum numbers of ½ and -½. *See*

Pauli principle, electron spin resonance, nuclear magnetic resonance.

spinel. Any of a class of minerals having the general formula $MO.M'_2O_3$, where M is a divalent metal and M' a trivalent metal. The *ferrates are types of spinel.

spiral galaxy. *See* galaxy.

spirits of salt. *See* hydrochloric acid.

spring balance. A simple instrument for measuring weight by the amount by which it stretches a coiled spring.

sputtering. A process in which atoms are ejected from a solid surface by the impact of high-energy ions. The effect can occur in gas discharges as a result of ions hitting the cathode. It is a technique of depositing thin films of the sputtered material on a nearby surface.

sputter-ion pump. *See* ion pump.

square. 1. A rectangle with all its sides equal.
2. The result of multiplying a number by itself: $x^2 = x \times x$.

square root. *See* root.

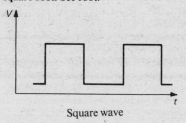

Square wave

square wave. A waveform that alternates between two steady values with (almost) instantaneous increases and decreases between these values and equal half periods (see illustration).

stabilizer. A substance added to another substance to prevent or retard its deterioration or chemical decomposition.

stable. 1. Denoting a chemical compound that does not decompose under normal conditions.
2. *See* equilibrium.

stainless steel. Any of a class of *steels that resist corrosion and are not attacked by weak acids. Stainless steels have a variety of compositions but all contain chromium (10-25%), carbon (0.1-0.7%), and sometimes small amounts of nickel and niobium. They are used for cutlery, household goods, turbine blades, industrial chemical equipment, and many other applications.

Stalloy. ® An alloy of iron and silicon used in transformer cores and similar applications requiring low hysteresis loss.

standard cell. Any cell with a constant reproducible e.m.f. that can be used as a standard. The *Weston cell is usually used for this purpose.

standard deviation. The square root of the average of the squares of the *deviations of a set of values.

standard solution. A solution of known concentration for use in a *titration.

standard temperature and pressure (S.T.P.). A temperature of 273.15 kelvins (0°) and 101 325 pascals (formerly 760 mmHg), used as standard conditions for properties of gases. The conditions are also called *normal temperature and pressure (N.T.P.).*

standing wave (stationary wave). *See* wave.

stannate. *See* tin hydroxide.

stannic. *See* tin.

stannic acid. *See* tin hydroxide.

stannite. *See* tin hydroxide.

stannous. *See* tin.

star. Any of the vast numbers of huge celestial objects that radiate light, heat, and other forms of electromagnetic radiation and are grouped into *galaxies. The stars, of which the sun is a typical example, are separated, even within the galaxies, by enormous distances and appear from earth only as points of light. In general, they fall into two classified groups, or *populations*. Stars of population I are young and are generally found in the outer arms of spiral galaxies; those of population II are much older and occur in the inner central region of galaxies and in globular clusters.

Stars range from *hot* (blue-white), with surface temperatures of up to 100 000°C, to *cool* (red), with surface temperatures of less than 2000°C. They can also be classified by their *spectral type. Within our Galaxy, stellar diameters show huge variation. At extreme ends of the scale lie the *pulsars and the *supergiants. Brightness varies considerably among the stars. *Canopus*, one of the brightest stars in the heavens, is 80 000 times as luminous as the sun, and lies about 650 light-years away; on the other hand, the red dwarf, *Wolf 359*, is only 1/30 000 as bright as the sun, though it is one of our nearest neighbours, only 7.7 light-years away.

Because of its position towards the outer rim of the Galaxy, the sun has relatively few near neighbours. Only eight stars lie within ten light-years of us. Proxima Centauri is the closest at a distance of only 4.3 light-years. Sirius, the brightest star in the sky, 26 times as bright as the sun, is only 8.7 light-years away. *See also* Hertzsprung-Russell diagram.

starch. A carbohydrate consisting of chains of glucose units, $(C_6H_{10}O_5)_x$. It has two forms (*amylose* and *amylopectin*) and is present in potatoes, rice, and cereals. In the body starch is hydrolysed to glucose.

The pure substance is a white powder, insoluble in cold water. It forms a bright blue compound with iodine.

stat-. Prefix placed in front of the name of a practical electrical unit to name the corresponding *electrostatic unit. Thus the e.s.u. of current is the statampere.

statcoulomb. The *electrostatic unit of charge, equal to $3.33\ 56 \times 10^{-10}$ coulomb.

state of matter. One of the three physical states—solid, liquid, or gas—in which matter may exist. The states of a particular substance change at definite transition temperatures. Plasmas, liquid crystals, glasses, colloids, and superfluids have all, at different times, been described as extra states of matter.

static electricity. *Electricity involving stationary electric charge. See* electrostatic, electrostatics.

statics. The branch of mechanics concerned with the properties of bodies in equilibrium: i.e. systems of forces in which no motion occurs.

stationary orbit. *See* synchronous orbit.

stationary point. A point on a curve at which the curve is neither increasing nor decreasing. At such a point the derivative is zero.

stationary state. One of the possible quantized states of a system as described by *quantum mechanics.

stationary wave. *See* wave.

statistical mechanics. The branch of physical science concerned with calculating the properties of matter from the statistical behaviour of large numbers of atoms and molecules.

statistics. The branch of mathematics concerned with collecting numerical data and making inferences from the data on the basis of probability.

stator. The stationary part of a *generator or *electric motor. It can be either the field winding or the armature.

steady-state theory. The theory that the universe has always existed and that its expansion is compensated for by the continuous creation of matter in such a way that the average density remains constant. To maintain such a state, matter has to be produced at a rate of about 10^{-43} kilogram per cubic metre per second. The theory would be supported by observations of radio waves from far in space showing that the density is constant at all points: such evidence has not yet been obtained. Many observations tend to support the alternative *big-bang theory. *See also* expanding universe.

steam. Water in the gaseous state at or above 100°C. Strictly, steam is an invisible gas but the term is also used for the white clouds which contain small droplets of liquid water.

steam distillation. A technique for separating immiscible liquids by bubbling steam through the hot mixture. The amount of a component present in the flow of steam is proportional to the product of its vapour pressure and its molecular weight. Thus, relatively involatile compounds can be distilled by this method if they have high molecular weights.

steam point. The temperature at which the liquid and vapour of water are in equilibrium at standard pressure: i.e. the boiling point of water (100°C). *See* temperature scale.

stearate. Any salt or ester of stearic acid.

stearic acid (octadecanoic acid). A crystalline water-insoluble fatty acid, $CH_3(CH_2)_{16}COOH$, present, in the form of its glycerides, in many natural *fats and oils. M.pt. 69.6°C; b.pt. 376°C (decomposes); r.d. 0.94.

stearin. *See* tristearin.

steel. Any of various forms of iron containing carbon (0.05–1.5%) and usually quantities of other elements. Two main types are distinguished: *carbon steels* contain mainly iron and carbon with only small amounts of other elements; *alloy steels* have larger amounts of other metals. If the carbon content is below 0.05% the material is iron (*see also* wrought iron). If it exceeds 1.5% it is usually described as *cast iron. Steel is made by reducing the carbon content of *pig iron in an *open-hearth furnace or a *Bessemer converter. Alloy steels are usually made from a *ferroalloy in an electric-arc furnace.

Many different types of steel exist depending on the amounts of carbon and other elements present and on the heat-treatment given to the steel. Carbon steels are classified as *mild steel* (0.1–0.2% carbon), *medium-carbon steel* (0.3–0.4% carbon), and *high-carbon steel* (above 0.5% carbon). Steels with low carbon content tend to be tougher and more ductile whereas a higher content of carbon produces a harder material. Alloy steels have a variety of properties depending on the elements present (*see* high-speed steel, stainless steel).

Steel has a complex variety of different phases. If molten iron containing carbon is solidified the material is a solid solution of carbon in gamma iron (*austenite*). As the temperature is reduced, *cementite* (Fe_3C) is formed. Below a certain temperature the gamma iron changes to alpha iron (*ferrite*) and a mixture of ferrite and cementite is produced (called *pearlite*). If the austenite is cooled quickly by quenching in water, the gamma iron changes to ferrite but the cementite does not precipitate. This yields a metastable structure (called *martensite*) consisting of a (supersaturated) solid solution of carbon in ferrite. Slower quenching produces a different structure (*troostite*) in which very fine grains of cementite and ferrite occur.

Stefan's law. The principle that the energy radiated per unit area per second from a black body is proportional to the fourth power of the thermodynamic temperature: i.e. $M = \sigma T^4$. The constant σ is *Stefan's constant*. It has the value 5.6697×10^{-8} W m^{-2} K^{-4}. [After Joseph Stefan (1835–93), Austrian physicist.]

St. Elmo's fire. *See* corona discharge.

steradian. Symbol: sr. The *SI unit of geometric solid angle, equal to the solid angle that cuts off an area on a sphere equal to the square of the sphere's radius, the vertex of the solid angle being at the centre of the sphere.

stere. A unit of volume equal to one cubic metre.

stereochemistry. The arrangement in space of the groups in a molecule and the effect this has on the compound's properties and chemical behaviour.

stereoisomerism. A type of *isomerism in which two or more compounds have the same functional groups but different arrangements of these groups in space. Such compounds are said to be *stereoisomers*. Two types of stereoisomerism exist: *geometrical isomerism and *optical isomerism.

stereoregular. Having a regular steric structure. The term is used for materials that are tactic *polymers. *Stereoregular rubbers*, for example, are synthetic rubbers with a structural order resembling that found in natural rubbers.

stereospecific. Involving particular steric arrangements of the atoms of molecules. *Stereospecific catalysis* is the use of catalysts to produce tactic *polymers. *See* Ziegler-Natta catalyst.

Stern-Gerlach experiment. An experiment in which a molecular beam of silver atoms was passed through a nonuniform magnetic field to demonstrate that the beam splits into two separate components. The experiment shows that the magnetic moments of the atoms can have only certain distinct orientations to the direction of the field (*see* spin).

stibine (antimony hydride). A colourless gas, SbH_3. It decomposes to antimony and hydrogen when heated. M.pt. $-88°C$; b.pt. $-17°C$.

stibnite. A grey mineral consisting of antimony sulphide, Sb_2S_3. It is the principal ore of antimony.

still. Any apparatus used for distillation.

stimulated emission. The process in which a photon colliding with an excited atom causes emission of a second photon with the same energy as the first. It is the phenomenon on which

depend the action of *lasers and *masers.

stochastic. Random. A *stochastic process* is one that involves random behaviour, so that it can be described in terms of probabilities.

stoichiometric. Involving chemical combination in exact simple ratios. A *stoichiometric compound* is a normal compound: i.e. one in which the elements have combined in small whole numbers. A *stoichiometric mixture* is one that would combine to give other substances with no excess of any reactant.

stoichiometry. The branch of chemistry concerned with the proportions in which compounds react together and the use of these in finding atomic weights, chemical equivalents, etc. The term is often taken to mean the proportions themselves.

stokes (stoke). Symbol: St. A CGS unit of kinematic *viscosity equal to the kinematic viscosity of a fluid with a dynamic viscosity of one poise and a density of one gram per cubic centimetre. [After Sir George Gabriel Stokes (1819-1903), British mathematician and physicist.]

Stokes' law. The principle that a small spherical particle falling under gravity in a viscous fluid reaches a terminal velocity given by $v = 2gr^2(\rho_1 - \rho_2)/9\epsilon$, where g is the acceleration of free fall, r the radius of the sphere, ρ_1 and ρ_2 the densities of the sphere and the fluid respectively, and ϵ the viscosity of the fluid. The equation can be used in experimental methods of determining the viscosities of liquids and gases.

stop. An aperture used to limit the width of a beam of light in an optical system.

storage battery. *See* accumulator.

storage ring. *See* accelerator.

store. The part of a *digital computer in which the information is held. Computer storage systems are classified into *direct-access* devices, in which any part of the stored file can be retrieved directly, and *sequential-access* devices, in which the file has to be scanned in sequence until the information is found. Direct-access devices have a much lower access time. In most digital computers more than one type of store is used. The *main store* (or *memory) is used for holding the information that is required directly by the central processing unit. It consists of a large number of cores, either ferrite rings or semiconductor devices. The *backing store* is used for holding large amounts of data. The information is stored by magnetized regions on a magnetic *tape, *drum, or *disk.

S.T.P. *See* standard temperature and pressure.

straight chain. A chain of atoms, usually carbon atoms, in a molecule with no side chains attached. Butane, for example, is a straight-chain hydrocarbon, whereas isobutane has a *branched chain.

strain. Deformation, either temporary or permanent, produced when a body is subjected to a *stress. The *tensile strain* (symbol: e) is the fractional increase in length of the body: i.e. $\Delta l/l$, where Δl is the increase in length and l is the original length. The *bulk* or *hydrostatic strain* (symbol: θ) is the fractional increase in volume, $\Delta V/V$. The *shear strain* is the angle (in radian measure) through which the twisting motion takes place. *See also* modulus.

strain gauge. A device for measuring strain. The term is commonly used for small devices attached to the surface of a body for sensing and measuring deformation at different points. The simplest type has a fine resistance wire attached to a piece of paper or thin plastic. Small changes in the length of the wire are followed by monitoring its resistance. Other types of strain gauge exploit changes in inductance or capacitance or make use of the piezoelectric effect.

strain hardening (work hardening). Hardening of a metal by subjecting it to strain, as by stretching or rolling. It results from the production of dislocations in the crystal structure.

strangeness. Symbol: *S.* A property of certain *elementary particles, originally assigned to account for the fact that they decay more slowly than would be expected. Kaons, for example, would be expected to decay in about 10^{-23} second whereas their actual lifetime is about 10^{-8}-10^{-10} second. To account for this, it is postulated that they have an additiional property—strangeness—that is described by a quantum number (called the *strangeness number*—or simply the strangeness).

Normal particles, such as the proton, have $S = 0$: strange particles have integral values of *S.* The strangeness number is conserved in *strong and *electromagnetic interactions but not in *weak interactions. A particle's strangeness number is its hypercharge minus its baryon number.

stratopause. *See* stratosphere.

stratosphere. The region of the earth's *atmosphere extending from a height of about 10 kilometres to about 80 kilometres. It lies between the troposphere and the mesosphere and has an increasing temperature. The temperature rises (up to $10°C$) near the top of the stratosphere—the maximum occurring at the *stratopause*.

streamline. A line in a fluid such that, at any point in the fluid, the tangent to the line is the direction of the fluid's velocity at that point. Fluid flow in which continuous streamlines can be drawn through the fluid is called *streamline flow*.

stress. The effect producing or tending to produce deformation of a body (*strain), equal to the applied force per unit area. There are several different types of stress.
Tensile stress (symbol: σ) is a stress tending to stretch a material in a parti-

cular direction. It is the force in that direction divided by the cross-sectional area. *Hydrostatic* or *bulk stress* is the stress tending to compress the whole body: it is simply the pressure on the body. In these examples the forces are acting at right angles to the surface—they are examples of *normal stress*.

When the forces tend to produce a twisting motion, the effect is a *shear stress* (symbol: τ) which is given by the tangential force per unit area. Although stress has the same units as pressure it is always given in newtons per square metre—not pascals. *See* modulus.

striations. *See* glow discharge.

stroboscope. An instrument for viewing objects that have fast cyclic or vibrating motion by illuminating or viewing them intermittently at the same frequency as that of their motion, thus making them appear stationary. For example, if a vibrating bar makes n vibrations per second and is illuminated by a lamp giving n short bursts of light per second, then every time the bar is seen it is in the same position. If the light frequency differs slightly from that of the bar, the bar appears to move slowly. A similar method of producing a stroboscopic effect is by viewing the object through a hole in a rotating disc.

Stroboscopes can be used for inspecting moving parts in operation and for measuring frequencies of vibration and speeds of rotation. Stroboscopic effects are also produced in other ways: perhaps the best known is the behaviour of wheels in films. The film camera takes a rapid series of still images. If the wheel makes one complete rotation between each frame, it will appear to be stationary.

strong acid or **base.** *See* acid, base.

strong interaction. A type of interaction between *elementary particles occurring at short range (about 10^{-15} m) and having a magnitude about 100 times greater than that of the electromagnetic interaction. Strong interactions occur between hadrons and account for the

forces that hold nucleons together in the atomic nucleus. They are thought to be due to the exchange of virtual mesons (*see* virtual particle) between the interacting hadrons.

strontia. *See* strontium oxide.

strontianite A mineral, white, grey, yellow, or green in colour, consisting of strontium carbonate, $SrCO_3$. It is a source of strontium salts.

strontium. Symbol: Sr. A soft pale yellow metallic element occurring principally as strontianite ($SrCO_3$) and celestine ($SrSO_4$). It is made by electrolysis of fused strontium chloride and used in some alloys and as a getter. Strontium is one of the *alkaline-earth metals. A.N. 38; A.W. 87.62; m.pt. 800°C; b.pt. 1300°C; r.d. 2.54; valency 2.

strontium-90. The radioactive isotope of strontium with mass number 90. It is a particularly hazardous constituent of radioactive fallout, being metabolized like calcium and included in bone: its half-life is 28 years.

strontium oxide (strontia). A greyish-white powder, SrO, prepared by decomposing strontium carbonate or hydroxide. It is a basic oxide, used in the manufacture of strontium salts, pyrotechnics, and soaps, and as a desiccant. M.pt. 2430°C; b.pt. 3000°C; r.d. 4.7.

structural formula. *See* formula.

structural isomerism. A type of *isomerism in which the isomers have different structural formulas. The compounds may be quite unrelated, as in the case of ethanol, C_2H_5OH, and dimethyl ether, CH_3OCH_3, which both have the formula C_2H_6O. Often structural isomers are similar, as in butane, CH_3-$CH_2CH_2CH_3$, and isobutane, CH_3CH-$(CH_3)CH_3$. Sometimes such compounds differ only slightly in physical properties. Another example is the existence of isomers with functional groups in different positions, as in propan-1-ol, $CH_3CH_2CH_2OH$, and propan-2-ol,

ortho   meta

para

Structural isomers

$CH_3CH(OH)CH_3$. Derivatives of benzene with two substituted groups can have three isomers depending on the relative positions of the groups on the ring: *ortho*, in which the groups are adjacent, *meta*, in which they have one unsubstituted carbon atom between them, and *para* in which the groups are at opposite sides of the benzene ring.

strychnine. A colourless crystalline highly poisonous alkaloid, $C_{21}H_{22}N_2O_2$, found in certain plants. M.pt. 270–80°C; r.d. 1.36.

styrene. A colourless oily liquid with an aromatic odour, $C_6H_5CH:CH_2$, prepared by dehydrogenation of ethylbenzene. It is polymerized to make *polystyrene. M.pt. -31°C; b.pt. 145°C; r.d. 0.9.

subatomic. Smaller than an atom. *Subatomic particles* are particles such as protons and electrons that are found in atoms.

subcritical. Involving a nuclear *chain reaction that is not self-sustaining.

sublimate. A solid deposited by *sublimation.

sublimation. The passage of certain substances from the solid state into the gaseous state and then back into the solid state, without any intermediate liquid state being formed. The term is often used to describe the direct change from solid to vapour without reference to subsequent change back to solid.

sub-shell. A subdivision of an electron *shell, for which all the electrons have the same azimuthal quantum number. The sub-shells of a particular shell are designated by the letters *s*, *p*, *d*, or *f*, according to the value of their quantum number. *See* atom.

subsonic. Relating to a speed less than Mach 1. *See* Mach number.

substantive dye. *See* dye.

substantivity. *See* dye.

substituent. An atom or group that replaces another atom or group in a *substitution reaction. The term is often used for an atom or group replacing hydrogen in a derivative of a compound.

substitution. A chemical *displacement reaction, especially one in which a hydrogen atom is displaced. An example is the chlorination of benzene, $C_6H_6 + Cl_2 = C_6H_5Cl + HCl$, in which the hydrogen atom is replaced by a chlorine atom. *Compare* addition.

substrate. A substance that is acted upon in some way. For example, the substrate is the compound acted on by a catalyst, the material to which a dye is attached, or the solid on which a compound is adsorbed.

subtend. To determine an angle at a given point; said of curves. Thus an arc AB subtends the angle AOB at O.

subtractive process. The process in which a colour is produced by combining coloured pigments, dyes, filters, or other absorbing materials. The final colour is produced by absorption of different wavelengths of light. Most colours can be produced by a suitable mixture of primary colours. *Compare* additive process.

succinate. Any salt or ester of succinic acid.

succinic acid (butanedioic acid). A colourless crystalline weak carboxylic acid, $HOOC(CH_2)_2COOH$, made by fermentation of ammonium tartrate and used as a sequestrant and raw material for the manufacture of dyes, lacquers, and perfumes. M.pt. 185°C; b.pt. 235°C; r.d. 1.5.

sucrose (sugar). A white sweet soluble crystalline disaccharide, $C_{12}H_{22}O_{11}$, obtained from sugar beet or sugar cane and used as a sweetener. M.pt. 185–6°C; r.d. 1.6. *See also* sugar, invert sugar.

Sugar

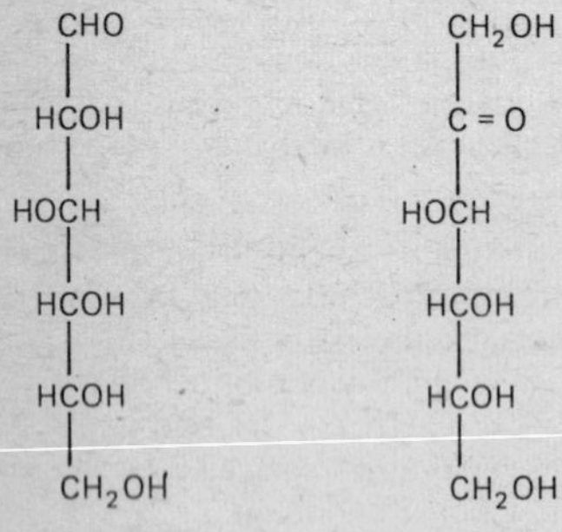

D-glucose   D-fructose

Fig. 1 Glucose and fructose

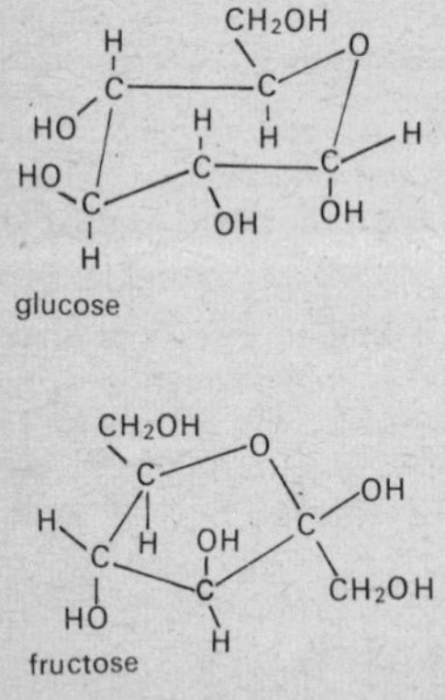

glucose

fructose

Fig. 2 Ring forms of glucose and fructose

sugar. 1. *See* sucrose.
2. Any of a class of simple carbohydrates. The *monosaccharides* or *simple sugars* have the general formula $(CH_2O)_n$ and their molecules contain hydroxyl groups and either an aldehyde group or a ketone group. Sugars containing an aldehyde group are *aldoses*: those containing a ketone group are *ketoses*. The simple sugars are also classified according to the number of carbon atoms in the molecule: *trioses* contain three carbon atoms, *tetroses* contain four, *pentoses* five, etc. The commonest types are the *hexose* sugars, with formula $C_6H_{12}O_6$. The different compounds with this formula differ in the relative arrangements of groups attached to the carbon atoms. Fig. 1 shows the formulas of glucose (an *aldohexose*) and fructose (a *ketohexose*).

In fact the monosaccharides do not normally exist in these straight-chain forms but in a ring form. The hydroxyl (alcohol) group of one end of the molecule can condense with the carbonyl group of the other to give a hemiacetal (*see* acetal). Compounds such as glucose, which have six-membered rings, are called *pyranoses*: those that have five-membered rings are *furanoses* (see Fig. 2).

In the solid state the sugar has a ring form and in solution this is in equilibrium with a small amount of the straight-chain form. The sugars all show *optical isomerism and if a pure optical isomer is prepared and dissolved in water the optical activity changes. Conversion occurs through the straight-chain form—the process is called *mutarotation. The presence of the straight-chain form is also responsible for the positive reaction of sugars with *Fehling's solution.

More complex sugars have molecules formed by condensation of two or more monosaccharide units and are known as *disaccharides*, *trisaccharides*, etc., according to the number of units involved. Carbohydrates with large numbers of units are *polysaccharides*. The condensation of two simple sugars can be

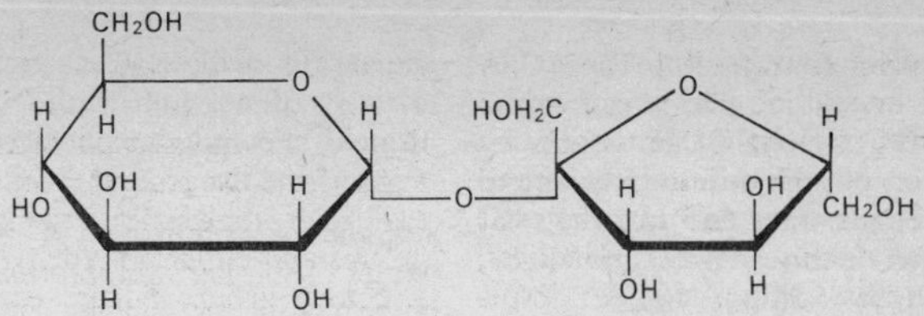

Fig. 3 Sucrose

visualized as elimination of H_2O between two -OH groups, giving an -O- bond between two rings. This is known as a *glycosidic bond*. Sucrose, for instance, has a glucose ring bound to a fructose ring (see Fig. 3).

Sucrose is linked in such a way that it cannot form a noncyclic molecule containing an aldehyde or ketone group and thus does not reduce Fehling's solution. Such compounds are said to be *nonreducing sugars* to distinguish them from *reducing sugars*.

sugar of lead. *See* lead acetate.

sulpha drug. *See* sulphonamide.

sulphate. Any salt or ester of sulphuric acid.

sulphation. The production of an insoluble layer of lead sulphate on the electrodes of a *lead-plate accumulator when it is left discharged.

sulphide. 1. Any compound of sulphur with a more electropositive group or element. Inorganic sulphides are salts of hydrogen sulphide and contain the ion S^{2-}. *Polysulphides* can also be produced containing ions of the type S_x^{2-}, in which there are chains of several sulphur atoms.
2. *See* thio ether.

sulphite. Any salt or ester of sulphurous acid.

sulphonamide. Any of a class of compounds that are amides of sulphonic acid: i.e. they contain the group -SO_2NH_2 or are derived from such a compound by replacement of one of the hydrogen atoms. An example is *sulphanilamide*, $H_2NC_6H_4SO_2NH_2$. The

sulphonamides are an important class of drugs (*sulpha drugs*) active against bacteria.

sulphonic acid. Any of a class of organic compounds of the general formula $Ar.SO_2.OH$, where Ar is an aryl group. The simplest example is benzenesulphonic acid, $C_6H_5SO_3H$. The sulphonic acids tend to be strong water-soluble acids. They are made from chlorosulphonic acid: $ArH + 2ClSO_3H = ArSO_3Cl + HCl + H_2SO_4$. The acid chloride then reacts with water to give the sulphonic acid.

sulphonyl group. *See* sulphuryl group.

sulphur. Symbol: S. A yellow nonmetallic element occurring in many minerals and also as the native element in large underground deposits in the U.S. It is obtained by the *Frasch process. Sulphur exists in several allotropic forms. At room temperature the stable form has a rhombic structure and above $95.6°C$ this slowly changes into a monoclinic form. Both these solids contain S_8 rings. Another form of sulphur, containing S_6 rings and having a hexagonal structure, is obtained by adding sodium thiosulphate to hydrochloric acid and crystallizing the sulphur from toluene. Plastic sulphur is made by pouring molten sulphur at about $160°C$ into cold water. It is an amorphous solid containing S_8 molecules with atoms in helical chains. *Flowers of sulphur* is a fine powder produced by condensing sulphur vapour. Liquid sulphur also has several forms. Just above the melting point it is a mobile yellow liquid containing S_8 rings. Above $160°C$ it becomes brown and increases in viscosity as the temperature is increased. This is probably due to the formation of chains of sulphur atoms and the reaction of these chains with each other and with rings to form polymers. Above 200°C the viscosity again falls. Sulphur vapour probably contains an equilibrium mixture of S_8, S_6, S_4, and S_2 molecules. The element is used for making sulphuric acid, the most important compound, and paper and for vulcanizing rubber. It is also an agricultural fungicide. Chemically, it is one of the *chalconide elements. A.N. 16; A.W. 38.06; m.pt. 112.8°C (rhombic), 119.0°C (monoclinic); b.pt. 444.674°C; r.d. 2.07 (rhombic), 1.96 (monoclinic); valency 2, 4, or 6.

sulphur chloride. Either of two chlorides of sulphur, both of which are readily hydrolysed liquids used as chlorinating agents. *Disulphur dichloride*, S_2Cl_2, is yellowish in colour and is prepared by passing chlorine into molten sulphur. M.pt. -80°C; b.pt. 138°C; r.d. 1.7. *Sulphur dichloride*, SCl_2, is reddish-brown. It is made by passing chlorine into disulphur dichloride at a low temperature. M.pt. -78°C; r.d. 1.6.

sulphur dichloride. *See* sulphur chloride.

sulphur dioxide. *See* sulphur oxide.

sulphuretted hydrogen. *See* hydrogen sulphide.

sulphuric acid (vitriol, oil of vitriol). A colourless hygroscopic oily liquid, H_2SO_4, made by the *contact process. It is a strong acid in aqueous solution and also an oxidizing and dehydrating agent. Sulphur trioxide can be dissolved in sulphuric acid to form *fuming sulphuric acid*, $H_2S_2O_7$, which is also known as *oleum* and *pyrosulphuric acid*. Sulphuric acid is one of the most important commerical chemicals, being used in the manufacture of fertilizers, paints, rayon, explosives, and many other products as well as in petroleum refining and lead-plate accumulators. M.pt. 10°C; b.pt. 315-338°C; r.d. 1.8.

sulphurous acid. A weak acid, H_2SO_3, made by dissolving sulphur dioxide in water. It is a reducing agent and is used in bleaching.

sulphur oxide. Either of two acidic oxides of sulphur. *Sulphur dioxide*, SO_2, is the colourless pungent gas produced by burning sulphur. It is made by roasting pyrites in air and used to make sulphuric acid and as a preservative for foods, a bleaching agent, and a fumigant. It is also a reducing agent. In solution it yields a mixture of sulphuric and sulphurous acids. M.pt. -76°C; b.pt. -10°C; r.d. 1.4 (liquid at 0°C). *Sulphur trioxide* is a white solid, SO_3, made from sulphur dioxide by the *contact process. It is the acid anhydride of sulphuric acid. Three crystalline forms exist with different melting points.

sulphur point. *See* temperature scale.

sulphur trioxide. *See* sulphur oxide.

sulphuryl chloride. A colourless liquid, SO_2Cl_2, with a pungent odour, made by reacting sulphur dioxide with chlorine in the presence of activated carbon as a catalyst. It is used for chlorination and dehydration. M.pt. -54°C; b.pt. 69°C; r.d. 1.7.

sulphuryl group (sulphonyl group). The divalent group $=SO_2$.

sun. The star around which orbit the earth, the planets, and all other bodies in the solar system. The sun is a yellow dwarf, halfway in energy output between the hottest and brightest bluish-white stars and the coolest and dimmest red ones. In space it occupies a position about one third of the way from the perimeter to the central core of the *Galaxy and revolves with the Galaxy at a velocity of 216 kilometres per second, completing one revolution in 200 million years. Lying at a mean distance of 149 million km from earth, its light and heat take eight minutes to reach us.

As with other evolving stars, its energy is provided by nuclear fusion reactions in its centre. Here, at a temperature of

15 000 000°C, hydrogen is turned first into deuterium and then helium (*see* carbon cycle). The surface is the *photosphere*. Its temperature is about 6000°C. Above it extends the solar atmosphere, the *chromosphere*, an incandescent gaseous envelope stretching outwards for some 15 000 km. Here *prominences and *solar flares occur. Above this the *corona*, split into the inner or K corona and the outer or F corona, extends into space. The corona, visible during a total eclipse as a halo, contains tenuous highly ionized gases at a temperature of about 1 000 000°C.

sunspot. A region of the sun's surface that is much cooler and therefore darker than the surrounding area, having a temperature of about 4000°C as opposed to 6000°C for the rest of the photosphere. Sunspots range from a few hundred kilometres to several times the earth's diameter and last from a few hours to a few months. A sunspot typically has a dark central core (the *umbra*) surrounded by a lighter region (the *penumbra*). The number of sunspots follows a pattern of fluctuation (the *sunspot cycle*) passing from maximum to maximum in about 11 years. High numbers of sunspots are associated with *solar flares and *prominences, giving out large numbers of charged particles that affect the earth's *ionosphere.

superconductivity. A phenomenon occurring in certain metals and alloys at temperatures close to absolute zero, in which the electrical resistance of the solid vanishes below a certain temperature (the *transition temperature*). A current induced in a ring of superconductor continues to flow when the magnetic field inducing it is removed—an effect used in high-current superconducting magnets.

supercooling. The cooling of a system to a temperature below that at which a phase change should occur without any change taking place. For example, a pure liquid can sometimes be cooled slowly to a few degrees below its freezing point without freezing. Similarly, a solution of a solid in a liquid might be cooled to a temperature at which the solvent contains more solute than a saturated solution at that temperature. The solution is then *supersaturated*. Supercooling of liquids and solutions results in the production of a *metastable state, in which the system is not at equilibrium. It occurs when no particles are present on which crystals may form. Crystallization may be induced by introducing a single crystal (a *seed*) on which the crystals can nucleate. Another method is to scratch the inner wall of the vessel—thus producing roughness on which the crystals can form. Supercooling can also be observed in vapours when they are free from dust or other small particles on which drops of liquid are produced. The vapour then exists at a temperature below its normal *dew point. A *supersaturated vapour* results. *See also* superheating.

supercritical. Involving a nuclear *chain reaction that is self-sustaining.

superdense theory. *See* big-bang theory.

superfluid. A fluid that flows without friction and has extremely high thermal conductivity, a phenomenon occurring in liquid helium at very low temperatures. Liquid helium makes the transition from the normal state to the superfluid state below 2.186 K.

supergiant. Any of a class of stars with higher absolute magnitude than ordinary giants, forming a small separate group above the giant branch on the *Hertzsprung-Russell diagram.

superheating. The process of heating a liquid to a temperature above its normal boiling point without boiling occurring. Superheating can happen when there are no small particles of matter in the liquid or roughness on the vessel walls on which bubbles can form. *See also* supercooling.

superheterodyne. Denoting the combination of radiofrequency signals by the *heterodyne principle to give a resultant signal with a frequency above the audiofrequency range. This effect is used in superheterodyne radio reception. The term is short for *supersonic heterodyne*.

super-high frequency (SHF). A *frequency in the range 3 gigahertz to 30 gigahertz.

superior planet. Any of the planets Mars, Jupiter, Saturn, Uranus, Neptune, and Pluto, whose orbits lie beyond that of the earth. *Compare* inferior planet.

supernova. A star that suffers an explosion, becoming up to 10^8 times brighter in the process and forming a large cloud of expanding debris (the supernova remnant). The Crab nebula is an example of the remnants of a supernova. The explosion is probably caused by a star exhausting its nuclear fuel and contracting under gravitation to an unstable condition. The outer layers are thrown off. *See also* pulsar, black hole.

superoxide. An inorganic compound containing the O_2 ion. Potassium, rubidium, and a few other electropositive metals form superoxides by direct combination.

superphosphate. A mixture consisting mainly of calcium hydrogen phosphate $(Ca(H_2PO_4)_2)$ and calcium sulphate $(CaSO_4)$, obtained by treating calcium phosphate with concentrated sulphuric acid. It is used as a fertilizer.

supersaturated. Denoting a solution that contains a higher concentration of solute than a *saturated solution at the same temperature. Similarly, the term describes a vapour that contains a higher pressure of vapour than a saturated vapour at the same temperature. Supersaturation is a consequence of *supercooling.

supersonic. Relating to a speed greater than Mach 1. *See* Mach number.

supplementary angle. An angle that together with a specified angle makes 180°. Thus, 30° is the supplementary angle of 150°.

supplementary unit. *See* SI units.

suppressor grid. The *grid between the screen grid and the anode in a *pentode valve.

surd. An *irrational root of a number. $\sqrt{2}$ and $\sqrt{3}$ are examples of surds; $\sqrt{4}$ (= 2) is not.

surface-active agent. *See* surfactant.

surface tension. Symbol: γ. The property of liquids that makes them act as if they had an elastic skin at their surface. The effect is caused by forces between the molecules—a molecule in the interior is attracted by other molecules on all sides whereas one at the surface experiences a force tending to pull it into the liquid. The phenomenon is responsible for the rise of certain liquids in vertical capillary tubes, the absorption of liquids by cloth, paper, and other porous substances, the ability of drops of liquid to stand on a surface without flowing, and the spherical shapes of soap bubbles, raindrops, etc.

Surface tension can be thought of quantitatively in two ways. It is the force acting tangential to the surface on one side of a line of unit length in the surface: i.e. its units are newtons per metre. Alternatively it can be thought of as the energy of the surface and defined by the work required to produce unit increase in surface area: its units are then joules per square metre. The two definitions are equivalent. The value of surface tension can be determined by *capillary rise, *Jaegar's method, or the Langmuir balance.

surfactant (surface-active agent). A substance used to increase the spreading or wetting properties of a liquid. Surfactants are often *detergents, which act by lowering the surface tension. They are used in many processes including froth-flotation of ores, the dispersion of insoluble powders in liquids, and the dyeing of textiles (to increase the rate of penetration).

susceptance. Symbol: B. The imaginary part of the *admittance of an electric circuit, equivalent to $-X/(R^2 + X^2)$, where R is the resistance and X the reactance.

susceptibility. 1. (**magnetic susceptibility**). Symbol: χ_m. A measure of the extent to which a substance is magnetized by an applied magnetic field, equal to the *magnetization (M) per unit *magnetic field strength (H): i.e. $M = \chi_m H$. The susceptibility is related to the relative *permeability (μ_r) by $\chi_m = 1 - \mu_r$. Diamagnetic materials have small negative susceptibilities, paramagnetic substances have small positive values, and ferromagnetic materials have large positive values. The susceptibility is a dimensionless ratio.
2. (**electric susceptibility**). Symbol: χ_e. A measure of the extent to which a substance is polarized by an applied electric field, equal to the *polarization (P) per unit *electric field strength (E): $P = \chi_e \epsilon_0 E$, where ϵ_0 is the *electric constant. The susceptibility is related to the relative permittivity (ϵ_r) by $\chi_e = 1 - \epsilon_r$. The electric constant is used to make the susceptibility a dimensionless ratio.

suspension. A mixture containing small particles of solid or liquid dispersed in a liquid or gas.

suspensoid. See sol.

symmetry. A property of geometric figures, mathematical expressions, patterns, etc., whereby certain operations leave the structure unchanged. A square, for example, has symmetry about its centre and about certain lines drawn across it. Each point on one side of a diagonal has a corresponding point on the other side, as if it had been reflected in a mirror placed along this line. Another way of describing the symmetry of shapes is by rotation. If the whole square were rotated for $180°$ about a diagonal it would be unchanged. This is said to be a two-fold axis of

symmetry—because two such rotations give a complete turn. The square also has a four-fold axis through its centre perpendicular to its plane. The symmetry of shapes can be described by such mirror planes and symmetry axes—known as *symmetry elements*. Patterns also have similar symmetry elements which involve translation operations. The properties of symmetry operations can be treated by *group theory and are important in many branches of science, in particular in the study of the shapes and energy levels of molecules, the structure of crystals, and the properties of elementary particles.

synchrocyclotron. A type of *cyclotron in which the frequency of the accelerating electric field is slowly decreased in such a way that the particles stay in phase with the field as their velocity increases. The decrease in frequency compensates for the relativistic increase in mass. Synchrocyclotrons are able to produce particles with energies in the range 400–500 MeV. Higher energies can be obtained in the *synchrotron.

synchronous motor. See electric motor.

synchronous orbit. The orbit of an artificial satellite that makes one revolution every 24 hours. If the orbit is in the same plane as the earth's equator the satellite remains stationary with respect to a point on the earth. Communications satellites placed in such orbits (*stationary orbits*) are used for reflecting radio waves.

synchrotron. A type of particle *accelerator in which the particles move in circular paths and are accelerated by alternating magnetic and electric fields. The *electron synchrotron* is similar to the *betatron, having a torus-shaped vacuum chamber and a pulsed magnetic field. In addition a radiofrequency electric field is applied across gaps in a metal cavity inside the chamber. The electric field has a fixed frequency, synchronous with the orbiting motion of the electrons. Electrons are travelling at close to the velocity of light at energies in the MeV range and the relativistic increase in mass can be compensated for by varying the magnetic field. In the *proton synchrotron* this cannot be done and the electric-field frequency must also be varied, just as in the *synchrocyclotron. Proton synchrotrons work at very high energies—up to 500 GeV.

synchrotron radiation. Electromagnetic radiation produced when electrons move at high velocities in circular paths in a magnetic field. The radiation is emitted in the direction of motion of the particle and is plane-polarized perpendicular to the magnetic-field direction. Its frequency depends on the velocity and can lie anywhere in the electromagnetic spectrum from radio waves to X-rays. Synchrotron radiation is produced in synchrotrons and is also thought to be the cause of many astronomical radiation sources.

syndiotactic. *See* polymer.

syneresis. The process in which a liquid separates from a gel.

synodic. Concerning two successive *conjunctions or *oppositions of the moon or a planet as viewed from the centre of the earth. *See also* lunar time.

synthesis. The preparation of chemical compounds, usually from simpler compounds.

synthetic. Denoting a substance or material that has been made artificially rather than naturally: i.e. a substance made by chemical reaction in a laboratory or in an industrial process.

T

tachometer. Any instrument for measuring speeds of rotation.

tachyon. A hypothetical elementary particle that has the property of travelling faster than the speed of light. According to the special theory of *relativity, this would be possible only if either the rest mass or the energy were *imaginary. No tachyons have been detected.

tactic. *See* polymer.

talc. A mineral consisting of hydrated magnesium silicate, $Mg_6Si_8O_{20}(OH)_4$. It is used as a lubricating agent.

tall oil. An oily resinous substance containing fatty acids, obtained in the processing of wood pulp and used in making soaps and emulsions.

tandem generator. An electrostatic *generator consisting of two *Van de Graaff generators in series, thus producing twice the voltage. Tandem generators are used in accelerators.

tangent. 1. *See* trigonometric function.
2. A line or plane touching a curve or surface at one point. The gradient of a tangent to a curve at a point is the derivative at that point.

tangent galvanometer. A type of galvanometer consisting of a vertical flat coil with a magnetic needle pivoted to swing horizontally and mounted at the centre of the coil. The plane of the coil is aligned in the direction of the earth's field and the current to be measured is passed through the coil, thus causing a deflection of the needle.

tannic acid (tannin). A yellowish amorphous powder, $C_{76}H_{52}O_{42}$, which may be extracted from powdered nutgalls. It is used in tanning, as a mordant, and in making pyrogallic acid. Decomposes at 210°C.

tannin. *See* tannic acid.

tantalite. *See* columbite-tantalite.

tantalum. Symbol: Ta. A heavy hard grey *transition element found principally in columbite-tantalite $(Fe,Mn)(Nb,-Ta)_2O_6$. It is separated and extracted in the same way as niobium. The element is used in alloys, incandescent filaments, and as a vacuum getter. Its chemistry is very similar to that of niobium. A.N.

73; A.W. 180.95; m.pt. 2996°C; b.pt. 5425°C; r.d. 16.65; valency 2-5.

tar. Any of various thick semi-solid substances produced by decomposition of organic substances. They contain mixtures of hydrocarbons, other organic compounds, and free carbon, which gives them a black colour. *See also* coal tar.

tartar (argol). A red-brown crystalline deposit of potassium hydrogen tartrate obtained from wine vats.

tartaric acid. A white crystalline powder, HOOC(CHOH)₂COOH, with an acid taste. In the preparation, potassium hydrogen tartrate from wine-lees is treated with calcium hydroxide, and the precipitated calcium tartrate is decomposed by sulphuric acid. Tartaric acid is used to make tartrates and as a sequestrant and constituent of baking powders. M.pt. 170°C; r.d. 1.8.

tartrate. Any salt or ester of tartaric acid.

tautomerism. A type of isomerism in which two isomers exist in equilibrium, the two forms (*tautomers*) being easily interconvertible. A common type is *keto-enol tautomerism.

tear gas (lachrymator). A substance with a vapour that causes irritation and watering of the eyes.

technetium. Symbol: Tc. A silvery-grey radioactive *transition element obtained from fission products of uranium and plutonium. Technetium was the first artificially produced element, made by bombarding molybdenum with deuterons. It is not found naturally on earth. The most stable isotope, ⁹⁹Tc, has a half-life of 5 × 10⁵ years. A.N. 43; m.pt. 2200°C; b.pt. 5030°C; r.d. 11.5; valency 2-7.

Teflon. Ⓡ *See* polytetrafluoroethylene.

tektite. Any of numerous small spheroidal glassy objects found in large quantities in certain parts of the world. They do not resemble terrestrial rocks and are thought to be of meteoric origin.

telecommunications. The communication of signals, speech, pictures, or other information over long distances, as by telegraphy, radio, television, etc.

telegraph. An apparatus for transmitting and receiving messages in the form of electrical pulses conducted along cables.

telemeter. Any instrument for making distant measurements of electrical quantities, as by radio.

telephoto lens. A compound lens consisting of a converging lens system followed by a diverging lens system, designed to replace the normal lens of a camera and produce a magnified real image at the same distance from this lens.

telescope. Any of several optical instruments used for producing magnified images of far objects. *Refracting telescopes produce their magnification by combinations of lenses: *reflecting telescopes use large concave mirrors.

television. A system for electrical transmission of moving pictures, either by cable (closed-circuit television) or by radio waves (broadcast television). The visual image is converted into electrical signals by a television camera. In broadcasting, this is amplified and used to modulate the transmitted radio wave. Television reception uses the *superheterodyne principle and the demodulated signal produces the image on the screen of a cathode-ray tube. In the receiver the spot on the screen is continually scanned across and down, forming a number of lines (625 in the British system). The received signal is used to control the intensity of the electron beam, thus forming the picture. Audio signals are transmitted along with the visual signals and the two are synchronized in the receiver.

In colour television three separate visual signals are transmitted corresponding to red, green, and blue components of the image. The cathode-ray tube used has three separate electron beams, one for each colour, and the screen is covered by groups of three coloured phosphor dots. The coloured picture is formed by an additive process involving different amounts of the three primary colours.

telluric. 1. Relating to the planet earth. **2.** *See* tellurium.

tellurium. Symbol: Te. A silvery-white metalloid element, usually obtained from the anode sludge produced in electolytic copper refining. The element is used in some alloys and semiconductors. Its chemistry is similar to that of sulphur: it forms *tellurous* compounds with low valency and *telluric* compounds with higher valency. Tellurium is one of the *chalconide elements. A.N. 52; A.W. 127.6; m.pt. 449.5°C; b.pt. 990°C; r.d. 6.24; valency 2, 4, or 6.

tellurous. *See* tellurium.

temperature. Symbol: *T*. A measure of the hotness of a body. If two bodies have different temperatures heat will flow from the higher temperature to the lower. Heat is a measure of the total kinetic energy of the atoms or molecules in the system and temperature is a measure of their average kinetic energy. It is measured in units of degrees on a scale such as the *centigrade, *Fahrenheit, or *Réaumur scale. The SI unit is the kelvin. *See also* thermometry, pyrometry, temperature scale.

temperature scale. A scale for the practical measurement of temperature. Scales such as the *centigrade, *Fahrenheit, *Réaumur, and *Rankine are defined by two *fixed points* with the interval between them divided into degrees. The fixed points used on these scales are the freezing and boiling points of water. *Thermodynamic temperature is measured in *kelvins and practical measurements are made by reference to the *International Practical Temperature Scale*, which has 11 fixed points (see

International Practical Temperature Scale

fixed point	TK	C
triple point of hydrogen	13·81	−259·34
temperature of hydrogen with vapour pressure 25/76 atmosphere	17·042	−256·108
b.pt. of hydrogen	20·28	−252·87
b.pt. of neon	27·102	−246·048
triple point of oxygen	54·361	−218·789
b.pt. of oxygen	90·188	−182·962
triple point of water	273·16	0·01
b.pt. of water	373·15	100
m.pt. of zinc	692·73	419·58
m.pt. of silver	1235·08	961·93
m.pt. of gold	1337·58	1064·43

table). Between these values temperatures are measured with a platinum *resistance thermometer or, above 1337.58 K, with a *pyrometer.

temporary hardness. *See* hardness.

temporary magnetism. *See* magnet.

tensile strength. The applied *stress necessary to break a material under tension. It is measured in newtons per square metre, etc.

tensimeter. An apparatus for measuring the vapour pressure of a liquid by comparing it with that of water at the same temperature.

tensor. An abstract mathematical entity used in transformations from one coordinate system to another. Tensor analysis is a generalization of vector analysis.

terbium. Symbol: Tb. A soft silvery-grey metallic element belonging to the *lanthanide series. A.N. 65; A.W. 158.925; m.pt. 1360°C; b.pt. 3041°C; r.d. 8.23; valency 3 or 4.

terephthalic acid. A white crystalline solid, $C_6H_4(COOH)_2$, made by the catalytic oxidation of xylenes. It is used to produce polyester resins and fibres such as Dacron and Terylene. Sublimes at 300°C; r.d. 1.5.

terminal. 1. A point on an electrical circuit, battery, etc., at which a connection can be made.
2. A piece of apparatus connected to a computer for the input or output of information.

terminal velocity. The constant velocity achieved by a body moving through a fluid under the action of a constant force, produced when this force is balanced by the resistance of the fluid. A body falling in air attains a terminal velocity as a result of air resistance. *See also* Stokes' law.

terminator. The line separating the illuminated portion from the dark portion on the moon or on a planet.

ternary. Denoting a chemical compound formed from three different elements. An example is sodium sulphate, Na_2SO_4.

terpene. Any of a class of naturally occurring unsaturated hydrocarbons found in the essential oils of many plants. They are built up of isoprene molecular units (C_5H_8), and have formulas of the type $(C_5H_8)_n$. *Monoterpenes* contain two units, $C_{10}H_{16}$, *diterpenes* contain four, $C_{20}H_{32}$, *triterpenes* six, $C_{30}H_{48}$, etc. *Sesquiterpenes* contain three isoprene units, $C_{15}H_{24}$.

terrestrial magnetism (geomagnetism). The magnetism of the earth. The earth has a magnetic field in which lines of force run from a point near the geographic true North Pole to one near the true South Pole, as if it contained a large bar magnet orientated at a small angle to its axis of rotation. The points on the earth at which the field is most intense are the *magnetic North Pole* and the *magnetic South Pole*. Their positions do not coincide with those of the geographic poles and also change slowly with time. At present they are lat. 76°N, long. 102°W and lat. 68°S, long. 145°E. The field at points on the earth's surface is characterized by three

properties: the *horizontal component, the angle of *dip, and the angle of *declination. Lines can be drawn on a map corresponding to equal values of these parameters: called *isodynamic*, *isogonal*, and *isoclinal lines* respectively. The earth's magnetic field is still not fully understood but the field probably arises from the iron core of the earth and is influenced by solar phenomena and the ionosphere. It is slowly changing with time and also shows periodic annual and daily variations. *See also* agonic line, magnetic equator.

terrestrial planet. Any of the planets Mercury, Venus, earth, Mars, and—according to some authorities—Pluto, whose solid constitution and relatively small size cause them to differ markedly from the *giant planets. The difference may result from the fact that the terrestrial planets are at a later evolutionary stage than the supposedly younger giants, though this is still uncertain.

terrestrial telescope. Any telescope suitable for use on the earth, as opposed to use in astronomy. *See* Kepler telescope, Galilean telescope.

tervalent. *See* trivalent.

Terylene. Ⓡ A type of *polyester resin used in synthetic fibres.

tesla. Symbol: T. An *SI unit of *magnetic flux density equal to a magnetic flux density of one weber per square metre. [After Nikola Tesla (1856-1943), Croatian-born U.S. electrical engineer.]

Tesla coil. An apparatus for producing high-frequency high voltages, consisting of a *transformer with its primary coil supplied through a spark gap. The high-frequency voltage will induce a discharge in a low pressure of gas within a glass vacuum system and Tesla coils are used for testing such systems for leaks.

tetrachloroethane. A colourless nonflammable toxic liquid, $CHCl_2CHCl_2$, made by chlorination of acetylene. It is a widely used solvent. M.pt. -43°C; b.pt. 146°C; r.d. 1.6.

tetrachloroethene (tetrachloroethylene).
A colourless nonflammable liquid,
$Cl_2C:CCl_2$, used as a solvent, particularly in dry-cleaning. M.pt. $-22\,°C$;
b.pt. $121\,°C$; r.d. 1.6.

tetrachloromethane. *See* carbon tetrachloride.

tetraethyl lead (lead tetraethyl). A
colourless poisonous oily liquid,
$Pb(C_2H_5)_4$, with a pleasant odour. It is
made by treating lead-sodium alloys
with chloroethane and is a petrol additive preventing knocking in internal-
combustion engines. M.pt. $-136\,°C$;
b.pt. $198\,°C$; r.d. 1.6.

tetragonal. *See* crystal system.

tetrahedron. A *polyhedron with four
faces. The faces are triangles, which are
congruent and equilateral in a *regular
tetrahedron*.

tetrahydrate. A solid compound with
four molecules of *water of crystallization per molecule of compound, as in
calcium tartrate tetrahydrate, $CaC_4H_4O_6$.
$4H_2O$.

tetrahydrofuran. A colourless liquid
heterocyclic compound, C_4H_8O, widely
used as a solvent. M.pt. $-65\,°C$; b.pt.
$66\,°C$; r.d. 0.9.

tetravalent. *See* quadrivalent.

tetrode. A *thermionic valve with four
electrodes. It is a *triode modified by
the inclusion of a second grid, the
screen grid, between the control grid
and the anode to reduce the large grid-
anode capacitance. It is held at a positive potential. *See also* pentode.

thallic. *See* thallium.

thallium. Symbol: Tl. A soft bluish-
white metallic element resembling lead,
present in small amounts in some zinc
ores. Thallium has two valencies, forming both thallium III (*thallous*) compounds and thallium I (*thallic*) compounds. A.N. 81; A.W. 204.37; m.pt.
$303.5\,°C$; b.pt. $1460\,°C$; r.d. 8.27.

thallous. *See* thallium.

theodolite. An instrument consisting of
a small telescope moving over angular
horizontal and vertical scales, used for
measuring angles in surveying.

theorem. A proposition that can be
proved logically from a set of basic
assumptions.

theory. 1. A proposition put forward to
explain observed facts but not yet
established as true.
2. A body or collection of principles,
methods, etc., used in explaining a fairly
wide set of phenomena, as in the theory
of relativity, quantum theory, etc.

theory of games. A mathematical theory
applied to situations in which there is a
conflict between interests, designed to
predict the optimum strategy.

therm. A unit of energy equal to 10^5
British thermal units. It is equivalent to
$1.055\ 056 \times 10^8$ joules. The therm is
used for measuring heat energy, as in
the calorific value of gas and other fuels.

thermal capacity. *See* heat capacity.

thermal conductivity. Symbol: λ. A
measure of the ability of a substance to
conduct heat, equal to the rate of flow
of heat per unit area resulting from unit
temperature gradient. It is given by the
equation $dQ/dt = \lambda A dT/dx$, where
dQ/dt is the rate of heat flow normal to
an area A and dT/dx is the temperature
gradient. Thermal conductivity is
sometimes defined as the heat flow per
unit time between two opposite faces of
a unit cube of the material when these
faces have unit temperature difference.
It is measured in joules per second per
metre per kelvin ($J\ s^{-1}\ m^{-1}\ K^{-1}$).

thermal diffusion. The phenomenon in
which a mixture of gases subjected to a
temperature gradient tends to separate
so that the heavier molecules concentrate in the cooler region and the lighter
molecules in the hotter region. The
effect is the basis of the *Clusius
column for separating isotopes.

thermal equilibrium. Equilibrium in which there is no net flow of heat between different parts of the system. The temperature is then uniform.

thermalize. To reduce the energies of neutrons with a *moderator so that they become thermal neutrons.

thermal neutron. A neutron with a low kinetic energy, of the same order of magnitude as the kinetic energies of atoms and molecules. Thermal neutrons have energies of the order kT, where k is the Boltzmann constant: i.e. about 0.025 electronvolts at normal temperature.

thermal reactor. See fission reactor.

thermionic emission. Emission of electrons from a hot solid. The effect occurs when significant numbers of electrons have enough kinetic energy to overcome the solid's *work function. The number of electrons increases with temperature. Thermionic emission is the process used for producing electrons in *thermionic valves, *X-ray tubes, and *cathode-ray tubes.

thermionic valve. An electronic device for controlling the flow of current in a circuit by regulating the flow of electrons produced by thermionic emission. Thermionic valves are often called *electron tubes*. They consist of an evacuated or gas-filled glass or metal container into which are sealed two or more electrodes. The main electrodes are the cathode, which is at a negative potential and is heated so as to emit electrons, and the anode, which is at positive potential. The cathode may be a metal filament heated to incandescence by an electric current. Alternatively, it may be a separate electrode heated by a nearby coil (the *heater*).
Electrons can only flow between the cathode and the anode when the anode has a positive potential with respect to the cathode. Extra electrodes (*grids*) can be introduced between the two to control this current. Types of valve are named according to the number of

electrodes they have (see diode, triode, tetrode, and pentode). In most applications they have been replaced by transistors although they are still used for special purposes (see thyratron) and for handling large voltages and currents.

thermistor. A small piece of semiconductor with an electrical resistance that decreases sharply with temperature. Thermistors can be used to measure temperature or compensate for temperature increases in electrical equipment.

thermite. A mixture of aluminium powder with a metal oxide, usually iron III oxide, mixed in the correct proportions for reduction of the oxide by aluminium. When ignited, aluminium oxide and metal are produced: $2Al + Fe_2O_3 = Al_2O_3 + 2Fe$. The reaction is strongly exothermic and the iron is molten. Thermite is used in a technique for *in situ* welding of steel (the *Thermit process*). It is also a constituent of some incendiary bombs. The reaction is the same as that in the *Goldschmidt process for extracting metals.

thermochemistry. The branch of chemistry concerned with the measurement and use of heats of reaction, heats of solution, etc. See Hess's law.

thermocouple. A junction between two different metals used for measuring temperature by the potential difference produced. In general, any two metals in contact develop a potential difference (the *contact potential) across the junction and this increases with increasing temperature. In practice thermocouples are made by welding together the ends of two pieces of wire: suitable pairs of metals are ones that show a large increase in potential difference for small changes in temperature. They include copper/constantan, iron/constantan, and platinum/platinum-iridium. Two junctions are often used connected in series: one to measure the temperature and the other maintained at a constant low temperature. This is an application of the *Seebeck effect. *See also* thermopile.

thermodynamics. The branch of science concerned with the relationship between *heat, *work, and other forms of *energy. The theory is based on three laws:

The *first law* states that the heat absorbed by any system is equal to the work done by the system plus its change in *internal energy: $\Delta q = \Delta w + dU$. It is equivalent to the idea that heat is a form of energy and an application of the law of *conservation of energy. The *second law* states that heat cannot pass from a colder to a hotter body without any other external effect occurring: i.e. that there can be no spontaneous self-sustaining heat flow against a temperature gradient. This is only one of several more-or-less equivalent statements of the law, which is generally concerned with the availablity of energy. Use of the second law and the *Carnot cycle leads to the idea of *entropy. The *third law* states that the *entropy of a substance approaches zero as its *thermodynamic temperature approaches zero. The third law is a modified form of the earlier Nernst heat theorem. Thermodynamics is used in studies of heat engines and also in studies of chemical reactions and chemical equilibrium. *See also* free energy, enthalpy.

thermodynamic temperature (absolute temperature). Symbol: T. Temperature that is a function of the internal energy possessed by a body, having a value of zero when the internal energy is zero. The unit of thermodynamic temperature is always the *kelvin, which is defined in such a way that the *triple point of water is 273.16 kelvins. In practice, measurements of thermodynamic temperature are made using the *International Practical Temperature Scale* (*see* temperature scale).

thermoelectric effect. Any of three effects involving direct interconversion of heat and electricity: the *Seebeck effect, the *Peltier effect, and the *Thomson effect.

thermoelectricity. Electricity produced by a thermoelectric effect.

thermoluminescence. *Luminescence produced when certain solid materials are heated. The effect arises when the solid contains electrons trapped at crystal defects, a condition produced by the impact of ionizing radiation. As the material is heated the electrons are freed and the energy released is emitted as photons. Thermoluminescence is used as a method of dating archeological finds, particularly pottery. The low intensity of light produced by heating a sample is measured, thus enabling an estimate to be made of the amount of natural ionizing radiation to which the object has been subjected. This is proportional to time that has elapsed since the pottery was fired.

thermometer. Any instrument for measuring temperature. Many different types of thermometer exist, all depending for their action on some easily measured physical property that varies with temperature. They include the *mercury thermometer, *alcohol thermometer, *gas thermometer, and *resistance thermometer. Other methods of measuring temperature involve the use of *thermocouples, *thermistors, and various types of *pyrometer.

thermonuclear bomb. *See* nuclear weapon.

thermonuclear reaction. *See* nuclear fusion.

thermopile. An instrument for measuring radiant heat, consisting of a number of *thermocouples connected in series with alternate junctions exposed to the radiation, the other junctions being shielded. Usually the thermocouples are made up of short metal rods. The current generated by the apparatus is detected by a galvanometer.

thermoplastic. *See* plastic.

thermosetting. See plastic.

Thermos flask. ® A commercial type of *Dewar flask.

thermostat. An apparatus for maintaining a constant temperature. Thermostats have a heater controlled by a temperature-sensing device such as a *bimetallic strip. The term is used both for the controlling device and for the constant-temperature enclosure.

thick lens. See lens.

thin film. A thin layer of solid deposited on the surface of another solid, usually by vacuum evaporation or sputtering. Thin films are used in many semiconductor devices.

thin-layer chromatography. See chromatography.

thin lens. See lens.

thio-. Prefix indicating the presence of sulphur in a chemical compound.

thiocarbamide. See thiourea.

thiocyanate. Any salt or ester of thiocyanic acid.

thiocyanic acid. An unstable gaseous compound, HSCN.

thio ether. Any of a class of organic compounds of formula RSR', where R and R' are hydrocarbon groups. They are sometimes called organic *sulphides.*

thionyl chloride. A yellowish fuming liquid, $SOCl_2$, prepared by oxidation of sulphur dichloride with sulphur trioxide. It is used in organic chemistry to replace hydroxyl groups with chlorine atoms, as in the preparation of acid chlorides. M.pt. $-105°C$; b.pt. $79°C$; r.d. 1.6.

thiosulphate. Any salt or ester of thiosulphuric acid.

thiosulphuric acid. An unstable dibasic acid, $H_2S_2O_3$, known only in the form of thiosulphate salts.

thiourea (thiocarbamide). A white bitter crystalline solid, $(NH_2)_2CS$, made by heating ammonium thiocyanate. It is used as a photographic fixing agent. M.pt. $180°C$; r.d. 1.4.

thixotropy. A phenomenon shown by some *non-Newtonian fluids in which the viscosity decreases as the rate of shear increases. In other words, the fluid becomes less viscous the faster it moves. Many colloidal mixtures are thixotropic, particularly *gels. A useful application is in non-drip paints, which are firm when on the brush but become thinner when worked. Some lubricating oils are also thixotropic, being viscous when the lubricated parts are at rest and more mobile when movement occurs. The opposite effect to thixotropy, *dilatancy, is less common.

Thomson effect. The phenomenon in which a temperature gradient along a conductor produces a gradient of electrical potential along it. A potential difference, depending on the substance, occurs between any two points at different temperatures. [After Sir William Thomson—Lord Kelvin.]

thoria. See thorium dioxide.

thorium. Symbol: Th. A soft silvery metallic element belonging to the *actinides, found in monazite ($ThPO_4$) and thorite ($ThSiO_4$). It is a radioactive element, the most stable isotope, ^{232}Th, having a half-life of 1.4×10^{10} years. Thorium is used in incandescent gas mantles, certain magnesium alloys, and as a source of nuclear power. A.N. 90; A.W. 232.04; m.pt. $1750°C$; b.pt. $3800°C$; r.d. 11.72; valency 4.

thorium dioxide (thoria). A heavy white powder, ThO_2, obtained by reducing thorium nitrate. It is used in high temperature ceramics, gas mantles, and nuclear reactors. M.pt. $3300°C$; b.pt. $4400°C$; r.d. 9.7.

thorium series. A *radioactive series beginnning with thorium-232 and ending with lead-82.

thoron. A radioisotope of radon, radon-220, produced by decay of thorium.

thou (mil). One thousandth of an inch.

three-body problem. The problem of calculating the motions of three objects moving under the influence of their mutual gravitational field. In general, no exact solution is possible although the motions can be obtained by numerical methods. The problem is the simplest case of the n-body problem, which applies, for example, to the motion of planets around the sun.

thulium. Symbol: Tm. A soft silvery-grey metallic element belonging to the *lanthanide series. A.N. 69; A.W. 168.9342; m.pt. 1545°C; b.pt. 1950°C; r.d. 9.3; valency 2 or 3.

thyratron. A type of gas-filled *triode valve used in particle counters and relays. The grid is given a negative potential just large enough to prevent current flow from the cathode. If a small positive potential is superimposed on the grid potential, it reduces its magnitude and a discharge starts. This current flows even when the grid potential is restored because of the presence of positive ions. It is switched off by reducing the anode potential.

timbre. See quality.

time. The duration between two events as determined by comparison with some changing standard. Time is usually measured against a periodic process, such as the swing of a pendulum, the rotation of the earth, or the vibration of electromagnetic radiation. The SI unit of time, the *second, is defined in terms of lines in the spectrum of caesium. Several astronomical measures of time also exist: *solar time, which is measured with reference to the sun, *sidereal time, which is measured with reference to the stars, and *lunar time, measured with reference to the moon. The *calendar year* (or *civil year*) has an average length equal to that of the solar year. To obtain an exact number of days, each year is 365 days with a leap year every fourth year containing 366 days. Century years (1800, 1900, etc.)

are not leap years unless they are divisible by 400.

time base. A circuit that generates a voltage that increases linearly with time at a controlled rate. Usually the process is continually repeated: i.e. it gives a *sawtooth waveform. Time bases are used in cathode-ray tubes to deflect the spot.

time dilation. See relativity.

time sharing. The simultaneous use of a computer by two or more separate terminals.

tin. Symbol: Sn. A silvery-white malleable metallic element found chiefly in cassiterite (SnO_2), from which it is extracted by reduction with carbon. Below 13.2°C this metallic form (*white tin*) changes slowly into a powdery grey allotrope (*grey tin*). It is used as a coating for steel plate (tinplate) and in alloys such as solder, Babbitt metal, brass, bronze, and pewter. Tin is a reactive element: it dissolves in hydrochloric acid to give tin II chloride, $SnCl_2$; it reacts directly with chlorine to give $SnCl_4$ and with oxygen to give SnO_2; it is dissolved by alkalis to give *stannates. It forms tin IV (*stannic*) compounds, which are covalent, and the more ionic tin II (*stannous*) compounds. A.N. 50; A.W. 118.69°C; m.pt. 231.89°C; b.pt. 2270°C; r.d. 7.31 (white).

tin chloride. Either of two chlorides of tin. *Tin II chloride* (stannous chloride) is a white crystalline solid, $SnCl_2$, obtained by the action of hydrochloric acid on the metal. It is used as a reducing agent. M.pt. 246.8°C; b.pt. 652°C; r.d. 3.95. *Tin IV chloride* (stannic chloride) is a colourless fuming liquid, $SnCl_4$, produced by direct reaction of the elements. M.pt. -33°C; b.pt. 114°C; r.d. 2.28.

tincture. A solution of a substance in alcohol.

tin hydroxide. A white substance precipitated from solutions of tin salts by alkali. The compound is amphoteric—it dissolves in excess alkali and is sometimes regarded as *stannic acid*, with formula H_4SnO_4 or H_3SnO_2. In fact it is hydrated tin IV oxide, $SnO_2.xH_2O$, which dissolves in alkali to give salts known as *stannates*: M_2SnO_3 or M_4SnO_4, where M is a monovalent metal. Hydrated tin II oxide also exists, $SnO.xH_2O$. Its salts are *stannites*.

tin oxide. Either of two oxides of tin. *Tin IV oxide* (stannic oxide) is a white crystalline substance, SnO_2, obtained by heating tin in air. It is used in ceramics and as a metal polish. M.pt. $1127\,°C$; sublimes at $1800\,°C$; r.d. 6.6 - 9. The lower oxide, *tin II oxide* (stannous oxide), is a black insoluble powder, SnO, obtained by precipitating the hydroxide from a solution of a stannous salt and heating at $100\,°C$. M.pt. $1080\,°C$ (600 mmHg); r.d. 6.3.

tin plague. The change of the white metallic form of *tin into its grey allotrope at low temperatures.

tinplate. Steel coated with a thin layer of tin to prevent corrosion.

tinstone. *See* cassiterite.

tin sulphide. Either of two sulphides of tin. *Tin IV sulphide* (stannic sulphide) is a yellow compound, SnS_2, precipitated from solutions of tin IV salts by hydrogen sulphide. It is used as a yellow pigment (*mosaic gold*). Decomposes at $600\,°C$; r.d. 4.5. The lower sulphide, *tin II sulphide* (stannous sulphide), is a grey substance obtained by direct combination of the elements. It slowly changes into tin IV sulphide and metallic tin. M.pt. $880\,°C$; b.pt. $1230\,°C$; r.d. 5.1.

tint. *See* colour.

titanate. Any of a number of oxy compounds of titanium having formulas of the type $MTiO_3$ or M_2TiO_5, where M is a monovalent metal. Titanates are not salts, in the sense that they do not contain ions of the type TiO_3^-. Instead, they are best considered as mixed oxides. Ilmenite and perovskite are naturally occurring examples.

Titania. *See* Uranus.

titanic. *See* titanium.

titanium. Symbol: Ti. A light strong lustrous white *transition element, ductile when free of dissolved oxygen, found in many minerals, notably rutile (TiO_2), perovskite ($BaO.TiO_2$), and ilmenite ($FeO.TiO_2$). The ore is treated with chlorine to give the tetrachloride and this is reduced by the *Kroll process. Titanium is extensively used in strong light corrosion-resistant alloys for aircraft, missiles, and ships. The element is unreactive at ordinary temperatures but combines with most nonmetals at high temperatures. It is not attacked by cold nonoxidizing acids. The most important compounds are the titanium IV (*titanic*) compounds, which are typically covalently bonded as in $TiCl_4$, and the *titanates. Titanium also forms *titanous* compounds with lower valencies, as in titanium III chloride, $TiCl_3$, titanium II chloride, $TiCl_2$, and many complexes. A.N. 22; A.W. 47.90; m.pt. $1675\,°C$; b.pt. $3620\,°C$; r.d. 4.54; valency 2, 3, or 4.

titanium chloride. Any of three chlorides of titanium, in which titanium has oxidation numbers 2, 3, and 4. The most important is *titanium IV chloride* (titanium tetrachloride), which is a colourless fuming liquid, $TiCl_4$, prepared by heating titanium IV oxide in a current of chlorine. It is used to make pure titanium and its salts, as a mordant, and as a polymerization catalyst. M.pt. $-30\,°C$; b.pt. $136\,°C$; r.d. 1.8.

titanium dioxide. A white crystalline solid, TiO_2, occurring naturally in *rutile and ilmenite. It is widely used as a pigment in paints and as a filler for paper, rubbers, and plastics.

titanous. *See* titanium.

titration. An operation in which a burette is used to add controlled amounts of one solution to a measured volume of another solution. Usually an *indicator is added to determine the point (called the *end point*) at which the reaction between the reagents is complete, so that neither reagent is in excess. If one solution has a known concentration (a *standard solution*) the concentration of the other can be calculated from the volumes of the two reagents. In some titrations the end point is found by continuously monitoring some physical property of the reaction mixture, such as its conductance or its colour, as solution is added from the burette. The technique is a form of *volumetric analysis.

TNB. *See* trinitrobenzene.

TNT. *See* trinitrotoluene.

Tollen's reagent. A solution of silver oxide in ammonia water used as a test for aldehydes, which reduce the oxide and deposit a bright mirror of silver on the wall of the test tube. Ketones do not cause this reduction.

toluene. A colourless flammable liquid, $C_6H_5CH_3$, obtained by catalytic reforming of petroleum. It is used in aviation fuels, in the production of phenol and other chemicals, and as a solvent for paints and resins. M.pt. -94°C; b.pt. 111°C; r.d. 0.9.

toluidene (aminotoluene). Any of three isomeric derivatives of toluene, $CH_3C_6H_4NH_2$, prepared by reduction of the appropriate nitrotoluene. The most widely used is *orthotoluidene* (2-aminotoluene), which is a pale yellow liquid used in the manufacture of dyes and saccharin. M.pt. -16°C; b.pt. 200°C; r.d. 1.0.

tolyl group. Any of the three isomeric groups $CH_3C_6H_4-$, derived by removing a hydrogen atom from the ortho-, meta-, or para- position of the benzene ring in toluene.

tone. 1. An audible note with no harmonics: i.e. a sound of a single frequency of the type produced by a tuning fork. A pure tone has a sinusoidal waveform.
2. *See* quality.

tonne (metric ton). A unit of mass equal to 1000 kilograms. It is equivalent to 0.9842 (short) tons.

topology. The branch of geometry concerned with general types of shape and pattern rather than with particular shapes or sizes. The basic notion is that of equivalence based on whether one figure or surface could be deformed into another without tearing it and without joining points together. Thus, a square and a circle are topologically equivalent but neither is equivalent to a figure eight.

toroidal. Having a torus shape.

torque. Symbol: *T*. A force or system of forces producing a turning effect, measured by its *moment.

torr. A unit of pressure equal to the pressure that would support one millimetre of mercury. It is equivalent to 133.322 newtons per square metre. The torr is used for measuring the low pressures used in vacuum systems. [After Evangelista Torricelli (1608-47), Italian scientist.]

Torricellian vacuum. The vacuum produced by filling a long tube, closed at one end, with mercury and inverting the open end in a reservoir of mercury, so that the mercury column is held up by atmospheric pressure on the reservoir. The space above the mercury contains the Torricellian vacuum, in which the pressure equals the vapour pressure of mercury (1.3×10^{-7} mmHg).

torsion balance. A device in which a torque is measured by the amount by which it twists a vertical wire or fibre. The angle of twist, which is usually measured by reflecting a beam of light off a small attached mirror, is proportional to the torque and depends on the rigidity *modulus of the fibre.

torsion pendulum. *See* pendulum.

torus. A surface or solid generated by rotating a circle about an axis that does not pass through the circle. A torus thus has the same shape as a doughnut or anchor ring.

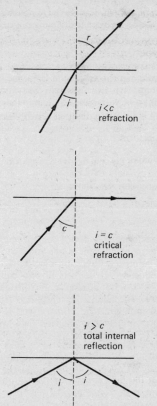

i < c
refraction

i = c
critical
refraction

i > c
total internal
reflection

Refraction and total internal reflection

total internal reflection. Reflection of an incident ray of light back into the initial medium at an interface between this medium and one that has a lower refractive index. A ray of light travelling from glass into air is refracted (see illustration) provided that the angle of incidence does not exceed the value at which the angle of refraction is 90°. Angles of incidence greater than this

value (the *critical angle*) give total internal reflection rather than refraction.

totality. The point at which the sun's disc is completely obscured during a total *eclipse.

total-radiation pyrometer. *See* pyrometer.

tourmaline. A blue or black crystalline naturally occurring borosilicate. Crystals of tourmaline show piezoelectricity and pleochroism.

Townsend discharge. A type of electrical discharge occurring between two electrodes in a low pressure of gas. The Townsend discharge differs from the *glow discharge in having a low current (a few microamps), which is limited by placing a high external resistance in the circuit. A fairly uniform luminous plasma is produced between the electrodes and the voltage falls uniformly with distance. The voltage drop increases with increasing discharge current. [After Sir John Townsend (1868–1957), Irish physicist.]

tracer. An isotope used for investigating chemical or physical changes. An example of a physical application is in studies of diffusion. A layer of a radioactive metal isotope deposited on the surface of the normal metal will slowly diffuse in, the extent of diffusion being determined by cutting thin slices of metal and testing their radioactivity. Tracer studies are also extensively used in finding mechanisms of chemical reactions. For instance in the hydrolysis of an ester, $RCOOR' + H_2O = RCOOH + R'OH$, it can be shown that the oxygen from the water molecule appears in the acid rather than the alcohol. This is done by including water made from the stable ^{18}O isotope in the reaction and demonstrating that $RCO^{18}OH$ is formed. The isotopic atom used in such studies is called a *label*. Radioactive-tracer studies are usually easier than those using stable isotopes due to difficulties in analysis.

trans. Indicating the position of a group or atom in a molecule opposite to the position of a specified group or atom: i.e. on the opposite side of a double bond or central atom. *See* geometrical isomerism.

transcendental. 1. Denoting a number that cannot be the root of a polynomial equation with rational coefficients. Transcendental numbers are a type of irrational number. Examples are π and e.
2. Denoting a mathematical function that cannot be written as a finite number of terms. Sin x, e^x, and log x are examples of transcendental functions.

transducer. A device that is supplied with the energy of one system and converts it into the energy of a different system, so that the output signal is proportional to the input signal but is carried in a different form. Thus, a microphone is a transducer for converting energy carried as sound waves into electrical energy. A record-player pick-up converts the vibrations of the stylus into an electrical signal. A loudspeaker converts electrical signals into sound waves.

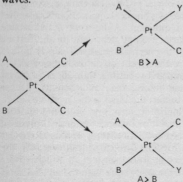

Trans substitution products

trans effect. An effect observed in the substitution of square-planar inorganic complexes, in which certain ligands have the power to direct the substituent into the *trans* position. For example, if the compound $PtABC_2$ is substituted by a ligand Y to give PtABCY, the position of Y depends on the relative *trans*-directing powers of A and B (see illustration). The cyanide ligand (CN^-), for example, is more efficient at causing *trans* substitution than the bromide ligand (Br^-). Some typical ligands, in order of their *trans*-directing power, are $CN^- > NO_2 > I^- > Br^- > Cl^- > NH_3 > H_2O$.

transformation. 1. A change of a mathematical expression from one form to another by replacing the variables with new variables, which are related to the old ones by defining equations. For instance, the equation $y = x^2$ can be transformed by substituting $y_1 = y + 3$ and $x_1 = x - 2$, to give $y_1 = x_1^2 + 4x_1 + 7$. This particular transformation corresponds to a translation of axes in Cartesian coordinates.
2. A change of one nuclide into another, as by emission of an alpha or beta particle.

transformer. A device for changing an alternating current of one voltage to one of another voltage by electromagnetic *induction. It consists of two coils of wire wound around a ferromagnetic core. The alternating current is passed into one coil (the *primary*) and it produces a changing magnetic field, which induces an alternating current of the same frequency in the other coil (the *secondary*). The core of the transformer links the magnetic field between the two coils: often it is laminated to reduce loss caused by *eddy currents. In a perfect system, the power input is equal to the power output and the ratio of voltage in the primary to that in the secondary is equal to the ratio of the number of turns of wire in the primary to the number in the secondary: $V_p/V_s = n_p/n_s$. If the number in the secondary is larger than in the primary the voltage is increased and the device is a *step-up transformer*. If the number in the secondary is less the device is a *step-down transformer*. If $n_p = n_s$, the voltage is unchanged. Transformers of this type are used for isolating a piece of equipment from its power supply so that there is no direct electrical connection.

transient. 1. A sudden short-lived disturbance in a system.
2. A short-lived chemical species, such as a free radical or excited atom or molecule.

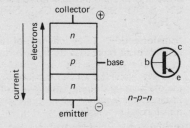

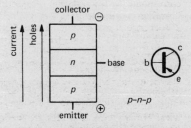

Fig. 1 Junction transistors

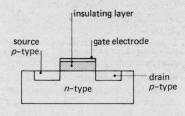

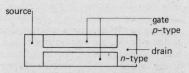

Fig. 2 Types of field-effect transistor

transistor. A device made of semiconducting material in which a flow of current between two electrodes can be controlled by a potential applied to a third electrode. In the most common type, known as a *junction transistor*, three pieces of *semiconductor are joined together, each with a separate attached electrode (see Fig. 1). Two forms exist, designated *n-p-n* and *p-n-p transistors* according to the arrangement of n-type and p-type material.

An n-p-n transistor has a potential difference applied so that one n-region is positive and the other negative. The central p-region (the *base*) has a positive potential with respect to the negative n-region (called the *emitter*). Electrons flow from the emitter into the base and holes from the base flow to the emitter. The electrons entering the base diffuse across and are carried through the p-n junction into the other n-region (the *collector*). Such a device can be used as an amplifier: a small current input to the base leads to a large current drawn from the collector. A p-n-p transistor works in a similar manner with the exception that the main carriers are holes and the emitter has a positive potential with respect to the collector. Holes have a lower mobility than electrons and p-n-p transistors operate more slowly. Transistors of this type have replaced *thermionic valves in all but the most specialized applications.

Other types of transistor exist. The original form had a single piece of semiconductor with one large metal contact and two other electrodes making contact at points. These *point-contact transistors* are now obsolete. Modern *field-effect transistors* are more complicated devices in which conduction occurs through a channel between a *source* and a *drain* (see Fig. 2) and is controlled by a potential applied to a *gate*.

The gate is either a doped semiconductor of opposite type to that of the source and drain, or a metal electrode acting through an insulating layer. Field-effect transistors (*FETs*) have a very high input impedance and are used in amplifiers and integrated circuits.

transition. 1. A change in *energy level in an atom, ion, or molecule.

2. A change in energy level of an atomic nucleus, causing either the emission of a gamma-ray photon or the emission of a beta or alpha particle. When a particle is ejected a different nucleus is formed: i.e. a transformation occurs.

3. A change of a substance from one physical state to another. Transitions include changes of phase (freezing, melting, etc.), the conversion of one crystal structure into another at a particular temperature (*see* enantiotropy), and the onset of *superconductivity in a metal at low temperature. The temperature at which a transition occurs is the *transition temperature*.

transition metal. Any of a class of elements characterized by an incomplete inner shell of electrons. In the *periodic table the elements of increasing atomic number have their electron shells filled in a regular way up to calcium, which has a configuration 2, 8, 8, 2. The next nine elements, from scandium to copper, have extra electrons in their penultimate shell: thus scandium has 2, 8, 9, 2, titanium has 2, 8, 10, 2, etc. The elements from titanium to copper form the first transition series: by convention scandium is not usually included in the transition metals. Two other transition series exist: from zirconium to silver and from hafnium to gold.

The transition elements are all metals, having a lustrous appearance and good electrical and thermal conductivity. They often have more than one valency and form many coordination *complexes. Compounds of transition metals are often coloured and, because of unpaired electrons, many of them are paramagnetic.

transition state. The state of highest energy occurring during the breaking and formation of bonds in a chemical reaction. In a simple case, the reaction

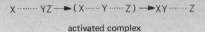

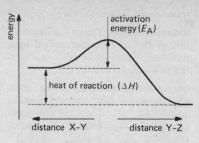

Energy profile

$X + YZ = XY + Z$, the atom X approaches as Z leaves (see illustration) and the total energy increases to a maximum and then decreases again. The partially bonded system of atoms corresponding to this transition state is the *activated complex* of the reaction.

translation. Motion of a body, molecule, etc., in which every point in the body moves an equal distance and the paths of the points are parallel.

translucent. Denoting a material that allows partial transmission of incident radiation, especially light. A substance that allows all the light to be transmitted is said to be *transparent* whereas one that absorbs all the incident light is *opaque*.

transmission coefficient. *See* transmittance.

transmission density. Symbol: D. The logarithm to base 10 of the reciprocal of the transmittance T. $D = \log_{10} 1/T$.

transmittance (transmission coefficient). Symbol: T. A measure of the extent to which a body transmits incident light or other electromagnetic radiation, equal to the ratio of the flux transmitted to the incident flux. The transmittance depends on the path length of the radiation and on its wavelength.

transmutation. The changing of one element into another. Transmutation occurs in radioactive decay or can be caused by bombardment of nuclei with particles.

transparent. *See* translucent.

transport number. Symbol: *t.* The fraction of the total current carried by a particular type of ion when an electrolyte conducts electricity.

transuranic element. Any of the *actinide elements with atomic number greater than 92. They do not occur naturally, being produced by nuclear reactions.

transverse wave. *See* wave.

trapezium. A quadrilateral that has one pair of opposite sides parallel. The area of a trapezium is one half of the sum ($a + b$) of the lengths of the parallel sides multiplied by the perpendicular distance between them (h): i.e. $h(a + b)/2$.

travelling wave. *See* wave.

triangle. A plane geometric figure with three sides. The area of a triangle is one half of the product of its base (b) and its height (h): i.e. $bh/2$.

triangle of forces. A triangle formed by representing the magnitudes of three forces acting at a point by the sides of a triangle. A triangle can only be formed for such forces when they are in equilibrium, a consequence of the rule for addition of *vectors.

triatomic. Having molecules composed of three atoms. Carbon dioxide, CO_2, and ozone, O_3, are both examples of triatomic gases.

triazine. Any of three isomeric azines, $C_3N_3H_3$.

triazole. Any of four isomeric heterocyclic compounds, $C_2N_3H_3$.

tribasic. Denoting a compound with three acidic hydrogen atoms in its molecules. Orthophosphoric acid is a common tribasic acid, forming the salts M_3PO_4, M_2HPO_4, and MH_2PO_4, where M is a monovalent metal.

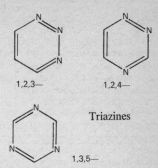

Triazines

triboelectricity. Static electricity produced by friction. When one material is rubbed against another it is possible for electrons to be transferred, so that one material acquires a positive charge and the other an equal negative charge. If a glass rod is rubbed with silk the glass becomes negatively charged and the silk positively charged.

tribology. The study of friction between solid surfaces, including the origin of frictional forces and the practical problems of wear and lubrication of moving parts.

triboluminescence. Light emission produced by friction between certain solids. Triboluminescence may be observed when some crystalline materials, such as sugar, are crushed. It results from a static electric charge generated by the friction.

trichloracetic acid. A corrosive deliquescent crystalline solid carboxylic acid, CCl_3COOH, made by oxidizing chloral hydrate with nitric acid. M.pt. 57°C; b.pt. 197°C; r.d. 1.6.

trichloroethylene. A stable toxic colourless liquid, $CHCl:CCl_2$, prepared by treating tetrachloroethane ($CHCl_2CHCl_2$) with alkali. It is used as a solvent and dry-cleaning agent. M.pt. -73°C; b.pt. 87°C; r.d. 1.4.

trichloromethane. *See* chloroform.

triclinic. *See* crystal system.

triglyceride. A *glyceride derived by substituting all three hydroxyl groups of glycerol.

trigonal. *See* crystal system.

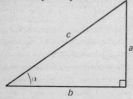

Right-angled triangle

trigonometric function. One of a number of functions of an angle defined as ratios of the lengths of the sides in a right-angled triangle containing the angle. The three most important are the *sine* (sin α = a/c), the *cosine* (cos α = b/c), and the *tangent* (tan α = a/b). Three other functions are the reciprocals of these: the *cosecant* (cosec α = c/a), the *secant* (sec α = c/b), and the *cotangent* (cot α = b/a). Certain other relationships apply between the trigonometric functions, notably $\sin^2 \alpha + \cos^2 \alpha = 1$, $1 + \tan^2 \alpha = \sec^2 \alpha$, and $1 + \cot^2 \alpha = \csc^2 \alpha$. The trigonometric function of a number is that of the angle equal in radians to the number. The sine and cosine functions can be expressed as series: $\sin x = x - x^3/3! + x^5/5! - x^7/7! \ldots$ and $\cos x = 1 - x^2/2! + x^4/4! - x^6/6! \ldots$ Trigonometric functions of complex numbers can also be expressed in terms of exponential functions: $\sin z = (e^{iz} - e^{-iz})/2i$ and $\cos z = (e^{iz} + e^{-iz})/2$.

trigonometry. The study of the properties of triangles with particular reference to the ratio of the sides of right-angled triangles. It is used in surveying, navigation, and astronomy.

trihydrate. A solid compound with three molecules of *water of crystallization per molecule of compound, as in potassium aluminate trihydrate, $K_2Al_2O_4.3H_2O$.

trihydric. Denoting an alcohol or phenol containing three hydroxyl groups per molecule.

trimer. A compound whose molecules are formed by addition of three molecules of a simpler compound (the *monomer*). A reaction leading to formation of a trimer is a *trimerization*. Acetaldehyde, for example, trimerizes to give paraldehyde.

trimethylamine. A colourless flammable gaseous tertiary *amine, $(CH_3)_3N$, with a fishy odour, prepared by catalytic reaction between methanol and ammonia. It is strongly alkaline and is used in the manufacture of disinfectants and synthetic resins. M.pt. $-117\,°C$; b.pt. $2.8\,°C$; r.d. 0.6 $(-5\,°C)$.

trinitrobenzene (TNB). A yellow crystalline solid, $C_6H_3(NO_2)_3$, prepared from trinitrotoluene. The compound is 1,3,5-trinitrobenzene: it is used as a high explosive, being more powerful than TNT. M.pt. $122\,°C$; r.d. 1.69.

trinitrotoluene (TNT). A pale yellow crystalline solid, $CH_3C_6H_2(NO_2)_3$, prepared by the nitration of toluene. The compound is 2,4,6-trinitrotoluene: it is used as a high explosive. M.pt. $81\,°C$; r.d. 1.6.

triode. A type of *thermionic valve containing three electrodes—an anode, a cathode, and a grid. The triode differs from the *diode by the interposition of the grid, usually made of a wire spiral or mesh, between the other two electrodes. A potential applied to this electrode is used to control the flow of current between anode and cathode—small changes in the grid potential cause large changes in the electron current. The triode valve can be used as an amplifier. If an alternating signal is applied between the grid and the cathode, a larger signal of the same frequency is produced between the anode and the cathode. The valve has the disadvantage that there is a large grid-to-anode capacitance, an effect that is reduced in the *tetrode valve.

triol. A trihydric *alcohol; an alcohol containing three hydroxyl groups per molecule.

triolein (olein). A simple *glyceride of oleic acid, $C_3H_5(OOCC_{17}H_{33})_3$, found in many natural fats and oils.

triose. A simple *sugar containing three carbon atoms per molecule.

tripalmitin (palmitin). A simple *glyceride of palmitic acid, $C_3H_5(OOCC_{15}H_{31})_3$, found in many natural fats and oils.

triphenylmethane. An aromatic hydrocarbon, $HC(C_6H_5)_3$. Its derivatives, obtained by substitution in the benzene rings, constitute an important class of dyes.

triple bond. See covalent bond.

triple point. The point at which the solid, liquid, and gas phases of a pure substance can all co-exist in equilibrium. It corresponds to definite values of pressure and temperature for each substance. The triple point of water (273.16 kelvins at 101 325 pascals) is used in the definition of the kelvin.

trisaccharide. A *sugar with molecules containing three monosaccharide units.

tristearin (stearin). A simple *glyceride of stearic acid, $C_3H_5(OOCC_{17}H_{35})_3$, found in natural fats.

triterpene. See terpene.

tritiated. Indicating that a chemical compound contains tritium atoms in place of hydrogen atoms, as in tritiated methane, CT_4.

tritium. Symbol: T, ^3H. The hydrogen isotope with one proton and two neutrons in its atomic nucleus. It is a radioactive substance produced by irradiation of lithium in an atomic pile. Tritium is not known to occur naturally. It is used in radioactive tracer studies. A.N. 1; A.W. 3.016; half-life 12.5 years.

triton. A nucleus of a tritium atom, containing two neutrons and one proton.

trivalent (tervalent). Having a valency of three.

trochoid. A curve that is the locus of a point on the radius of a circle as that circle rolls along a straight line. If the point is at the end of the radius, i.e. on the circumference, then the curve is a *cycloid.

troostite. A dispersion of cementite Fe_3C in ferrite (alpha-iron) produced in steels by tempering martensite below $500°C$.

tropical year. See solar time.

troposphere. The lowest part of the earth's *atmosphere in which the temperature decreases with increasing height. It extends to a height of about 10 kilometrès and is separated from the stratosphere by a boundary region, the *tropopause.*

tropylium ion. The positive ion $C_7H_7^+$, found in certain salts. It has a seven-membered symmetrical aromatic ring of carbon atoms.

Trouton's rule. The principle that the ratio of the molar heat of vaporization of any liquid to its boiling point is a constant. This value, the *Trouton constant*, is about 88 joules per mole per kelvin. The rule is equivalent to the statement that the entropy of vaporization is constant. It is not always followed, especially by liquids such as water in which hydrogen bonding occurs between the molecules.

troy weight. A system of weights used for gems and precious metals. *See Appendix.*

Tryptophan

tryptophan. A sweet white crystalline essential amino acid, $C_{11}H_{12}N_2O_2$, obtained synthetically or by hydrolysis of protein and used as a dietary supplement. M.pt. 275-90°C.

tungstate. Any of a number of oxy compounds of tungsten formed by dissolving tungsten trioxide in an alkali. The simplest tungstates have formula M_2WO_4 (where M is a monovalent metal) and contain WO_4^{2-} ions. Formally, they are salts of the hypothetical *tungstic acid*, H_2WO_4. If alkaline solutions of simple tungstates are acidified, *polytungstates* are produced in solution. They contain large anions of the type $(W_{12}O_{46})^{20-}$, and it is possible to obtain crystalline salts and acids (*polytungstic acids*) from such solutions.

tungsten (wolfram). Symbol: W. A grey or white usually brittle *transition element found in wolframite ((Fe,Mn)-WO_4) and scheelite ($CaWO_4$). The ores are dissolved in sodium hydroxide and WO_3 is precipitated by acidification and reduced with hydrogen. Tungsten is used in alloys, especially high-speed steels, and in electric-lamp filaments. Its chemical properties are similar to those of molybdenum and, like molybdenum, it has a complicated chemistry, forming many compounds, particularly oxides and oxy compounds (*see* tungstate). A.N. 74; A.W. 183.85; m.pt. 3410°C; b.pt. 5927°C; r.d. 19.3; valency 2-6.

tungsten carbide. A hard grey powder, WC, obtained by direct combination of the elements at about 1600°C. It is almost as hard as diamond and is used in making tools and in abrasives. M.pt. 2780°C; b.pt. 6000°C; r.d. 15.6.

tungstic acid. *See* tungstate.

tuning fork. A metal fork with two prongs that vibrate to produce a pure tone of known pitch.

tunnel diode (Esaki diode). A type of highly doped semiconductor *diode that has a negative resistance over part of its operating range: i.e. the current flowing decreases with increasing voltage. This occurs under forward bias conditions and is caused by electrons tunnelling

through the junction from the conduction band to a level of equal energy in the valence band (*see* tunnel effect). At higher forward-bias voltages the device behaves like a normal p-n junction. Tunnel diodes are used in ultra-high-frequency circuits and in switching circuits.

tunnel effect. The passage of particles through a potential barrier that they do not have sufficient energy to pass according to classical mechanics. The effect is explained by *wave mechanics in terms of the probability of the particle moving through the barrier.

turbulent flow. Flow of a fluid in which the motion is irregular and the velocity at any point can vary with time in magnitude and direction. *See* Reynolds number.

turpentine. An oily liquid extracted from pine resin. It consists chiefly of pinene, $C_{10}H_{16}$, and is used as a solvent.

tweeter. A loudspeaker for use at high audiofrequencies. *Compare* woofer.

twin. A crystal that has grown in two branches, both crystal lattices sharing a common plane so that one branch of the crystal is the mirror image of the other. The process of growing in this way is called *twinning*.

Tyndall effect. The scattering of light by small particles in its path. The Tyndall effect causes a light beam to be visible in a dusty atmosphere, colloidal solution, etc. [After John Tyndall (1820-93), English physicist.]

type metal. An alloy of lead (60%), antimony (30%), and tin (10%). It expands when it solidifies, a property that makes it suitable for casting type and other intricate work.

U

UHF. *See* ultra-high frequency.

ultracentrifuge. A centrifuge designed to work at very high angular speeds, so that the centrifugal force produced is large enough to cause sedimentation of colloids. The rate at which the sedimentation occurs depends on the size of the particle: the apparatus can be used to investigate sizes of colloidal particles and molecular weights of proteins and other macromolecules.

ultra-high frequency (UHF). A *frequency in the range 0.3 gigahertz to 3 gigahertz.

ultra-high vacuum. See vacuum.

ultramicroscope. A microscope in which colloidal particles can be observed by using *dark-field illumination.

ultrasonic. Above audible frequencies: i.e. above 20 kHz.

ultrasonics. The study of the properties and applications of ultrasound.

ultrasound. Propagated vibrations with a frequency above that of audible *sound: i.e. above 20 kHz. Ultrasound has many applications including flaw detection of metals, medical diagnosis, depth finding, cleaning, dispersion of one liquid in another, and coagulation of some dispersions.

ultraviolet radiation. Electromagnetic radiation with wavelengths between 13 and 397 nanometres. Ultraviolet radiation lies between light and X-rays in the electromagnetic spectrum. It is produced by emission from excited atoms and ions and can be generated by gas-discharge tubes. Large amounts are produced by the sun, although the atmosphere absorbs radiation with wavelengths below 200 nm.
Ultraviolet radiation causes ionization of matter and also induces certain chemical reactions. It can split some molecules into free radicals and initiate polymerization processes. In the human body it converts ergosterol into vitamin D and in plants it induces photosynthesis.
The ultraviolet region is arbitrarily divided into the *near ultraviolet* (397-200 nm) and the *far ultraviolet* (200-13 nm). Radiation with wavelengths shorter than about 200 nm lies in the *vacuum ultraviolet.

ultraviolet spectrum. An emission or absorption *spectrum in the ultraviolet region. Ultraviolet emission occurs from gas discharges and is caused by electron transitions in atoms, molecules, and ions. It gives information on energy levels and molecular structure. Ultraviolet absorption is used in *Rydberg spectroscopy. Most ultraviolet radiation is absorbed by glass so the radiation is dispersed by quartz prisms or by reflection gratings.

umbra. 1. A region of complete *shadow. **2.** The central darker portion of a *sunspot.

Umbriel. See Uranus.

uncertainty principle. See Heisenberg uncertainty principle.

underdamped. See damping.

uniaxial crystal. A crystal with one optic axis. See double refraction.

unified-field theory. A theory that seeks to explain *gravitational and *electromagnetic interactions and the *strong and *weak nuclear interactions in terms of a single set of equations. The approach is an extension of the general theory of *relativity in which the gravitational field arises as a consequence of the curvature of space-time. A unified-field theory would involve discovering a geometry of space-time from which the electromagnetic and nuclear interactions also arise naturally.

unimolecular. Relating to one molecule. A *unimolecular layer* is an adsorbed layer one molecule (or atom) thick. A *unimolecular reaction* is one in which single molecules of reactant undergo a change: i.e. the transition state does not involve more than one molecule. Such reactions must be distinguished from first-order reactions (see order).

unit. A magnitude of a physical quantity used for measurements of other magnitudes of the physical quantity by comparison. In defining a unit, a standard is adopted which, by agreement, is taken to have a specified value. For example, a particular piece of platinum has been adopted as the SI unit of mass—the kilogram. This is an example of a *primary standard* of a unit: i.e. the standard that defines the unit. Such a standard does not have to be an actual object: the *metre for example is defined in terms of the wavelength of light from a particular spectral transition. Standards of this type, based on atomic constants or on the properties of substances, are preferable as they cannot change with time and can be reproduced in different places. Any measurement is effectively a comparison of the object or system measured with the standard. Masses, for example, can be compared by weighing with a balance. Measurements are usually not made against the primary standard but against *secondary standards* of predetermined value.

Units defined by reference to primary standards are *base units*, and other units—*derived units*—can be defined by reference to these base units. For example, the SI units of length and mass—the metre and the kilogram—are base units, whereas the unit of density—the kilogram per cubic metre—is a derived unit. In making physical measurements, systems of units are employed, determined by a choice of certain base units. *Imperial, *CGS, and *MKS units are examples. These have now been replaced for most scientific work by *SI units.

unitary symmetry. A relationship between sets of elementary particles whereby they show a special symmetry between their *isospin quantum numbers and their *hypercharge. These quantum numbers are related by the operations of a certain *group (called SU3). Using this idea, particles can be classified into *multiplets that can be thought of as different states of the same

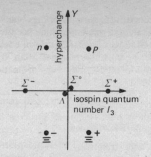

Octet of baryons

particle. All particles in the same multiplet have the same parity and spin. The relationship between particles in a multiplet is shown on a graph (see illustration). Unitary-symmetry theory was used to predict the existence of the *omega-minus particle. The extension of unitary-symmetry theory to a larger symmetry group (SU4) leads to the concept of *charm.

unit cell. The smallest repeated unit in a crystal *lattice. In sodium chloride, for example, the unit cell is a cube with eight ions at its corners. *See also* crystal system.

unit pole. A *magnetic pole of unit strength. The unit pole in *electromagnetic units is defined as the pole that repels an identical magnetic pole with a force of one dyne when placed one centimetre away in vacuum.

unit vector. A *vector with unit magnitude and specified direction.

univalent. *See* monovalent.

universal gas constant. *See* gas constant.

universe. The whole of space and all the matter (galaxies, etc.) contained in it. The age of the universe is estimated as about 5×10^9 - 50×10^9 years and it is thought to contain 10^{41} kilograms of matter. Its origin is still under discussion. *See* big-bang theory, steady-state theory.

unsaturated. Denoting a chemical compound that is not *saturated: i.e. one that contains double or triple bonds. Unsaturated compounds, such as olefines and acetylenes, tend to form compounds by *addition rather than by substitution unless they are *aromatic compounds.

unstable. 1. Denoting a chemical compound that is not stable.
2. *See* equilibrium.

upper atmosphere. The atmosphere from a height of about 30 kilometres upwards: i.e. that part investigated by rockets and artificial satellites rather than by balloons.

uranium. Symbol: U. A heavy ductile malleable silvery metallic element belonging to the *actinides, found in several ores, notably pitchblende (UO_2). The element is a mixture of three isotopes: ^{234}U, ^{235}U (0.7%), and ^{238}U (99%). The most stable isotope, ^{238}U, has a half-life of 4.5×10^9 years and is used in breeder reactors as a source of the plutonium isotope, ^{239}Pu. Uranium enriched with ^{235}U (half-life 7.1×10^8 years) is used as a fuel in *fission reactors. A.N. 92; A.W. 238.03; m.pt. 1150°C; r.d. 18.68.

uranium dioxide. *See* uranium oxide.

uranium hexafluoride. A volatile white crystalline solid, UF_6, obtained by successive hydrofluorination and fluorination of uranium dioxide. It is used in gas-diffusion processes for separating the isotopes of uranium. M.pt. 64.5°C; r.d. 5.1.

uranium oxide. Any of several oxides of uranium. The most important is *uranium IV oxide* (uranium dioxide), a black crystalline solid, UO_2, made by oxidation of uranium ore to *uranium VI oxide* (UO_3) followed by reduction with hydrogen. It is used to pack nuclear fuel rods. M.pt. 3000°C; r.d. 10.9.

uranium series. A *radioactive series beginning with uranium-238 and ending with lead-82.

Uranus. The seventh planet in order of succession outwards from the sun; the third of the *giant planets with a dense atmosphere of hydrogen and methane. Orbiting the sun once in 84.01 years at a mean distance of 2870 million kilometres, the planet has a diameter of 47 100 km, a mass 14.5 times that of the earth, and an average relative density of 1.7. Its period of axial rotation is about 10 hours 45 minutes and its temperature is probably as low as –200°C. One of its most interesting features is that, unlike any other planet, the inclination of its equator to the plane of its orbit is 98°, giving earthbound observers a view of the planet as if it were lying on its side: its lines of latitude cross the disc vertically. Uranus has five satellites: *Oberon*, *Titania*, *Ariel*, *Umbriel*, and *Miranda*.

uranyl. Containing the ion UO_2^{2+} or the divalent group UO_2:. The uranyl ion is a greenish-yellow colour and is present in such salts as uranyl nitrate, $UO_2(NO_3)_2$.

urea (carbamide). A white powder, $CO(NH_2)_2$, manufactured by heating ammonia with carbon dioxide at high pressure and allowing the ammonium carbamate ($NH_4CO_2NH_2$) formed to decompose. It is present in urine. Urea is used as fertilizer, in animal feeds, and in urea-formaldehyde resins. M.pt. 133°C; r.d. 1.3. *See also* Wöhler's synthesis.

urea-formaldehyde resin. A synthetic *aminoresin made by reacting urea with formaldehyde. The process yields a water-soluble syrup which is used as an adhesive. The syrup can be dried to a powder which, when mixed with filler, is used in moulding a strong hard white thermosetting plastic.

urethane. A white crystalline solid, $CO(NH_2)(OC_2H_5)$, made by heating ethanol with urea nitrate. M.pt. 49°C; b.pt. 180°C; r.d. 0.9. *See also* polyurethane.

uric acid. A colourless odourless tasteless powder, $C_5H_4N_4O_3$, occurring in

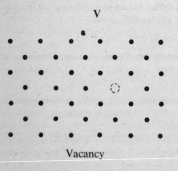

Uric acid

small quantities in urine. Decomposes on heating; r.d. 1.87.

V

● ● ● ● ● ●
● ● ● ● ● ●
● ● ● ● ● ●
● ● ● ○ ● ●
● ● ● ● ● ●
● ● ● ● ● ●
● ● ● ● ● ●

Vacancy

vacancy (Schottky defect). A type of *defect in crystals in which a position that is normally occupied by an atom, ion, or molecule is unoccupied.

vacuum. A region containing gas at a pressure below atmospheric pressure. A vacuum system is a sealed apparatus evacuated by a pump. The ultimate pressure achieved depends on the pumping speed and on the presence of adsorbed gas, volatile materials, leaks, and other sources of gas in the system. Grades of vacuum are classified into *soft* (or *low*) vacuum (down to about 10^{-4} mmHg) and *hard* (or *high*) vacuum (below 10^{-4} mmHg). An *ultra-high vacuum* has a pressure below about 10^{-9} mmHg. Vacuum technology is important in the manufacture of light bulbs, thermionic valves, cathode-ray tubes, and similar devices and is also used in industrial processes such as freeze drying and vacuum evaporation.

Vacuum systems are also necessary in many branches of scientific research.

vacuum distillation. Distillation of liquids under reduced pressure, thus lowering the boiling point of the liquid and allowing a lower temperature to be used. The technique is valuable for distilling substances that would decompose or explode at higher temperatures.

vacuum evaporation. A process for depositing thin films of solid on a surface by evaporating the material at high temperature in a vacuum. Under these conditions atoms leave the hot material and travel directly to the cool surface without colliding with other molecules in the gas. They condense on the cool surface, gradually building up a layer of solid.

vacuum flask. *See* Dewar flask.

vacuum ultraviolet. The part of the ultraviolet region of the spectrum in which the radiation is absorbed by air. Experiments using this radiation have to be performed in evacuated apparatus. The region includes ultraviolet radiation with wavelengths less than about 200 nm.

valence. *See* valency.

valence band. The energy band of a solid that is occupied by the valence electrons of the atoms forming the solid. *See* band theory.

valence bond. A single linkage between two atoms in a molecule, considered to be a shared pair of electrons between the atoms. The valence-bond approach to the theory of chemical bonding is based on the idea that a molecule is held together by definite linkages between pairs of atoms. In the alternative approach (*see* molecular orbital) the molecule is thought of as a collection of atomic nuclei with the electrons moving in the resultant field of these nuclei.

valence electron (valency electron). An electron in the outer shell of an atom that participates in the chemical bonding when the atom forms compounds.

valency (valence). The combining power of an element, atom, ion, or radical, equal to the number of hydrogen atoms that the atom, ion, etc., could combine with or displace in forming a compound. For instance, carbon has a valency of four, as displayed in methane, CH_4. Many elements exhibit more than one valency; phosphorus has a valency of three in the trichloride, PCl_3, and five in the pentachloride, PCl_5. The valency of an ion is equal to its charge: the ammonium ion, NH_4^+, has a valency of one. The valency of a radical is the number of hydrogen atoms with which it is capable of combining: the methyl radical, CH_3-, has a valency of one. In general, the valency of a chemical entity is the number of single covalent or electrovalent bonds it can make. See also oxidation state.

valeric acid. See pentanoic acid.

valine. A white crystalline essential amino acid, $(CH_3)_2CHCH(NH_2)COOH$, occurring in proteins. It is used as a dietary supplement.

valve. An *electron tube for controlling a flow of current by motion of electrons between electrodes. Valves are either evacuated or gas-filled. See also thermionic valve, klystron.

valve voltmeter. A type of *voltmeter in which the low input voltage, which may be either direct or alternating, is amplified by a valve circuit. The output current is proportional to the input voltage and is registered by a meter calibrated in millivolts. The instrument draws very little current because of its high input impedance.

vanadate. Any of a large number of oxy compounds of vanadium in its +5 oxidation state. Simple compounds with formulas of the type M_3VO_4 (orthovanadates), MVO_3 (metavanadates), and $M_4V_2O_7$ (pyrovanadates) can be prepared. They are formally salts of the hypothetical acids orthovanadic acid, H_3VO_4, metavanadic acid, HVO_3, and pyrovanadic acid, $H_4V_2O_7$. However, the behaviour of vanadates is quite complicated. When vanadium V oxide, V_2O_5, is dissolved in alkali, the solution contains simple orthovanadate ions VO_4^{3-} and more complicated polymeric anions. The $[V_3O_9]^{3-}$ ion, for example, has a ring containing three vanadium atoms and three oxygen atoms. A large number of different polymeric vanadates (polyvanadates) can be precipitated from such solutions. Polyvanadates are also formed by dissolving V_2O_5 in acid solutions: they contain ions of the type $V_{10}O_{28}^{6-}$. Any vanadate can be regarded as derived from a corresponding vanadic or polyvanadic acid.

vanadic acid. See vanadate.

vanadinite. A red, orange, brown, or yellow mineral, $Pb_5Cl(VO_4)_3$, used as an ore of lead and vanadium.

vanadium. Symbol: V. A silvery-white ductile *transition element found in many minerals, especially patronite (VS_4), carnotite $(K_2(ClO_2)_2(VO_4)_2.$ $3H_2O)$, and vanadinite $(Pb_5Cl(VO_4)_3)$. The pure metal can be obtained by the *van Arkel-de Boer process or by reduction of the oxide with calcium. Usually it is produced alloyed with iron for use in steels.

Chemically the element resists corrosion, although it combines with oxygen and other nonmetals at high temperature. It does not dissolve in alkalis or nonoxidizing acids. Vanadium forms compounds with several valencies, the higher oxidation states being more important. Vanadium IV compounds tend to be covalently bonded: in particular they include the *vanadyl compounds and complexes. Vanadium V compounds are also covalent, as in the molecular halides and oxyhalides and the *vanadates. Numbers of vanadium II and III complexes are also known, including the hydrated cations $[V(H_2O)_6]^{3+}$ and $[V(H_2O)_6]^{2+}$. A.N. 23; A.W. 50.94; m.pt. 1890°C; b.pt. 3380°C; r.d. 6.1; valency 2–5.

vanadium oxide. Any of three oxides, *vanadium III oxide*, V_2O_3, *vanadium IV oxide*, V_2O_4, or *vanadium V oxide* (vanadium pentoxide), V_2O_5. Vanadium V oxide is a yellow to brown crystalline solid, extracted from vanadium minerals. It is a starting material for other vanadium compounds and a widely used catalyst, particularly for the oxidation of sulphur dioxide. M.pt. 690°C; r.d. 3.4.

vanadyl. Containing the ion VO^{2+} (*vanadyl ion*) or the divalent group $=VO$ (*vandyl group*). The vanadyl ion is a bright blue colour.

Van Allen belts (radiation belts). Two belts of charged particles surrounding the earth in the region of the equator at heights of 2000-5000 and 13 000-19 000 kilometres. They are mainly protons and electrons, originating in the *solar wind and in *cosmic rays, trapped by the earth's magnetic field. [After James Alfred Van Allen (b. 1914), U.S. physicist.]

van Arkel-de Boer process. A process for obtaining pure specimens of certain elements by forming a volatile iodide and then decomposing the vapour on an incandescent tungsten filament.

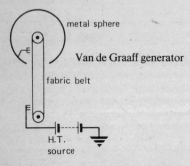

Van de Graaff generator

Van de Graaff generator. A type of electrostatic *generator in which charge is accumulated on a metal sphere from an endless belt of fabric. The belt is mounted vertically with its top end inside a hollow metal sphere and charge is applied to the belt from points at the bottom, which are connected to an external source of high potential. This charge is collected by other points attached to the sphere. The device is capable of developing potentials of millions of volts. It is often used as a high voltage source for a linear accelerator (*Van de Graaff accelerator*). [After R. J. Van de Graaff (1901-67), U.S. physicist.]

van der Waals equation. A modified form of the *ideal gas law: $(P + a/v^2)(v -b) = RT$, where a and b are constants for a particular gas. The equation was designed to apply to real gases. The term a/v^2 corrects for a reduction in pressure caused by attraction between molecules and the b term is a correction for the size of the molecules, being equal to about four times their actual volume. [After J. D. van der Waals (1837-1923), Dutch physicist.]

van der Waals forces. The attractive forces existing between molecules. Van der Waals forces are responsible for the cohesion of molecules in liquids and nonionic crystals. They act over short distances, being proportional to the seventh power of the distance, and are responsible for the correction to the pressure in van der Waals' equation. They are caused by attraction between *dipoles and, in nonpolar molecules, by *dispersion forces.

van't Hoff factor. Symbol: i. The ratio of the total number of entities (ions, molecules, etc.) of solute in a solution to the number that would be present if the solute consisted of undissociated molecules. The factor is used in the equation for *osmotic pressure when the solute is an electrolyte: $\Pi V = iRT$. It reflects the fact that osmotic pressure depends on the number of entities present in solution. [After J. H. van't Hoff (1852-1911), Dutch physical chemist.]

van't Hoff's isochore. A relation between the equilibrium constant (K) of a reaction and the temperature (T):

$$\frac{d \log K}{dT} = \frac{\Delta H}{RT^2}$$

where H is the enthalpy of the reaction.

van't Hoff's law. The principle that the osmotic pressure of solution is equal to the pressure that the solute would exert if it were an ideal gas at the same temperature as the solution and occupying the same volume. Thus the osmotic pressure (Π) is given by the equation $\Pi V = RT$.

vapour. A substance in the gaseous state at a temperature below its *critical temperature, so that it could be liquefied by pressure alone.

vapour density. The density of a vapour under specified conditions. The vapour density is often taken as a relative density—the ratio of the weight of a given volume of a gas or vapour to the weight of the same volume of hydrogen measured under the same conditions. The vapour density of a substance is equal to one half of its molecular weight and measurement of vapour density is a method of determining molecular weights. The principle techniques are Victor Meyer's method and Dumas' method.

vapour pressure. The pressure of vapour produced by a solid or liquid. In particular, the term is used for the *saturated vapour pressure*, which is the pressure of vapour in equilibrium with the solid or liquid. Thus, if a solid is placed in an enclosed space the pressure of its vapour increases until a steady state is reached in which the number of molecules leaving the solid is equal to the number returning to the solid from the gas phase. The pressure at which this occurs is the saturated vapour pressure. It depends on the temperature.

variable. A quantity that can take a range of values in a mathematical expression. A *dependent variable* is one whose values are determined by other *independent variables*. For example, in

$y = 7x^3 + 2x$, y is the dependent variable and x the independent variable.

Variac. Ⓡ A device for producing a variable alternating voltage from a fixed input voltage. It is essentially a toroidal winding on a core, tapped by a moving brush.

variance. The square of the mean *deviation of a statistic.

variation. *See* declination.

variometer. A variable inductor, usually consisting of two coils connected in series and arranged so that one can be rotated inside the other, thus changing the self-inductance of the pair of coils.

varistor. A resistor, usually of semiconductor material, that does not obey Ohm's law: i.e. the current is not directly proportional to the voltage.

Vaseline. Ⓡ A colourless or pale yellow semisolid containing a mixture of petroleum hydrocarbons, used as an inert protective coating, particularly in pharmacy.

vat dyes. Any of a class of dyes that can be reduced to a soluble *leuco form and then oxidized to the coloured dye after application to the material. *Indigo is an example of a vat dye.

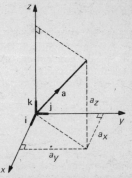

Components of a vector

vector. A quantity that is specified both by its magnitude and its direction. Velocity, acceleration, and force are examples of vector quantities. Mass and electric charge, on the other hand, are *scalars, having only magnitude. Usually symbols for vectors are printed in italic boldface type or written with an arrow or a bar above them. A vector can be represented by a line with the same direction as the quantity and a length equal to its magnitude.

Multiplication of a vector by a scalar gives another vector with the same direction but with a magnitude equal to its original magnitude multiplied by that of the scalar. It is often convenient to express a vector as a *unit vector*, specifying the direction, multiplied by a number that is its magnitude. In three dimensions, a vector a can be represented as the sum of three vectors directed along the axes of a *Cartesian coordinate system (see diagram). Thus, $a = i\,a_x + j\,a_y + k\,a_z$, where i, j, and k are unit vectors. The *modulus of a is $(a_x^2 + a_y^2 + a_z^2)^{1/2}$. Vectors can be added together (see parallelogram rule) and multiplied together in two ways (see vector product, scalar product).

vector boson. See virtual particle.

vector product (cross product). The product of two vectors yielding a vector whose magnitude is the product of their magnitudes and the sine of the angle between them and whose direction is perpendicular to their plane. The vector product is written: $a \times b$ or $a \wedge b$. This is read as "a cross b". Thus, if $a \times b = c$, then c has a magnitude of $|\,a\,|\,|\,b\,|\sin\theta$. By convention, its direction is taken to be such that a, b, and c form a right-handed system: i.e. one in which a right-handed screw turning from a towards b would move in the direction of c. (Note that $b \times a$ would have the opposite direction). A physical example of a vector product is that involving the force F on a charged particle moving in a magnetic field B. The force is perpendicular to the plane of motion and the field: $F = Q\,v \times B$, where Q is the particle's charge and v its velocity. If the vectors are represented in Cartesian coordinates, their vector product has the form: $a \times b = i(a_y b_z - a_z b_y) + j(a_z b_x - a_x b_z) + k(a_x b_y - a_y b_x)$. Two vectors can also be multiplied together to give a *scalar product.

velocity. Symbol: v. A measure of rate of motion, equal to the distance moved per unit time. Velocity is a vector quantity: its value at a point is determined both by its magnitude (dx/dt) and its direction. The velocity of a particle travelling in a curved path has the direction of the tangent to the path at the point considered. See also speed.

velocity modulation. See klystron.

velocity ratio. See machine.

Venetian white. A white pigment consisting of a mixture of lead carbonate and barium sulphate.

Venus. The second planet in the solar system in order of succession outwards from the sun. Its mean distance from the sun is about 108.2 million kilometres. Of all the planets, Venus approaches closest to the earth: at inferior conjunction it comes as near as 42 million km. Moving with a mean velocity of some 35 km per second, it takes 225 earth days to orbit the sun. In size, Venus is comparable with our own planet. Its mass and gravity are about four-fifths those of the earth, its diameter (excluding atmosphere) is about 12 100 km, and its mean relative density is 5.1. Venus supports an extremely dense atmosphere consisting mainly of carbon dioxide (about 95%) and minute amounts of nitrogen, water vapour, carbon monoxide, oxygen, hydrochloric and hydrofluoric acids, and ammonia. The atmosphere has an outer cloud covering of 100% and is impenetrable to normal optical instruments. The period of axial rotation is 243 earth days, Venus's motion being *retrograde. The outer clouds, perhaps made of ice, have a temperature of $-43\,°C$, but this temperature rises quickly to $+330\,°C$ just below the clouds. The temperature increases enormously on descending through the atmosphere and it is believed to reach 480°C at the surface. It is believed that the terrain of Venus is a rocky desert that probably glows dimly by virtue of its own red heat. Atmospheric pressure is believed to be about 95 atmospheres. The eccentricity of Venus's orbit is 0.0067 and its inclination is about 3° to the plane of the ecliptic. The brightest object in the sky after the moon, Venus is observable as a morning or evening star.

verdigris. Any of various greenish basic salts of copper. True verdigris is basic copper acetate, a green or blue solid of variable composition used as a paint pigment. The term is often applied to green patinas formed on metallic copper. Copper cooking vessels can form a coating of basic copper carbonate, $CuCO_3.Cu(OH)_2$. The verdigris formed on copper roofs and domes is usually basic copper sulphate, $CuSO_4.3Cu(OH)_2.H_2O$, while basic copper chloride, $CuCl_2.Cu(OH)_2$, may form in regions near the sea.

vermicide. A substance used for killing worms.

vermiculite. A hydrated silicate of magnesium, aluminium, and iron belonging to the *micas. When heated quickly it can expand by up to twenty times its original volume. This exfoliated form is used in insulation, packing, and lightweight concrete, and as a filler for paints and plastics, a soil conditioner, and a compost for plants.

vermilion. *See* mercury sulphide.

vernal equinox. *See* equinox.

vernier. A short scale moving alongside the main scale of a measuring instrument, used to obtain a more accurate value of the measurement. The vernier scale is constructed so that it has ten divisions that correspond in distance to nine divisions on the main scale. The pointer on the instrument is the zero on the vernier scale. When this falls between two graduations of the main scale, the next decimal place in the reading is obtained by noting the number of the graduation on the vernier that coincides with a graduation on the main scale. [After Pierre Vernier (1580-1637), French mathematician and soldier.]

versed sine. One minus the cosine of an angle: vers $\theta = 1 - \cos \theta$.

vertex. A point of intersection of two sides of a plane figure.

very high freqency (VHF). A frequency in the range 30 megahertz to 300 megahertz.

very low frequency (VLF). A frequency in the range 3 kilohertz to 30 kilohertz.

vesicant. Producing blisters.

VHF. *See* very high frequency.

vibrating-reed electrometer. A type of *electrometer in which a low steady potential difference is converted to an alternating potential by connecting it to a capacitor, one of the plates of which is mechanically vibrated. The alternating potential can then be amplified.

vibration. Rapid oscillatory motion, as of a tuning fork, stretched string, etc. *See* simple harmonic motion.

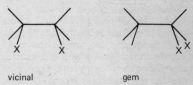

vicinal   gem

Vicinal and gem positions in molecules

vicinal. Denoting positions on adjacent atoms in a molecule (see illustration). In the names of compounds the abbreviation *vic-* is used. *Compare* gem.

Victor Meyer's method. A technique for measuring the *vapour density of volatile liquids and solids. A weighed amount of the sample is put in a small stoppered tube, which is dropped through a long tube into a bulb surrounded by a constant temperature bath at a temperature above the boiling point of the sample. The tube is then sealed. The sample is vaporized and the vapour

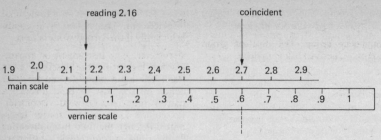

reading 2.16   coincident

Vernier scale

forces out the stopper of the sample bottle and displaces air from the tube, out of the side arm and into a graduated collection tube. Thus, the volume V of vapour produced by a mass m of sample at the temperature of the bulb is obtained. The density of vapour at this

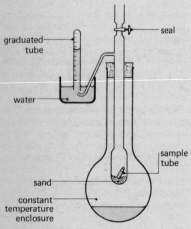

Victor Meyer apparatus

temperature is given by $mp/V(p-p')$, where p is the pressure at which the volume is measured and p' is the vapour pressure of water at the temperature at which the air is collected. [After Victor Meyer (1848–97), German chemist.]

video frequency. A frequency in the range used for transmitting television signals, lying between 10 hertz and 2 megahertz.

vidicon. A type of television *camera tube in which the image is produced on an insulating screen coated with a photoconducting layer. A positive potential is applied and each part of the *mosaic acts as a small capacitor storing charge. The amount of charge stored depends on the resistance of the screen at that point, which in turn depends on the amount of light falling on the photoconductor. The back of the screen is scanned with an electron beam, thus releasing the charge and causing a current to flow. For each element of the screen the discharged current depends on the amount of incident light.

vinegar. A dilute impure solution of *acetic acid made by fermentation of ethanol, particularly in cider, wine, or malted barley liquor. It is used as a preservative and food flavouring.

vinyl acetate. A colourless liquid $CH_3COOCH:CH_2$, manufactured by catalytic reaction between acetic acid and ethylene or acetylene. It is the starting material for polyvinyl acetate resins: the pure liquid must be stabilized as it polymerizes on exposure to light. M.pt. $-100°C$; b.pt. $73°C$; r.d. 0.9.

vinyl chloride. *See* chloroethene.

vinyl cyanide. *See* acrylonitrile.

vinyl ether. A colourless volatile flammable liquid, $CH_2:CHOCH:CH_2$, prepared by treating dichloroethyl ether with alkali and used as an anaesthetic. B.pt. $28°C$; r.d. 0.8.

vinyl group. The monovalent group CH_2:CH–, derived from ethylene.

vinylidene group. The divalent group CH_2:C =, derived from ethylene.

Vinyl polymer

vinyl resin. Any of a class of synthetic resins made by polymerizing a vinyl compound (see illustration). The polymerizations are usually carried out in the pure liquid phase or in emulsions. The most important vinyl polymers are polyvinyl chloride, polyvinyl acetate, and various copolymers of vinyl chloride with vinyl acetate. The vinyl resins belong to the class of *ethenoid resins. The term *vinyl* is used for any vinyl resin or material produced from such a resin.

virgin neutron. A neutron that has not experienced a collision.

virial equation. An equation of state representing real gases: $pV = A + Bp + Cp^2 + Dp^3 + ...$, where A, B, C, D, etc., are empirical constants (the *virial coefficients* of the gas).

virtual image. *See* image.

virtual particle. A particle thought of as existing for a very brief period of time in an interaction between two other particles. Thus, an *electromagnetic interaction between two charged particles can be described by emission and absorption of *virtual photons.* The *strong interaction between nucleons in the nucleus is described in terms of the exchange of *virtual pions.* Another particle, the *W-particle* or *intermediate vector boson,* has been postulated as the virtual particle responsible for the *weak interaction between particles.

virtual work. The work that would be done if a body or system subjected to a set of forces were given an infinitesimal displacement. The system is in equilibrium only if the virtual work is zero.

viscometer. Any instrument or apparatus for measuring the viscosity of a fluid. The main types depend for their action on the flow of fluid through a narrow tube (*see* Ostwald viscometer, Poiseuille's formula); the torque transmitted through the fluid from a rotating cylinder to a concentric stationary cylinder (*see* Searle's visometer); the fall of a sphere through the fluid (*see* Stokes' law); and the damping of a vibrating body by the surrounding fluid.

viscose. The viscous brown solution of cellulose xanthate in sodium hydroxide that is used in making *rayon.

viscose rayon. *See* rayon.

viscosity. The property of liquids and gases of resisting flow. It is caused by forces between the molecules of the fluid. In *streamline flow the fluid can be thought of as containing parallel layers, which move at different rates. The resistance caused by the viscous forces is a tangential force opposing the motion of the layer.
Newton's law of viscosity states that this force per unit area is proportional to the velocity gradient between layers: i.e. $F = \eta A dv/dx$. The shear stress is proportional to the rate of shear strain. Fluids that show this behaviour are called *Newtonian fluids* (*compare* non-Newtonian fluid). The constant η is the *coefficient of viscosity* of the fluid, usually called the *dynamic viscosity* or simply the *viscosity*. It is measured in newton-seconds per square metre or in *poise. The *kinematic viscosity* is the dynamic viscosity divided by the density of the fluid. It is given the symbol ν (= η/ρ) and is measured in square metres per second or in *stokes.

viscous. Having a high viscosity.

visible spectrum. 1. The spectrum of wavelengths in visible *light.
2. An emission or absorption spectrum in the visible region. Light emission is

caused by electron transitions in atoms and molecules or by vibration of atoms in incandescent solids. It gives information on the energy levels and on molecular structure.

visual binary. *See* binary.

visual-display unit. A device on the output of a computer that can produce a display on a cathode-ray tube in the form of characters or diagrams. It may be fitted with a *light pen.

vitamin. Any of a group of chemical compounds that are essential constituents of the diet of animals. In general, vitamins are not synthesized in the body, although vitamin D is produced in humans by the action of ultraviolet light on ergosterol. Chemically, the compounds are unrelated and they have a variety of functions in metabolism—many of them act as coenzymes. Their absence from the diet produces certain deficiency diseases: vitamin C prevents scurvy, vitamin D prevents rickets, etc.

vitamin C. *See* ascorbic acid.

vitreous. Glass-like; having the appearance or structure of a glass.

vitreous humour. *See* eye.

vitriol. 1. Any of several sulphate salts, such as copper sulphate (*blue vitriol*), iron II sulphate (*green vitriol*), or zinc sulphate (*white vitriol*).
2. *See* sulphuric acid.

VLF. *See* very low frequency.

volatile. Capable of vaporizing easily. Volatile substances, such as ether and camphor, have high vapour pressures.

volt. Symbol: V. The *SI unit of electric potential, potential difference, or electromotive force, equal to the potential difference between two points on a conductor carrying a steady current of one ampere when the power dissipated between the points is one watt. [After Count Alessandro Volta (1745-1827), Italian physicist.]

voltage. Symbol: *V*. Potential difference or electromotive force as measured in volts.

voltage divider. *See* potential divider.

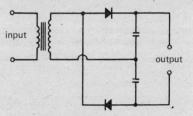

Voltage-doubler circuit

voltage doubler. An electrical circuit for producing an output voltage twice as large as the input voltage. The usual form has two rectifiers.

voltaic cell. *See* cell.

voltameter (coulombmeter). An electrolytic cell used for measuring a quantity of current electricity by electrodeposition of a metal, usually copper or silver. The cell contains a salt of the metal and, on passing the current, metal is deposited on the cathode. The total quantity of electricity passed can be calculated from the increase in weight of the cathode and the *electrochemical equivalent of the metal. *See also* ampere, Faraday's laws.

Volta's pile. An early battery invented by Volta, consisting of a pile of cells, each having a zinc and copper disc separated by a cloth disc soaked in sulphuric acid. *See also* Daniell cell.

voltmeter. Any of several instruments used to measure voltage. The term is usually used for meters that give a direct reading on a scale calibrated in volts, millivolts, etc. All voltmeters are designed so that they draw very little current from the circuit. When the required accuracy is not too great, *moving-coil and *moving-iron instruments can be used to measure potential difference. Both types have a high resistance incorporated so that they do not draw very much current, the moving-coil voltmeter being used for direct voltages and the moving-iron voltmeter being suitable for both direct and alternating voltages. Various types of *electrometer are also used as voltmeters. The *electrostatic electrometer is calibrated in the kilovolt range. Other types work by electronic amplification. The *valve voltmeter has a thermionic-valve circuit with a measuring instrument in its output: most modern electrometers have solid-state amplifiers and often a digital output (see digital voltmeter). Potential differences can also be measured by *potentiometer and by a *cathode-ray oscilloscope.

volume. Symbol: V. A measure of the amount of space occupied by an object or solid figure. It is measured in cubic metres, cubic feet, etc. See also Appendix.

volumetric analysis. Any of various forms of quantitative analysis, all of which depend on measurements of volumes. Volumetric analysis of gases depends on volume changes measured in a *eudiometer. A mixture of gases can be analysed by selectively absorbing different components and measuring the volume decrease: oxygen, for example, can be absorbed by pyrogallol. Another method of gas volumetric analysis is *Dumas' method of finding the percentage of nitrogen in organic compounds. Volumetric analysis of solutions and liquids is performed by finding the amount of unknown solution that reacts chemically with known amounts of standard solution. See titration.

vulcanite (ebonite). A hard black insulating material made by vulcanizing rubber with sulphur.

vulcanization. The process of improving the qualities of natural or synthetic rubber by heating it with sulphur. Temperatures of about 150°C are used and the reactions are promoted by *accelerators. Carbon black is also usually added.

vulgar fraction. See fraction.

W

wall effect. Any effect caused by the inside wall of a container in an experimental apparatus or other device. An example of a wall effect is adsorption and recombination of free radicals on the inside wall of the reaction vessel in flash-photolysis experiments.

warfarin. A colourless odourless tasteless crystalline solid, $C_{19}H_{16}O_4$. It is widely used as a rodenticide. M.pt. 161°C.

washing soda. See sodium carbonate.

water. A colourless odourless tasteless liquid, H_2O. Water is the commonest hydrogen compound, covering almost three-quarters of the earth's surface and found in all living matter and in many minerals. It is used as a standard for several units and physical quantities, in particular for the fixed points of the *centigrade scale, the *relative densities of liquids and solids, and the definitions of the *calorie and the *litre. Water has several important properties. The solid form (ice) is less dense than the liquid at 0°C: the densities are 916.8 kg/m³ and 999.8 kg/m³ respectively. This is why ice floats on water and why frozen water pipes often burst when they begin to thaw. Water also has a maximum density (1000 kg/m³) at 4°C, unlike most liquids, which have their maximum density at their melting point.
The molecules of water are polar and the liquid is held together by *hydrogen bonds. The polar nature of water makes it an excellent solvent for many compounds, particularly for salts, bases, and other ionic solids. See also hydrolysis, ionic product.

water equivalent. Symbol: w. The mass of water that would have the same *heat capacity as a given object.

water gas. A mixture of hydrogen and carbon monoxide made by passing steam over hot coke: $C + H_2O = H_2 + CO$. It is used as a fuel and as a raw material for the manufacture of some organic chemicals. The reaction is endothermic and water gas is sometimes made at the same time as *producer gas, either by using a mixture of steam and air or alternate blasts of steam and air to cool and heat the coke.

water glass. *See* sodium silicate.

water of crystallization. Water combined in the form of molecules in definite proportions in the crystals of many substances. The crystalline solid is a *hydrate* and can be further described as a *monohydrate*, *dihydrate*, etc., according to the number of water molecules present. The molecules are usually bound by coordination or by hydrogen bonding. Copper sulphate, for example, forms a pentahydrate, $CuSO_4.5H_2O$, in which four water molecules are coordinated to the copper II ion to give $[Cu(H_2O)_4]^{2+}$ and the fifth is held by hydrogen bonds to sulphate ions. The four molecules are lost at $100°C$ but the fifth is not removed until the solid is heated to $250°C$. Strongly bound water of crystallization of this type is sometimes called *water of constitution*.

watt. Symbol: W. The *SI unit of power, equal to one joule per second. In an electric circuit the power dissipated in watts is equal to the current in amperes multiplied by the potential difference in volts. [After James Watt (1736–1819), Scottish engineer.]

wattage. Symbol: *W*. Power as measured in watts.

wattmeter. An instrument for measuring electrical power directly in watts.

wave. A periodic change in some property or physical quantity through a medium or space, occurring in such a way that there is a regular periodic variation of the property with direction and usually with time. The varying property can be a physical displacement, such as that of a stretched string or the surface of water. Alternatively, it can be a quantity, such as a temperature, pressure, or strength of electromagnetic field.

Sound, for instance, is propagated by changes in pressure (or density) of a gaseous medium. At any instant of time the pressure will vary along a particular direction. A graph of pressure against distance would have the form of a *sine curve, with pressure increasing and decreasing regularly. If the behaviour of the medium is considered over a period of time, the pressure at each point along this direction will also rise and fall periodically with time. The overall effect is of the periodic displacement with distance travelling through the medium with constant velocity. In this way energy is carried through the medium.

This is an example of a *travelling wave*, in which the whole periodic displacement moves. Under certain conditions a *standing* or *stationary wave* can be obtained. In this case, the property does not change with time at any particular point in the medium.

Sound moves as a *longitudinal wave*, i.e. one in which the local displacement is in the direction of motion of the wave. A vibrating stretched string and electromagnetic radiation are both examples of a *transverse wave*, in which the displacement is perpendicular to the direction of motion. *See also* harmonic motion.

wave equation. The partial differential equation:

$$\nabla^2 U = \frac{1}{c^2} \frac{\partial^2 U}{\partial t^2}$$

relating the displacement U of a *wave to time t and to distances x, y, and z in three dimensions. c is the velocity of the waves.

waveform. The graph obtained by plotting the value of a periodic quantity against time.

wave front. A surface that is the locus of points having the same phase in a wave motion. Usually, the normal to the wave front at any point has the direction of propagation of the wave at that point.

— **wave function.** Symbol: ψ. The amplitude of the wave associated with a particle as obtained from the *Schrödinger equation. The wave function is a mathematical expression: a function of the position of the particle. Its physical significance is that the square of the value of ψ at any point, $|\psi|^2$, is proportional to the probability of finding the particle at that point. *See* wave mechanics.

waveguide. A hollow conducting tube of rectangular cross section used to guide microwaves propagated along it.

wavelength. Symbol: λ. The distance between adjacent points with the same phase in a wave: i.e. the distance between adjacent peaks. The wavelength of a wave is related to its velocity, c, and frequency, ν, by $\lambda = c/\nu$.

wave mechanics. A form of *quantum mechanics in which particles (electrons, protons, etc.) are regarded as *de Broglie waves, so that any system of particles (such as the electrons orbiting the nuclei of atoms) can be described by a wave equation—the *Schrödinger equation. The basic problems in wave mechanics are those of formulating and solving the Schrödinger equation for the system of interest. In general, only certain solutions are allowed, corresponding to standing waves. This leads to the existence of allowed values of the energy of each particle (*eigenvalues*) and associated wave functions (*eigenfunctions*) that give the probabilities of finding the particle at different points in space.

wavemeter. An instrument for measuring the wavelength of radio waves.

wave number. Symbol: σ. The number of single waves in unit distance, equal to the reciprocal of the wavelength.

wave theory. *See* light.

wax. Any of various naturally occurring solid or semisolid water-insoluble substances that are soft and can be melted or softened at fairly low temperatures. Substances secreted by plants or animals, such as beeswax, are mixtures of esters of high-molecular weight fatty acids. Mineral waxes, such as paraffin wax, are usually mixtures of high-molecular weight aliphatic hydrocarbons, especialy *alkanes.

weak acid or **base.** *See* acid, base.

weak interaction. A type of interaction between *elementary particles, occurring at short range and having a magnitude about 10^{10} times weaker than the electromagnetic force. Weak interactions occur in certain decay processes, in which a particle is thought of as breaking into *virtual particles that interact by weak interaction and then separate as true particles. In such decays, strong interactions cannot be responsible since *strangeness would not be conserved. It has been postulated that weak interactions are caused by a heavy *virtual particle—the intermediate vector boson.

weber. Symbol: Wb. The SI unit of *magnetic flux, equal to the flux linking a circuit of one turn that produces an e.m.f. of one volt when reduced uniformly to zero in one second. It is equal to 10^8 maxwells. [After Wilhelm Edward Weber (1804-91), German physicist.]

weight. 1. Symbol: W. The force exerted on a body by gravitational attraction by the earth. The weight of a body always acts towards the earth's centre and depends on its distance from the earth's surface and on the local value of g—the *acceleration of free fall. Weights can be expressed in the usual units of force

such as newtons: the weight of a mass
m is then given by *mg*. Sometimes units
such as the *gram-weight and *pound
weight are used. The term *weight* is
often used loosely for *mass
2. A number expressing the reliability
assigned to a measurement. If a series of
experimental values, x_1, x_2, x_3, ..., has
been obtained for a quantity, all the
values will not necessarily be of equal
worth. A *weighted mean* can be used,
$(w_1x_1 + w_2x_2 + ...)/(w_1 + w_2 + ...)$, in
which w_1, w_2, etc., are the weights given
to each value. The weight of a value is a
measure of the confidence that can be
placed in it and is either obtained from
the probable error or assigned intui-
tively.

weight thermometer. A small glass (or
silica) bulb with a narrow tube attached,
used for determining the coefficients of
expansion of liquids. The bulb is filled
with the sample and weighed so that the
mass of liquid can be determined. It is
then heated to a known temperature
and the mass of liquid expelled by
expansion is found by weighing. This
enables the coefficient of apparent
expansion of the liquid to be calculated.

Weston cell (cadmium cell). A type of
primary cell used as a standard of e.m.f.
It has a mercury anode covered with a
paste of mercurous sulphate and a
cadmium-amalgam cathode. The electrolyte is a saturated solution of cadmium sulphate and crystals of cadmium
sulphate are present in the cell. Its e.m.f.
is 1.0186 V at 20°C. [After Edward
Weston (1850–1936), American electrical engineer.]

wet- and dry-bulb hygrometer. A type of
*hygrometer consisting of two thermo-
meters placed side by side, one having
its bulb covered with moist cloth. The
wet thermometer indicates a lower
temperature because it is cooled by
evaporation of the water vapour, a
process that depends on the amount of
moisture present in the air. Thus the
difference in the readings of the two
thermometers gives an indication of the

relative *humidity, which can be found
from special tables. *See also* psychrometer.

wetting agent. A substance that causes a
liquid to spread when in contact with a
solid. Wetting agents act by lowering
surface tension.

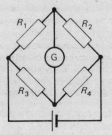

Wheatstone bridge

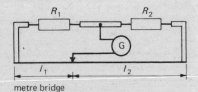

metre bridge

Forms of Wheatstone bridge

Wheatstone bridge. A network of
resistances used for the measurement of
resistance. The resistances are adjusted
until the galvanometer has a zero
reading. The bridge is then balanced
and $R_1/R_2 = R_3/R_4$. If three of the
resistances are known the value of the
fourth can be measured. A common
form of the bridge has a length of
resistance wire with a sliding contact
attached to the galvanometer. R_1 is a
standard resistance and R_2 is the
unknown one. At balance, $R_2/R_1 =
l_2/l_1$. [Named after Sir Charles Wheat-
stone (1802–75), English physicist.]

wheel and axle. A simple *machine
consisting of a wheel and an axle joined
together. The effort is applied to a rope
attached to the rim of the wheel and the
load is carried by a rope attached to the
axle. The mechanical advantage is equal

to the radius of the wheel divided by that of the axle.

whistler. A whistling noise of descending pitch interfering with radio reception. It is caused by radio waves produced by lightning flashes and reflected off the ionosphere.

white dwarf. Any of a class of dwarf *stars with low magnitudes that have diameters on the scale of that of the earth but masses commensurate with that of the sun. Matter in them is closely packed and totally ionized, consisting of electrons and atomic nuclei. Thus the densities of such stars are enormously high. It is generally believed that white dwarfs represent the closing stage in the life cycles of certain stars.

white lead. See lead carbonate.

white light. Light that contains all wavelengths in the visible range. Strictly speaking, the wavelengths should all have equal intensities but the term is used to describe light containing different intensities while still appearing white to the eye.

white noise. Noise that has a continuous wide range of frequencies of uniform energy.

white spirit. A mixture of hydrocarbons boiling in the range 150 - 200°C, obtained by distillation of petroleum. It is used as a solvent, particularly for paint, and as fuel for cigarette lighters.

white vitriol. See zinc sulphate.

Wiedemann-Franz law. The principle that the ratio of thermal conductivity to electrical conductivity has the same value for all metals at a given temperature. The ratio is directly proportional to the thermodynamic temperature for pure metals at normal temperatures.

Wien displacement law. The principle that, for the radiation emitted from a *black body, the product of the wavelength at which maximum emission occurs and the thermodynamic

temperature is a constant: i.e. $\lambda_{max}T = C$. The maximum of a curve of wavelength against temperature is displaced towards shorter wavelengths as the temperature is increased. [After Wilhelm Wien (1864-1928), German physicist.]

Wigner effect. A change in the crystal structure of a solid as a result of irradiation. It is observed in graphite bombarded with neutrons: the material changes size due to displacement of carbon atoms from their lattice positions. [After Eugene Paul Wigner (b. 1902), U.S. physicist born in Hungary.]

Wigner nuclide. Either of a pair of nuclides that are *isobars for which each nuclide has an odd mass number and a neutron number differing by one from its proton number. Tritium (^{3}H) and helium-3 (^{3}He) are a pair of Wigner nuclides.

will-o'-the-wisp (ignis fatuus). A pale flickering moving flame that is sometimes seen at night over marshy ground. It is caused by combustion of marsh gas (methane) produced by decomposing vegetation.

Wilson cloud chamber. See cloud chamber. [After C. T. R. Wilson (1869-1959), Scottish physicist.]

Wimshurst machine. A type of electrostatic generator having two circular discs of glass or other insulating material with radial strips of tinfoil stuck to their sides. The discs rotate in opposite directions and the charge is produced by friction with tinfoil brushes and collected by metal points. [After J. Wimshurst (1836-1903), English scientist.]

window. 1. A range of wavelengths over which a given material tansmits electromagnetic radiation. The atmosphere, for example, has three windows: an optical window, letting through light and ultraviolet radiation in the range 760-200 nm; an infrared window in the range 8-11 μm; and a radio window in the range 8 mm-20 m.

2. A sheet of material allowing the passage of electromagnetic or particle radiation. Thin sheets of metal are used as windows in X-ray tubes and Geiger counters.

witherite. A white or greyish mineral consisting of barium carbonate, $BaCO_3$, used as a source of barium compounds.

Wöhler's synthesis. The production of urea by evaporating a solution of ammonium cyanate $(NH_4)NCO$, first performed in 1828 by the German chemist Friedrich Wöhler (1800–82). The reaction was the first laboratory synthesis of an organic compound from inorganic materials.

wolfram. See tungsten.

wolframite. A naturally occurring tungstate of iron and manganese, (Fe,-Mn)WO_4, used as an ore of tungsten.

Wollaston prism. A prism for obtaining plane-polarized light. It is made from two triangular prisms of quartz or calcite, each cut so that their optic axes are at right angles to one another. The prism is designed to be used for ultraviolet radiation. [Named after W. H. Wollaston (1766–1828), English physicist.]

Wollaston wire. Very thin wire (diameters as small as 1 μm) made by encasing platinum wire in a layer of silver, drawing the wire out, and then dissolving the silver with acid.

wood alcohol. See methanol.

Wood's metal. A low-melting alloy of bismuth (50%), lead (25%), tin (12.5%), and cadmium. M.pt. 71°C.

wood sugar. See xylose.

woofer. A loudspeaker for use at low audio frequencies. Compare tweeter.

word. A number of bits processed as a single unit in a computer. Typical word lengths are 12, 32, 48, and 64 bits.

work. Symbol: w. Energy transferred to or from a body or system, thus causing a change in its energy. Work always involves an applied force moving a certain distance and is measured by the product of the force and the distance it moves along its line of action. If the force is perpendicular to the direction of movement, no work is done. Work, like other forms of energy, is measured in joules.

work function. Symbol: ϕ. The minimum energy necessary to remove an electron from a metal at absolute zero. See also band theory, photoelectric effect.

work hardening. See strain hardening.

W-particle. See virtual particle.

write. To enter information into the store of a computer.

wrought iron. A form of iron containing a very small amount of carbon, produced by melting pig iron with iron oxide to remove the carbon and hammering the hot solid metal to force out liquid slag. Wrought iron contains inclusions of slag with a fibrous structure produced by the hammering. It is less liable to rust than carbon steels and can easily be forged and welded. It has now largely been replaced by various types of *steel.

wurtzite. A naturally occurring form of zinc sulphide, ZnS.

Wurtz reaction. A chemical reaction for the synthesis of hydrocarbons by the action of sodium on alkyl halides: $2RCl + 2Na = 2NaCl + RR$. The similar reaction with aryl halides or with a mixture of an alkyl and an aryl halide to give an aromatic hydrocarbon is called the *Wurtz-Fittig reaction*. [After Adolphe Wurtz (1817–84), French chemist.]

X

xanthate. Any salt or ester of a xanthic acid. Cellulose xanthate is used in making *viscose rayon.

xanthic acid. Any of a class of organic acids with the general formula CS(OR)-(SH), where R is usually an alkyl group. The term is often used to denote *ethylxanthic acid*, $CS(OC_2H_5)(SH)$.

xenon. Symbol: Xe. A colourless odourless tasteless nonflammable gaseous element present in the atmosphere, from which it is extracted as a by-product in the liquefaction of air. Xenon is used in light bulbs, fluorescent lamps, and in lasers. It is one of the *inert gases. A.N. 54; A.W. 131.3; m.pt. -111.8°C; b.pt. -108°C; r.d. 4.53 (air = 1).

xenon fluoride. Any of a group of crystalline solids obtained by reacting xenon and fluorine. The compounds XeF_2, XeF_4, and XeF_6 are readily hydrolysed powerful fluorinating agents.

xerography. A process for copying documents by an electrostatic method. An electrostatically charged plate is used, usually coated with selenium. Ultraviolet light is passed through the document onto the plate, which it discharges. In this way an electrostatic image is produced depending on the amount of radiation falling on different parts of the plate. A dark powder with opposite charge to that of the plate is then applied. It sticks to the charged areas from which it can be transferred to paper. The powder, which consists of a mixture of graphite and a thermosetting resin, is fixed on the paper by heat.

xi particle. Symbol: Ξ. An *elementary particle with either zero charge and a mass 2572 times that of the electron or a negative charge and a mass 2585 times that of the electron. The two xi particles are classified as *hyperons.

X-radiation. Electromagnetic radiation composed of X-rays.

X-ray astronomy. The study of sources of X-rays in space. The radiation is absorbed by the earth's atmosphere and investigations in X-ray astronomy are carried out by means of counters mounted in space probes. The results show numbers of discrete X-ray sources together with a continuous *X-ray background. The discrete sources, many of which cannot be detected by optical or radio astronomy, may be caused by high-temperature emission from plasmas or by synchrotron radiation. *See* pulsar, black hole.

X-ray background. A fairly isotropic continuous background of X-rays observed in space. Its origin is still uncertain: one possible mechanism is the interaction of fast electrons in cosmic rays with low-energy photons in the *microwave background, producing X-ray photons by an inverse *Compton effect. Contributions may also come from numbers of discrete X-ray sources.

X-ray crystallography. The determination of the structure of crystals and molecules by use of *X-ray diffraction. Several experimental techniques exist applicable to powders, fibres, single crystals, etc. The sample is placed in a beam of monochromatic X-rays and the diffraction pattern recorded on a photographic plate. A typical pattern from a single crystal consists of a symmetrical array of spots, corresponding to diffraction from different crystal planes. The crystal structure can be calculated from the positions of the spots. It is also possible, by measuring the intensities of diffracted X-rays, to compute a map of electron density within molecules in the crystal.

X-ray diffraction. Diffraction of X-rays by crystals. The wavelengths of X-rays are the same order of magnitude as the interatomic distances in crystals and the periodic crystal lattice can act like a grating and produce a diffraction pattern from the X-rays. *See also* Bragg's law.

X-rays (Röntgen rays). Electromagnetic radiation with wavelengths in the range $10^{-7} - 10^{-10}$ metre. X-rays lie between ultraviolet radiation and gamma rays in the electromagnetic spectrum. They are produced by bombarding a solid target with a beam of high-energy electrons. In this process an incident electron knocks one of the inner electrons from an atom, forming a vacancy in an inner shell. An electron from an outer orbit fills this vacancy, the excess energy being emitted as an X-ray photon.

Similar emission can occur when atoms are hit by other X-ray photons or by gamma rays or when inner electrons are lost by nuclear *capture. The *X-ray spectrum produced by this process (*X-ray fluorescence*) contains lines with wavelengths characteristic of the target substance. In X-ray tubes, these are superimposed on the continuous *bremsstrahlung created by deceleration of the incident electrons.

X-rays affect photographic plates and cause some ionization in matter. Their most useful property is their ability to penetrate matter and they are extensively used for medical diagnosis and treatment and for the detection of flaws in metal parts.

X-ray spectrum. A spectrum of X-rays emitted by bombardment of a sample with electrons or photons. Each element emits X-rays with characteristic wavelengths (*characteristic X-rays*) and these can be used as a method of analysis. The spectrum is taken by using a *crystal spectrometer.

X-ray star. A star that emits X-rays.

X-ray tube. A tube for producing X-rays, typically consisting of an evacuated container in which an electron gun is used to direct high-energy electrons onto a metal anode (the *target* or *anti-cathode*). The X-rays leave the tube through a thin metal window.

X unit (Siegbahn unit). Symbol: XU. A unit of length equal to $1.002\ 02 \times 10^{-13}$ metre. It is used for expressing wavelengths of X-rays and gamma rays.

xylene (dimethyl benzene). Any of three colourless toxic flammable isomers, $C_6H_4(CH_3)_2$, obtained by fractional distillation of petroleum. The mixture of isomers is used as aviation fuel and as a solvent. Individual isomers are used to

D-xylose

make many organic compounds. B.pt. 135-145°C (mixture); r.d. 0.9 (mixture).

xylose (wood sugar). A white crystalline monosaccharide, $C_5H_{10}O_5$, extracted by hydrolysis of wood or straw. The naturally occurring substance is dextrorotatory. M.pt. 144°C; r.d. 1.53.

xylyl group. The univalent group $CH_3C_6H_4CH_2-$.

x-y recorder. A type of *pen recorder having two inputs, one displacing the pen in the x direction and the other in the y direction, so that a graph can be drawn of one varying electrical signal against another.

Y

yard. An imperial unit of length equal to 0.9144 metres. It was formerly defined as the distance, at 62°F, between lines scribed on two gold plugs in a standard bronze bar.

year. The time taken for the earth to move around the sun. It is measured in various ways (*see* time).

yield point. The stress beyond which a material undergoes a relatively large plastic deformation for small increases in stress. Beyond this point, the material no longer obeys *Hooke's law.

ylide. A type of compound in which the molecule has both a positive and negative charge, as in $R_2C-PR'_3$, where R and R' are organic groups.

Young's modulus. Symbol: E. A *modulus of elasticity applied to the stretching or compression of a material, equal to the ratio of the applied stress to the strain produced. [After Thomas Young (1773-1829), English physicist.]

Young's slits. An experiment to demonstrate *interference of light by placing a screen with two fine slits cut in it in front of a monochromatic light source. The slits act as coherent sources of light of equal intensity and produce interference fringes.

ytterbium. Symbol: Yb. A soft silvery metallic element belonging to the *lanthanide series. A.N. 70; A.W. 173.04; m.pt. 824°C; b.pt. 1193°C; r.d. 7; valency 3.

yttrium. Symbol: Yt. A silvery ductile metallic element. Yttrium falls into group IIIb of the periodic table but is usually classified with the *lanthanide elements. A.N. 39; A.W. 88.905; m.pt. 1523°C; b.pt. 3337°C; r.d. 4.46; valency 3.

Yukawa potential. A potential used to explain the short-range *strong interactions occurring between nucleons in the atomic nucleus. It is assumed that the potential energy is proportional to $[\exp(-\mu r)]/r$, where r is the distance apart and μ is a constant. [After Hideki Yukawa (b. 1907), Japanese physicist.]

Z

Zeeman effect. The splitting of single lines in a spectrum into two or more components with slightly different wavelengths when the sample is subjected to a magnetic field. The *normal Zeeman effect* involves splitting into three components and is caused by changes of the energy levels of the orbiting electrons due to interaction of the magnetic moment of the orbit with the applied field. The *anomalous Zeeman effect* is more complex and involves the *spin of electrons. In the Zeeman effect, the magnitude of the splitting is proportional to the strength

of the field. It can be used in determining the magnetic field associated with stars or with sunspots. [After Pieter Zeeman (1865-1943), Dutch physicist.]

Zener diode. A semiconductor *diode that gives a sudden increase in the reverse current when the reverse voltage increases above a certain value. Zener diodes can be used for regulating the voltage in a circuit. [After Clarence Melvin Zener (b. 1905), U.S. physicist.]

zenith. The point on the *celestial sphere directly above an observer.

zeolite. Any of a class of natural or synthetic *silicates of sodium, potassium, and calcium, having a three-dimensional crystal structure containing water molecules held in cage-like cavities in the lattice. When heated they lose water and the framework of the crystal can then trap other molecules of suitable size. Zeolites are used for separating mixtures by selective adsorption: for this reason they are known as *molecular sieves*.

zero point energy. The energy of vibration of atoms at absolute zero of temperature, equal to $h\nu/2$, where h is the Planck constant and ν the frequency of oscillation.

ZETA. An experimental fusion reactor at Harwell. It has a toroidal reaction chamber in which the plasma is contained by a magnetic field produced by a solenoidal winding. The name is an acronym for *Zero Energy Thermonuclear Apparatus*.

Ziegler-Natta catalyst. Any of a class of related catalysts that promote the *stereospecific polymerization of ethylene and other olefines. Typically, they consist of mixtures of transition-metal halides, such as titanium II chloride, with nontransition-metal alkyls, especially aluminium alkyls. [After Karl Ziegler (b. 1897), German chemist, and Giulio Natta (b. 1903), Italian chemist.]

zinc. Symbol: Zn. A bluish-white lustrous brittle metallic element found in zincite (ZnO), zinc blende (ZnS), smithsonite ($ZnCO_3$), and hemimorphite ($Zn_4Si_2O_7(OH)_2.H_2O$). The metal is obtained by roasting the ore and reducing the oxide by heating with carbon. It is used in many alloys, such as brass, bronze, nickel silver, German silver, and solder, and in *galvanizing iron.

Zinc is a fairly reactive electropositive element, reacting with oxygen, sulphur, the halogens, and other nonmetals. It is amphoteric, dissolving in nonoxidizing acids and also in alkalis to form *zincates. Most of its compounds contain Zn^{2+} ions although some are covalently bonded, as in the zinc *alkyls. A.N. 30; A.W. 65.37; m.pt. 419.6°C; b.pt. 907°C; r.d. 7.13; valency 2.

zincate. A salt containing the ion ZnO_2^{2-}, produced by dissolving zinc in an alkali. Zincates are formally salts of amphoteric zinc hydroxide, $Zn(OH)_2$, and are probably more correctly regarded as hydrated zinc hydroxide anions.

zinc blende. Naturally occurring zinc sulphide, ZnS.

zinc chloride (butter of zinc). A poisonous deliquescent white crystalline solid, $ZnCl_2$, prepared by adding hydrochloric acid to zinc or zinc oxide. It is used for galvanizing iron, as a catalyst, as a dehydrating agent, and as a soldering flux. M.pt. 290°C; b.pt. 732°C; r.d. 2.9.

zinc hydroxide. A white amphoteric solid, $Zn(OH)_2$, precipitated by addition of alkali to zinc salts.

zincite. A red or orange naturally occurring form of zinc oxide, ZnO, used as an ore of zinc.

zinc oxide (Chinese white, philosopher's wool). An odourless white powder, ZnO, which turns yellow when hot. It is made by vapour-phase oxidation of pure zinc or by roasting zinc oxide ore with coal followed by oxidation in air. Zinc oxide is a filler and pigment in plastics and paints, and is a widely used

mild antiseptic in ointments. M.pt. 1975°C; r.d. 5.5.

zinc sulphate (white vitriol). A colourless efflorescent crystalline solid, $ZnSO_4.7H_2O$, manufactured by roasting zinc blende with subsequent leaching. It is used as a styptic, a mordant, and as a preservative for wood. Loses water above 200°C; decomposes above 500°C; r.d. 1.9.

zinc sulphide. A yellow-white solid, ZnS, occurring naturally in wurtzite, sphalerite, and zinc blende. It is used as a pigment for paints and glass and as a phosphor in cathode-ray tubes. Sublimes at 1180°C; r.d. 3.98 (wurtzite).

zircon. A brown, red, or colourless naturally occurring silicate of zirconium, $ZrSiO_4$, used as a gemstone and a source of zirconium and zirconium oxide.

zirconium. Symbol: Zr. A greyish-white lustrous metallic element found principally in zircon ($ZrSiO_4$) and baddeleyite (ZrO_2), from which it can be extracted by the *Kroll process. Zirconium is used in nuclear reactors as a neutron absorber, as a vacuum getter, and in alloys. Its chemistry is similar to that of titanium, its main compounds being covalently bonded. A.N. 40; A.W. 91.22; m.pt. 1852°C; b.pt. 4377°C; r.d. 6.51; valency 2, 3, or 4.

zodiac. A band on the *celestial sphere containing the paths of the planets, the sun, and the moon. It is about 18° wide (i.e. 9° each side of the ecliptic) and is divided into 12 equal parts named after constellations.

zodiacal light. Faint light seen in the west just after sunset or in the east just before sunrise. It is caused by reflection of sunlight from small meteoric particles around the sun.

zone refining. A method of purifying solid materials, used especially for semiconductors. A long bar of the sample is heated at a point near one end so that a small region of the bar is melted. The molten phase is held in position by surface tension and impurities in the neighbouring regions tend to concentrate in the melt. By moving the heater, this molten zone is slowly moved along the bar, so that the impurities are transferred to one end. Impurity levels as low as 1 part in 10^{10} can be achieved by using several passes.

zoom lens. An adjustable compound lens used in cinematic or television cameras to change the magnification continuously without losing the sharpness of the image.

zwitterion (ampholyte ion). An ion that has both a positive and negative charge. Examples are found in the amino acids: glycine, (CH_2NH_2COOH), for example, can form the zwitterion $CH_2NH_3^+COO^-$ by ionization of the carboxyl group and protonation of the amino group.

APPENDIX

Table 1: Units of Length

1 inch		0·0254 m
1 foot	12 inches	0·3048 m
1 yard	3 feet	0·9144 m
1 mile	1760 yards	1·609 34 km
1 fathom	6 feet	1·8288 m
1 nautical mile	1·150 78 miles	1·8520 km
1 light-year		$9·4607 \times 10^{15}$ m
1 astronomical unit		$1·495 \times 10^{11}$ m
1 parsec		$3·0857 \times 10^{16}$ m
1 metre	1·093 61 yards	
1 kilometre	0·621371 miles	

Table 2: Units of Volume

1 minim	1/9600 pint	
1 fluid drachm	60 minims	
1 fluid ounce	8 fluid drachms	
1 gill	5 fluid ounces	
1 pint	4 gills	$0·568 26$ dm^3
1 quart	2 pints	
1 gallon	4 quarts	$4·546 09$ dm^3
1 gallon	0·160 544 cubic feet	
1 cubic foot	6·228 82 gallons	$0·028 316 8$ m^3

1 cubic decimetre (1000 cc) is 1 *litre*, although this name is not used in precision measurement.

1 gallon (U.S.) is equivalent to 0·832 68 gallon (U.K.).

Table 3: Avoirdupois Units of Mass

1 grain	1/7000 pound	0·0647 99 g
1 dram	1/256 pound	
1 ounce	16 drams	113·398 g
1 pound	16 ounces	0·453 592 kg
1 stone	14 pounds	
1 quarter	2 stones	
1 hundredweight	4 quarters	50·802 kg
1 ton	20 hundredweights	1016·047 kg
1 gram	0·0353 ounces	
1 kilogram	2·204 62 pounds	

The avoirdupois hundredweight and ton are sometimes called the *long hundredweight* and *long ton* to distinguish them from the U.S. measures, the *short hundredweight* (100 pounds) and *short ton* (2000 pounds).

Table 4: Troy Units of Mass

1 grain	1/7000 pound (avoirdupois)
1 carat	4 grains
1 pennyweight	6 carats
1 ounce (tr)	20 pennyweights
1 pound (tr)	12 ounces (tr)
1 hundredweight (tr)	100 pounds (tr)
1 ton (tr)	20 hundredweights (tr)

Table 5: Apothecaries' Units of Mass

1 grain	1/7000 pound (avoirdupois)
1 scruple	20 grains
1 drachm	3 scruples
1 ounce (apoth)	8 drachms
1 pound (apoth)	12 ounces (apoth)

The grain (1/7000 pound avoirdupois) has the same value in the avoirdupois, troy, and apothecaries' systems.

Table 6: Fundamental Constants

velocity of light	c	$2{\cdot}997\,925\,0 \times 10^8$	$\mathrm{m\,s^{-1}}$
magnetic constant	μ_0	$1{\cdot}256\,64 \times 10^{-6}$	$\mathrm{H\,m^{-1}}$
electric constant	ϵ_0	$8{\cdot}854\,16 \times 10^{-12}$	$\mathrm{F\,m^{-1}}$
electron charge	e	$1{\cdot}602\,192 \times 10^{-19}$	C
electron mass	m_c	$9{\cdot}109\,558 \times 10^{-31}$	kg
proton mass	m_p	$1{\cdot}672\,614 \times 10^{-27}$	kg
Avogadro constant	N	$6{\cdot}022\,52 \times 10^{23}$	$\mathrm{mol^{-1}}$
Boltzmann constant	k	$1{\cdot}380\,622 \times 10^{-23}$	$\mathrm{J\,K^{-1}}$
Planck constant	h	$6{\cdot}626\,196 \times 10^{-34}$	$\mathrm{J\,s}$
Loschmidt's number	L	$2{\cdot}687\,19 \times 10^{25}$	$\mathrm{m^{-3}}$
Faraday constant	F	$9{\cdot}648\,670 \times 10^4$	$\mathrm{C\,mol^{-1}}$
Gas constant	R	$8{\cdot}314\,34$	$\mathrm{J\,K^{-1}\,mol^{-1}}$
standard acceleration of free fall	g	$9{\cdot}806\,65$	$\mathrm{m\,s^{-2}}$
standard atmospheric pressure	p	$1{\cdot}013\,25 \times 10^5$	Pa